Theoretical concepts in physics

AN ALTERNATIVE VIEW OF THEORETICAL REASONING IN PHYSICS FOR FINAL-YEAR UNDERGRADUATES

M.S.LONGAIR

Astronomer Royal for Scotland, Regius Professor of Astronomy, University of Edinburgh, and Director, Royal Observatory, Edinburgh

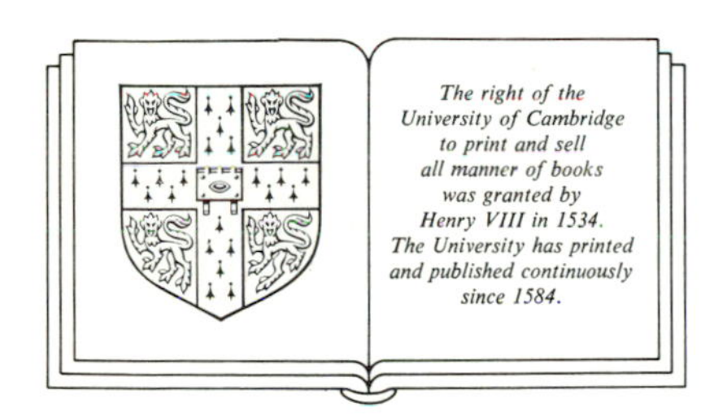

CAMBRIDGE UNIVERSITY PRESS

Cambridge

London New York New Rochelle

Melbourne Sydney

Published by the Press Syndicate of the University of Cambridge
The Pitt Building, Trumpington Street, Cambridge CB2 1RP
32 East 57th Street, New York, NY 10022, USA
296 Beaconsfield Parade, Middle Park, Melbourne 3206, Australia

First published 1984

Printed in Great Britain at the University Press, Cambridge

Library of Congress catalogue card number: 83-18928

British Library cataloguing in publication data

Longair, M.S.
Theoretical concepts in physics.
1. Mathematical physics
I. Title
530.1'5 QC20

ISBN 0 521 25550 3 hard covers
ISBN 0 521 27553 9 paperback

AN

CONTENTS

ACKNOWLEDGEMENTS

I am very grateful to the many people who helped in the preparation of the typescript for publication. The bulk of the text was expertly typed by Janice Murray. Susan Hooper kindly typed Chapters 14 and 15 which were needed for undergraduate teaching. The line drawings were beautifully drawn by Marjorie Fretwell. The reduction of these diagrams to a size suitable for publication and the production of all the photographs in the book are the work of Brian Hadley and his colleagues of the Photolabs, Royal Observatory, Edinburgh. The staff of the Royal Observatory library were very helpful in locating references and also in releasing for photographing the many treasures in the Crawford Collection of old scientific books. The production of this book would have been impossible without the splendid help of all those mentioned above.

I must also acknowledge the stimulation provided by the three generations of undergraduates who attended this course when it was given several years ago. Their comments and enthusiasm are largely responsible for the fact that this book has appeared at all.

PREFACE

The origin of this book was a course of lectures which I gave between 1977 and 1980 to undergraduates about to enter their final year in Physics and Theoretical Physics at Cambridge. The course was delivered in the summer term prior to the final year of the physics course. The aim of this lecture course was originally to provide students with a broad outline of the nature of theoretical physics which would put them in a receptive frame of mind for the very intense courses of lectures on all aspects of physics in the final year. As our ideas evolved, it became apparent that the material was of benefit to all physics students and the course was entitled 'Theoretical Concepts in Physics'.

An important feature was that the course was entirely optional being held at 9 a.m. every Monday, Wednesday and Friday for four weeks and was strictly non-examinable. I confess that this was the course of lectures I most enjoyed giving of all my Cambridge lectures. I was very gratified by the positive response from the students. Despite the timing of the lectures, the fact that the course was not examinable and the alternative attractions of the summer months in Cambridge, the course was very well attended throughout. This encouraged me to think of producing a published version of the course because I was aware of no other book which covered this material in the same way. In addition, I felt very strongly about the problems that I identify in my first lecture. Many students go through an undergraduate course in physics without gaining the basic insights, attitudes and techniques which are the tools of the professional physicist, let alone an impression of the intellectual excitement and beauty of the subject.

Because of other pressures, I laid aside the idea of putting the course into book form. In addition, I was very wary of getting the history of science wrong and I wanted more time to produce a more integrated view of theoretical ideas in physics.

I had the benefit of giving other courses with a strong theoretical content in Cambridge, in particular a second-year Examples Class in Mathematical Physics in which we went through many basic parts of physics where the mathematics can provide illumination of basic concepts. I have included some of that material in the present book. Finally, I have always felt that thermodynamics is an especially difficult subject for undergraduates. Having given an introductory course on the subject in my last year in Cambridge, I saw a way of integrating this topic into my grand plan.

The book which has resulted is a highly individual approach to physics and theoretical physics. I emphasise that it is in no way a substitute for the systematic exposition of physics and theoretical physics as they are taught in standard courses. It should be regarded as a complementary, non-examinable volume which illuminates the subject from the view point of real physics and theoretical physics. If I succeed in even marginally improving students' appreciation of physics as professional physicists know and love it, I will consider the book worthwhile.

I have purposely maintained the first person singular to a much greater degree than is normal. This is very important because it emphasises the individuality of every physicist's approach to the subject. I also feel free to express my own opinions (and experiences) of how physics is actually done. I do not expect everyone to agree with my points of view but that is part of the fascination of the subject. No matter how we as individuals think about the subject, we eventually have to quantify our ideas and then, no matter how we arrive at it, we all ought to get the same answer.

The views expressed in the text are obviously all my own but many of my Cavendish colleagues played a major part in formulating and clarifying my ideas. The idea of the basic course came from discussions with Alan Cook, Volker Heine and John Waldram. I inherited the Examples Class in Mathematical Physics from Volker Heine and the late J.M.C. Scott. Working on that class helped enormously in clarifying many of my own ideas. In later years Brian Josephson helped with the course and provided many startling insights. The course in thermodynamics was given in parallel with one by Archie Howie and I learned a great deal from discussions with him. Two committees provided valuable insight into physics. First, there was the Department of Physics Teaching Committee of which I was a long-standing member. I have often thought that a video recording of some of the heated discussions about how to teach physics and theoretical physics would have taught students more about physics than a whole course of lectures. Second, I was chairman (or fall-guy) of the Staff–Student Consultative Committee for Physics, where I faced a highly intelligent set of consumers at all stages in their physics education.

Perhaps the biggest debts I owe in my education as a physicist are to Martin Ryle and Peter Scheuer who supervised my research work in the Radio Astronomy Group. I learned more from them about real physics than from anyone else. Another powerful influence was Brian Pippard whose penetrating understanding of physics was a profound inspiration. Although he and I have very different views of physics, there is virtually no aspect of physics which we have discussed in which his insight has not added immensely to my understanding.

As in all my work, my debt to my wife, Deborah, and our children, Mark and Sarah, is incalculable.

March 1983 Malcolm Longair
Edinburgh

1

INTRODUCTION

1.1 An explanation for the reader

This lecture course is for those students who love physics and theoretical physics. It arises from a dichotomy which I believe pervades most attempts to teach the ideal course in physics. On the one hand, there is the way in which university teachers present physics and theoretical physics in lecture courses and examples classes. On the other hand, there is the way in which we actually practise physics as professional research workers. In my experience, there is often very little relation between these two activities. This is obviously a great misfortune because students are rarely exposed to their teachers when they are practising their profession as physicists.

There are, of course, very good reasons why the standard lecture course has evolved into its present form. First of all, physics and theoretical physics are not particularly easy subjects and it is important to set out the individual elements of the disciplines in as clear and systematic a manner as possible. It is absolutely essential that all students acquire a very firm grounding in the basic techniques and concepts of physics. But we should not confuse this process with doing real physics. Lecture courses in physics and theoretical physics are basically 'five-finger' exercises and five-finger exercises bear little relation to a performance of the Hammerklavier sonata at the Royal Festival Hall. You are only doing physics and theoretical physics when the answers *really* matter – in other words, when your reputation as a scientist hangs upon being able to do physics correctly in a research context or, in more practical terms, when your ability to reason correctly determines whether you are employable or whether your research grant is renewed. This is quite different from working through drill exercises for which answers are available at the back of the book.

Second, there is just so much material which lecturers feel they have to get through that all physics syllabuses are absolutely crammed full and there is little room for sitting back and asking 'What is all this about?' Indeed, one becomes so preoccupied with the technical aspects of the subject which are themselves fascinating that one generally leaves it up to the students to find out for themselves many essential truths about physics.

Let me give a list of the things which can be missed in our teaching and which I believe are essential aspects of the way in which physicists and theoretical physicists actually do physics.

(i) A course of lectures is by its nature a modular exercise. It is only too easy to lose *a global view* of the whole subject. Professionals use the whole of physics in tackling problems and there is no artificial distinction between thermal physics, optics, mechanics, electromagnetism, quantum mechanics, etc.

(ii) A corollary of this is that in physics, any problem can normally be tackled in a variety of different ways. There is *no single 'best way' of doing a problem.* You obtain a much deeper insight into how physics works if you approach a problem from totally different stand points, e.g. from thermodynamics, from electromagnetism, from quantum theory, etc.

(iii) Again, how one tackles a problem and talks about physics is a highly individual business. No two physicists think in exactly the same way about physics although, when they come to write down the relevant equations and solve them, they should come to the same answers. The *individual lecturer's response to the subject* is an integral part of the way physics is taught to a much greater extent than students or the teachers themselves would like to believe. But it is the diversity of different lecturers' approaches to physics which provides insight into the nature of the mental processes by which physicists understand their subject.

(iv) Another potential victim of the standard lecture course is an appreciation of what it feels like to be involved in *research science at the frontiers of knowledge.* Lecturers are always at their best when they reach the part of the course where they can slip in the things which excite them in their research work. For a few moments, the lecturer is transformed from a teacher to a research scientist and then the students see the real physicist at work.

(v) Again, it is often difficult to convey the sheer *excitement of the processes of research and discovery in physics* and yet this is the very reason that most of us get so enthusiastic about our research. It is the reason why many of us spend more hours pursuing our research work than would be expected in any normal 'job'. The well-known caricature of the 'mad' scientist is not wholly a myth in that, in doing frontier research, it is almost essential to become totally absorbed in the problem to the exclusion of the cares of normal life. A really important problem can absorb one totally. I recently had a note from a colleague who had just discovered the most distant object in the Universe. In it she stated that the thing which kept her awake at night was the question of whether or not there are even more distant objects and how to find them. When we are involved in issues of this nature, they become all absorbing and it is only later on looking back, that we regard these as among our best research experiences. Yet some students can complete a physics course without even being aware of what it is that drives us on.

(vi) Much of this can be conveyed through examples selected from the *history* of some of the great scientific discoveries. Yet, this seldom appears in our courses. First of all, there is just not time to include the material. Second, it is not so easy to be sure of all the relevant historical facts. Third, the history and philosophy of science is taught as a wholly separate discipline from physics and theoretical physics. A modest appreciation of some historical case studies can provide valuable illumination on the processes of discovery and of the intellectual background to them. In these historical case studies, we can recognise parallels with our own research experience.

(vii) Key factors that come through in these historical examples, which are familiar to all professional physicists, are the central roles of *hard work, experience* and, perhaps most important of all, *intuition.* Many of the most successful physicists depend very heavily upon their wide experience of a great deal of hard work in physics and theoretical physics. It would be marvellous if experience could be taught but I am convinced it is something that students can only discover for themselves by dedicated hard work. We all remember our mistakes and they teach us more about physics than many of our successes. As for intuition, I regard this as the distillation of all our experience as physicists. It is a very dangerous tool because one can make some very bad blunders by relying on it too much in a frontier area of physics. Yet it is certainly the source of many of the greatest discoveries in physics. These were not achieved by using the five-finger exercise techniques but involved leaps of the imagination which transcended known physics.

(viii) We are coming very close to what I regard as the central core of our experience as physicists and theoretical physicists. There is an essential element of *creativity* which is not so different from creativity in the arts. The creative leaps of the imagination involved in discovering the laws of motion, Maxwell's equations, relativity, quantum theory, etc. are no different in essence from the creations of the greatest artists, musicians, etc. The basic difference is that physicists must be creative within a very strict set of rules and that their theories must be testable by confrontation with experiment and observation. Now very few of us indeed attain the almost superhuman level of intuition involved in discovering a new physical theory. Most of our work is on a rather more mundane level but we are all driven by the same basic creative urge. Each small step we make contributes to the sum of our understanding of the nature of our physical universe. All of us in our own way tread in regions where no one else has trodden.

(ix) The result of this creativity is unquestionably a real sense of *beauty* about the greatest of physics and theoretical physics. Some of the great achievements of physics evoke in me, at least, the same sort of response that one finds with great works of art. I suspect that many of us feel the same way about physics but are normally too embarrassed to admit it. It is a pity because the achievements of physics and theoretical physics rank among the very peaks of human achievement. I think that it is very important to tell students when I find a piece of physics particularly beautiful – and there are lots of examples of such pieces. I find that I go through the same process of rediscovery that I go through when rehearing a familiar piece of classical music – my hundredth performance of the Eroica symphony of Beethoven or of Stravinsky's 'Sacre du Printemps'. I am sure that students should know about this.

(x) Finally, physics is *great fun.* The standard course with its concentration on technique can miss so much of the enjoyment and stimulation of the subject. Students should realise that physicists actually enjoy physics and that it is a very rewarding pursuit.

This book is an alternative approach to theoretical reasoning in physics in that it emphasises items (i) to (x) rather than attempts a systematic exposition of theoretical physics. The history of how I came to give this course is of some interest in that it exposes other aims which are wholly consistent with the above objectives.

1.2 The origin of the present lecture course

The origin of this lecture course arose from a feeling among a number of the members of the Physics Department at Cambridge who were involved in teaching undergraduate courses which were theoretically biased that the syllabuses lacked coherence from the theoretical point of view and that students were not quite clear exactly what theoretical physics is. What exactly is the scope of 'physics' as opposed to 'theoretical physics'? Are they really so different?

As our ideas evolved, it became apparent that a discussion of these ideas would be of value to all final-year students. The course which was entitled 'Theoretical Concepts in Physics' was given in the summer term in July and August to students entering their final year. The course was strictly non-examinable and entirely optional. Students obtained no credit from having attended the course beyond an increased awareness of physics and theoretical physics. I was very fortunate in being invited to give this course of lectures for the first time.

The basic aims of the course were initially the following:

(*a*) *The interaction between experiment and theory.* Particular stress was to be laid upon the importance of new technology in leading to theoretical advances.

(*b*) The importance of having available the appropriate *mathematical tools for solving a theoretical problem.* Sometimes the mathematics runs ahead of the physics, sometimes the mathematics is not yet available and has to be invented.

(*c*) *The theoretical background of the basic concepts of modern physics.*

(*d*) *The role of approximations and models in physics.*

(*e*) *The analysis of real scientific papers in theoretical physics.*

(*f*) *The underlying themes of theoretical physics – symmetry, conservation, invariance, etc.*

In pondering how to achieve these aims, I decided to approach the topics through a series of case studies designed to illuminate different aspects of physics and theoretical physics. The choice was entirely personal but I designed it so that I would achieve another aim which was:

(*g*) *The consolidation and revision of all the basic physical concepts which I expect all final-year undergraduates to have at their fingertips.*

Finally, I wanted:

(*h*) *To convey my own personal enthusiasm for physics and theoretical physics.* Although I am now professionally employed as an astronomer, I remain a physicist at heart and, to be entirely honest, I regard astronomy and astrophysics as a sub-set of physics but applied to the Universe on a very large scale. My own enthusiasm results from being involved in astrophysical and cosmological research problems at the very limits of our understanding of the Universe. Physics is not a dead, pedagogic subject, the only object of which is to provide examination questions for students. It is an active subject in a robust state of good health and indeed is at present enjoying another of those exciting epochs when fundamental new understandings are being gained.

In producing a published version of this lecture course, I have expanded the contents somewhat in order to provide more complete coverage of undergraduate physics. I have included material from examples classes in mathematical physics and a course I gave on thermodynamics.

1.3 A warning to the reader

The reader should be warned of two things. First, this is an *entirely personal view of the subject.* It is quite intentionally designed to emphasise items

(i) to (x) and (*a*) to (*h*) – in other words, to emphasise all those aspects which tend to be squeezed out of physics courses because of lack of time.

Second, and even more important, this set of lectures is not a text book on physics and theoretical physics. It is *in no way* a substitute for the systematic development of these subjects through the standard physics courses. You should regard this book as an optional extra but one which I hope may contribute something towards your understanding and appreciation of physics.

1.4 What is theoretical physics?

Let me begin by making a formal statement about the basis of all our work. The natural sciences aim to give a logical and systematic account of natural phenomena and to enable us to predict from our past experience to new circumstances. Theory is the formal basis for such arguments.

Theory need not be mathematical but mathematics is the most powerful and general method of reasoning which we possess. Therefore we attempt to secure *data* wherever possible in a form that can be handled *mathematically*. There are two immediate consequences for theory in physics.

The basis of all physics and theoretical physics is *experimental data* and the necessity that these data be in *quantified form.* Some people would like to believe that the whole of theoretical physics could be produced by pure reason. They are doomed to failure from the outset. The great achievements of theoretical physics have been solidly based upon the achievements of experimental physics. Experiment provides the only constraint upon physical theory. Every theoretical physicist should therefore have a firm understanding of the methods of experimental physics not only so that he can check his theory but also so that he can propose experiments which will be realistic and which can discriminate between rival theories.

The second consequence is that we must have adequate *mathematical tools* with which to tackle the problems we wish to solve. Historically, the mathematics and the experiments have not always been in step. Sometimes the mathematics is available but the experimental situation is unclear. In other cases, the opposite has been true – new mathematical tools have been necessary to produce a properly quantified theory.

The mathematics is of course central to reasoning in theoretical physics but we must beware of treating it as being the whole content of theory. Let me reproduce some words from the reminiscences of Dirac about his attitude to mathematics in theoretical physics. First of all, it should be noted that Dirac sought mathematical beauty in all his work. For example, he says

> Of all the physicists I met, I think Schrodinger was the one that I felt to be most closely similar to myself. . . . I believe the reason for this is that

> Schrodinger and I both had a very strong appreciation of mathematical beauty and this dominated all our work. It was a sort of act of faith with us that any equations which describe fundamental laws of Nature must have great mathematical beauty in them. It was a very profitable religion to hold and can be considered as the basis of much of our success.[1]

On the other hand, earlier he writes the following:

> I completed my course in engineering and I would like to try to explain the effect of this engineering training on me. Previously, I was interested only in exact equations. It seemed to me that if one worked with approximations there was an intolerable ugliness in one's work and I very much wanted to preserve mathematical beauty. Well, the engineering training which I received did teach me to tolerate approximations and I was able to see that even theories based upon approximations could have a considerable amount of beauty in them . . . .
>
> There was this whole change of outlook and also another which was perhaps brought on by the theory of relativity. I had started off believing that there were some exact laws of Nature and that all we had to do was to work out the consequences of these exact laws. Typical of these were Newton's laws of motion. Now we learned that Newton's laws of motion were not exact, only approximations and I began to infer that maybe all the laws of nature were only approximations . . . .
>
> I think that if I had not had this engineering training, I should not have had any success with the kind of work I did later on because it was really necessary to get away from the point of view that one should only deal with exact equations and that one should deal only with results which could be deduced logically from known exact laws which one accepted, in which one had implicit faith. Engineers were concerned only in getting equations which were useful for describing nature. They did not very much mind how the equations were obtained . . . .
>
> And that led me of course to the view that this outlook was really the best outlook to have. We wanted a description of nature. We wanted the equations which would describe nature and the best we could hope for was usually approximate equations and we would have to reconcile ourselves to an absence of strict logic.[2]

These are very important and profound sentiments and ones which I hope are now familiar to you. There is really no strictly logical way in which we can formulate theory – we are continually approximating and using experiment to keep us on the right track. You should note of course that Dirac was talking about theoretical physics at its very highest level – concepts like Newton's laws of motion, special and general relativity, Schrodinger's equation and the Dirac equation are *the very summits of achievement of theoretical physics* and very few of us can work at that sort of level. However, the same sentiments apply at the somewhat less elevated level where we are all burrowing away in our own little corner of the woods.

Most of us are concerned with *applying known laws to physical situations* in which so far their application has not been possible or foreseen and very often we make myriads of approximations to make the problem tractable at all. The essence of our training as physicists is to build up our confidence in our understanding of physics so that, when we are faced with a completely new problem, we can use this understanding to recognise the most fruitful ways of approaching it.

1.5 The influence of our environment

It is important to realise not only that all physicists are individuals but also that their approaches are strongly influenced by the tradition within which they themselves have studied physics. This applies to individual physics departments as well as to different countries where one can identify particular scientific traditions and approaches. I have personally had experience of working in a number of different countries, especially in the USA and the USSR, and I can recognise the distinctive flavour of the way in which physics and theoretical physics are practised. I have found that this has added greatly to my understanding and appreciation of physics.

If we talk simply about theoretical physics, there is a very wide range of opinion as to what constitutes 'theoretical physics' as opposed to 'physics'. In fact, in the physics department in Cambridge where this course was first given, most of the courses were very strongly theoretically biassed. By this I mean that most courses aimed to provide students with the basic theory and its development and paid relatively little attention to matters of experiment. If experiments were alluded to, the emphasis was upon the results rather than the experimental ingenuity by which the physicists came to their answers. This is somewhat unfortunate because as soon as you come to the real world, you have to take experimental data very seriously and you must be able to judge from a scientific paper whether or not you can trust the result as stated.

On the other hand, members of departments of theoretical physics or applied mathematics believe that they teach much more 'theoretical' theoretical physics than we did. In their undergraduate teaching, I believe that this is certainly true. There is by definition a very strong mathematical bias in the teaching of these departments and they are often much more rigorous in their use of mathematics than we were – or rather, they worried much more about rigour than we did. In other physics departments, the bias was often towards experiment rather than theory. It was rather amusing that a number of members of staff of the Cambridge Physics Department who were considered to be 'experimentalists' within the department were regarded as 'theoreticians' by all the other physics departments in the country!

The reason for discussing this question of environment is that it can produce a somewhat distorted view of what we mean by physics and theoretical physics. My own view is that physics and theoretical physics should not be considered as separate subjects. They are only different ways of looking at the same body of material and there are great advantages in developing one's mathematical models in the context of the experiments or at least in an environment where one can have day-to-day contact with those involved in the experiments.

A corollary of the fact that there are generally a number of ways of looking at physical problems is that different physicists 'understand' physics in different ways. You will have to decide how you best understand the concepts of physics. Among my colleagues there are some who have absolutely no intuitive feeling for physics but have to write down equations to understand anything. Others work entirely in terms of physical insight and are able to produce general answers to complicated problems in a few lines. There is not any real sense in which one is better than the other. They both arrive at the solution in the end. It is just that their methods of approaching physics are completely different. Remember that, when you hear different lecturers describing what sounds like material you have heard before, they are looking at the subject from their own points of view. Every separate presentation will add something to your understanding of physics.

An example of a distinctively British feature of physics is the tradition of model building to which we will return on a number of occasions. Model building seems to have been an especially British trait during the 19th and early 20th centuries and something of this certainly remains today. I confess that when I think about physics I generally have some picture in my mind of what I am thinking about rather than an abstract or mathematical idea.

The works of Faraday and Maxwell are full of models and at the turn of the present century, the variety of models for atoms was quite bewildering. The 'plum-pudding' model of the atom, which is perhaps one of the more vivid examples of model building, is just the tip of the iceberg. J.J. Thomson was quite straightforward about the importance of model building:

> The question as to which particular method of illustration the student should adopt is for many purposes of secondary importance provided that he does adopt one.[3]

This is splendidly illustrated by the set of lecture notes by Heilbron on *Lectures on the History of Atomic Physics 1900–1920.*[4] This approach is very different from the continental European tradition of theoretical physics – we find Poincaré remarking that in his experience all Frenchmen were oppressed by a 'feeling of discomfort, even of distress' at their first encounter with the works of Maxwell.[5] According to Hertz, Kirchhoff was heard to remark that he found it painful to see atoms and their vibrations wilfully stuck in the middle of a theoretical discussion.[6] This indicates clearly differences in opinion as to what does or does not constitute theoretical physics.

When I was at the Cavendish Laboratory, we attached great importance to the development of *physical insight.* I believe that this is part of the British model-building tradition. What we aimed to develop was the ability to guess what will happen in a given physical situation without having to write down all the mathematics. This is a very important talent and most of us learn it with time. However, it is mostly a matter of experience and this you can only acquire by hard work. It is also important to emphasise that having physical insight is no substitute for producing precise answers. If you wish to claim to be a theoretical physicist, you must be prepared to give the correct quantitative analysis as well.

1.6 The contents of the lecture course

The case studies through which I have endeavoured to accomplish my aims are each designed to cover major advances in theoretical physics, and, in the process, to look at the physics that you have already met from a totally different viewpoint. This is why the subtitle of this book is 'an alternative view of theoretical reasoning in physics . . .'. What I am aiming to do is to recreate the intellectual processes by which some of the greatest discoveries in theoretical physics came about and to draw from these historical case studies some insight into how real physics and theoretical physics are done. It will, I hope, convey some of the excitement and intense intellectual struggle which is involved in achieving new physical understanding. The seven major case studies are as follows.

1. The origins of Newton's law of gravitation.
2. Maxwell's equations.
3. Mechanics and dynamics.
4. Thermodynamics and statistical mechanics.
5. The origins of the concept of quanta.
6. Special relativity.
7. General relativity and cosmology.

In most cases, we will go through the processes of discovery by the same route which was followed by the scientists themselves. We will use only mathematical techniques available to scientists at the time. For example, we cannot cut corners by assuming we can represent electromagnetic waves by photons until after Case Study 5. In this process, we will revise many of the basic concepts of physics with which you should be familiar. I will also make a number of personal comments about the material, largely drawn from my own experience of research and teaching.

1.7 Apologies and words of encouragement

Let me make it clear that I am *not* a historian or philosopher of science. I am using examples from the history of science very much for my own purposes which are to illuminate my own experience of how real physicists think and behave. The use of historical case studies is simply a device for conveying something of the reality and excitement of physics. Maybe one day, I will have the time to rewrite the historical sections from a more erudite point of view but it might lack the conviction which impels me to set these ideas down on paper. I therefore apologise unreservedly to historians and philosophers of science for abusing subjects for which I have the most profound respect. It may be that students will gain a healthier respect for the works of professional historians of science from what they read in this book.

Establishing the intellectual processes by which scientific discoveries were made is a hazardous business and, even in the recent past, it is often difficult to disentangle the true story. In my background reading, I have relied heavily upon standard biographies and histories. For me, they provide vivid pictures of how science actually works and I can relate them to my own research experience. If I have erred in some places, my exculpation can only be by the words attributed to Giordano Bruno, 'Si non e vero, e molto ben trovato' (if it is not true, it is a very good invention).

This book was originally intended for prospective theoretical physicists, but is really for all undergraduates in physics. All of you should be able to get something from this course, even if you hold the discipline of theoretical physics in low esteem. An interesting example of someone who took little notice of theoreticians is Stark who made it a point of principle to reject almost all theories on which his colleagues had reached a consensus. Contrary to their view, he showed that spectral lines could be split by an electric field, the Stark effect, for which he won the Nobel prize.

Finally, you are meant to enjoy this. I am trying to put you into the proper frame of mind for appreciating all the theory which will be thrown at you during the final year of an undergraduate lecture course. I particularly want to get across an appreciation of the really great discoveries of physics and theoretical physics. These are achievements as great as any in any field of human endeavour.

Case Study 1

THE ORIGINS OF NEWTON'S LAW OF GRAVITATION

Tycho Brahe (1546–1601)[1]

Johannes Kepler (1571–1630)[2]

Isaac Newton (1642–1727)[3]

[1] From an old lithograph belonging to the Royal Observatory, Edinburgh.
[2] From the frontispiece of *Johannes Kepler – Gesammelte Werke, Vol. 1,* ed. M. Caspar, 1938, Beck, Munich.
[3] From the frontispiece of *The Correspondence of Isaac Newton, Vol. 1,* ed. H.W. Turnbull, 1959, Cambridge University Press.

2

TYCHO BRAHE, KEPLER AND NEWTON – THE ORIGINS OF NEWTON'S LAW OF GRAVITATION

2.1 Introduction

We begin our story with a classical example of the strong interaction between technological innovation and consequent advances in theoretical understanding. It also illustrates how the appropriate mathematical techniques must be available before the full content of the theory can be understood. We have to look into the history of the century preceding Newton's great discovery and it is, of course, very difficult to look back to these heroic, pioneering days without the dubious benefits of hindsight.

The period around 1600 is conventionally regarded as the epoch which marks the beginnings of modern science as we know it today. Probably more than any other development, it was Galileo's understanding of the nature of acceleration and inertia that marks the beginning of mathematical physics as applied to motion. This is not to underestimate the achievements of the Greek astronomers and mathematicians who devised models of ingenuity and elegance to account for the motions of the planets on the sky. These remarkable models, however, did not lead to any deeper understanding of the origin of these motions.

As far as quantitative science was concerned, observation was generally much more accurate than experiment. This is exemplified by the fact that there existed tables, called the Alphonsine and Prutenic Tables, which enabled eclipses and the positions of the planets to be predicted. It was the inaccuracies which he found in these tables in predicting a conjunction of Saturn and Jupiter in August 1563 that has been regarded as the turning point in the career of Tycho Brahe. His determination to improve their accuracies led ultimately to the great discoveries of Kepler and Newton.

2.2 Tycho Brahe and the observatory on the island of Hven

In 1543, Copernicus published his great work *De Revolutionibus Orbium Celestium.* On the basis of observations of the Sun and the planets available at that time, he proposed his revolutionary concept that the planets move in circular orbits around the Sun (Figure 2.1). With hindsight, we recognise that Copernicus had stumbled across the right answer but, at that time, his ideas did not gain any measure of acceptance and indeed met with strong opposition and hostility from the church authorities.

Tycho Brahe was born in 1546 and developed an early interest in astronomy and the motions of the planets. In 1572, he observed the bright supernova, or exploding star, which bears his name. These early visual observations were of sufficient accuracy for it to be possible to deduce the rate of decrease of the brightness of the supernova with time from Tycho's records. Throughout his youth and early manhood, he constructed instruments to measure accurately the positions of the stars and planets. His great scientific importance resulted from the fact that he eventually found himself in a position to be able to measure the positions of the Sun, planets and stars more accurately and systematically than anyone before.

Figure 2.1. The Copernican universe from Copernicus' treatise *De Revolutionibus Orbium Celestium*, 1543, opposite p. 10, Nurnberg. (From the Crawford Collection, Royal Observatory, Edinburgh.)

currens, In medio uero omnium residet Sol. Quis enim in hoc

pulcherimo templo lampadem hanc in alio uel meliori loco po
neret, quàm unde totum simul possit illuminare? Siquidem non
inepte quidam lucernam mundi, alij mentem, alij rectorem uo-

It was fortunate that his abilities as a scientist and astronomer were recognised early and that King Frederick II of Denmark was passionately interested in the sciences and literature. In 1576, he provided Tycho with 'our land of Hven with all our and the Crown's tenants and servants who thereon live, with all the rent and duty which comes from that . . . as long as he lives and likes to continue and follow his *studia mathematica* . . . '[1] What this meant was that he was granted the Island of Hven by Frederick, an island which lies off the coast of Denmark and Sweden. Besides the material and agricultural resources of the island, Tycho was granted essentially unlimited financial support with which to build an observatory and equip it with the most advanced instruments of its time. In Tycho's own words, he believed that the total enterprise had cost Frederick 'more than a tun of gold'.[2] Some estimates suggest this must have amounted to about 5–10% of the gross national product of Denmark at that time. He employed a sizeable staff to help make the observations and then reduce them to a set of tables of planetary, lunar and solar motions.

Tycho's methods will be recognised by any modern research worker as a model of how to undertake a major research project in the observational or experimental sciences. He designed and had built the most accurate instruments of the period, including a great mural quadrant of radius $6\frac{3}{4}$ feet (Figure 2.2). The sights moved round the arc of the quadrant and stars were observed precisely as they transitted through the hole in the wall which was located at the centre of the circular sector of the quadrant. This was by far the most important instrument because, being large and firmly fixed in one place, it was capable of the greatest accuracy. In the background of the picture, behind Tycho, are seen some of the other instruments of his observatory, as well as the basements in which chemical experiments were carried out. Tycho, in fact, built two observatories, one named Uraniborg and the other Stjerneborg. The assistants at the two observatories had to make independent measurements of the positions of the stars and planets and only Tycho himself was allowed to compare their results.

Technically, he made a number of notable advances in his procedure for measuring the positions of objects on the sky. He devised methods for taking account of the flexure of his instruments when they were pointed in different parts of the celestial sphere. He took out the effects of refraction when stars are observed at different angles above the horizon. Thus, he understood the importance of eliminating systematic errors in his observations. Even more important was the fact that, not only did he come up with by far the most accurate observations of his time, he also quoted the error on his observations. The last feature of his programme which was vital to its success was the fact that the observations were made systematically over a period of 20 years. We recognise in Tycho's approach to observational science all the very best attributes of modern experimental procedures.

The net result was a catalogue of star positions with much improved precision over anything which had been available before. Positions were measured for 777 stars with an accuracy of about 1–2 arcmin, roughly an order of magnitude

Figure 2.2. Tycho Brahe with the instruments he constructed for observation of accurate positions of the stars and planets. He is seated within his 'great mural quadrant' which produced the most accurate measurements of the positions of stars and planets at that time. (From *Astronomiae Instauratae Mechanica*, 1602, p. 20, Nurnberg, the Crawford Collection, Royal Observatory, Edinburgh.)

improvement over what was previously available. This amounts to about $\frac{2}{3}$ of all the stars visible to the naked eye from Denmark. In addition, he accumulated more or less daily observations of the Sun and a complete set of positions for the Moon and planets.

There is no question but that his greatest achievements were as a technologist and experimenter. He had a real interest in theory and developed his own model for the Solar System which apparently reconciled the geocentric and Copernican pictures of the kinematics of the Sun, Moon and planets (Figure 2.3). In this, the planets indeed move in circular orbits about the Sun but the Sun itself moves in a circular orbit about the Earth which maintains its fixed central position in Tycho's cosmology.

In 1597, he was forced to leave Hven after a major dispute with Frederick's successor and took up residence in Benatek close to Prague. There, just before his death in 1601, he employed as an assistant one of the most brilliant young

Figure 2.3. Tycho's universe from his book on the comet of 1577, *De Mundi Aetherei Recentioribus Phaenomenis – De Cometa Anni 1577*, 1610, p. 191, Frankfurt. (From the Crawford Collection, Royal Observatory, Edinburgh.)

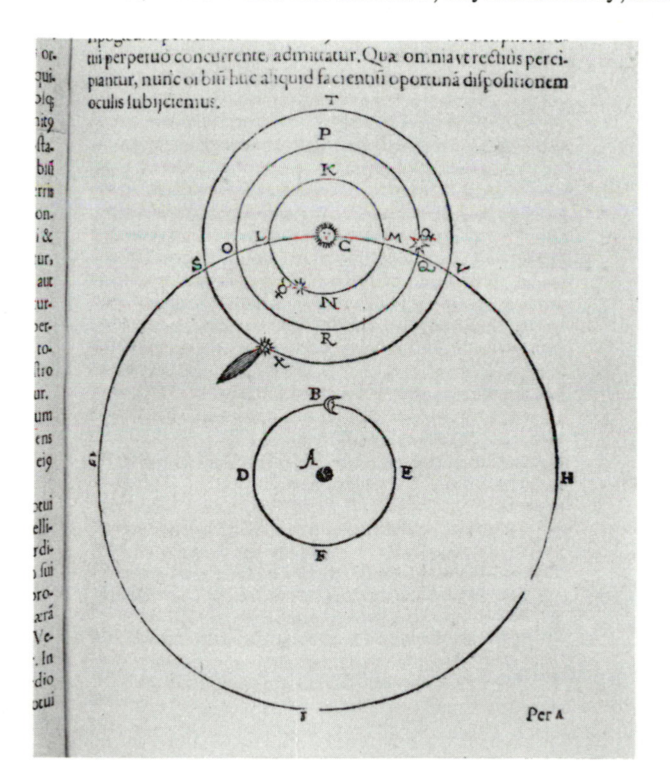

mathematicians in Europe to work on the analysis of his observations which he had ensured travelled with him to Prague. His name was Johannes Kepler.

2.3 Kepler's laws of planetary motion

Even during his university studies, Kepler was an enthusiast for the Copernican model of the Solar System under the instruction of Maestlin who was himself a rather cautious Copernican. By the time he came to work for Tycho Brahe in 1600, he had already made significant contributions to mathematics and astronomy through the publication of his short treatise *Mysterium Cosmographorum* in which the radii of the planetary orbits were related to regular geometric figures. Shortly before Tycho died in late 1601, he set Kepler to work on the orbit of the planet Mars, and following his death, Kepler succeeded him in the post of Imperial Mathematician.

To begin with, Kepler assumed circular orbits and soon established that the motion of Mars on the sky could not be explained if the motion were referred to the centre of the Earth's orbit. Rather, the motion had to be referred to the true position of the Sun. The arduous analysis proceeded with the fitting of circular orbits to the motion of Mars with the Sun displaced from the centre of the circle. After a great deal of trial and error, the best that could be achieved was agreement within about 6 to 8 arcmin with the observations of Tycho. In Kepler's own words, 'Divine Providence granted us such a diligent observer in Tycho Brahe that his observations convicted this Ptolemaic calculation of an error of 8 arcmin; it is only right that we should accept God's gift with a grateful mind . . . . Because these 8 arcmin could not be ignored, they alone have led to a total reformation of astronomy'.[3] In other words, this measure of disagreement was unacceptable and he had to start again. Kepler's rejection of circular orbits is a key step in this study – we note that Kepler's rejection was based on a disagreement which is only about four times the standard error of Tycho's observations. The crucial importance of the knowledge of the precision of Tycho's observations and the fact that they could be trusted is apparent.

Kepler's next attempts were based upon the use of ovoids in conjunction with a magnetic theory for the origin of the forces which drive the planets in their orbits. According to his model, the force should decrease inversely with distance and, in the same way, the velocity of the planet should be inversely proportional to distance. Kepler proposed to use this law as a measure of the velocity of the planet in its orbit. This led to difficulties in working out the apparent motion of the planets on the sky and consequently he adopted, intuitively, an alternative idea that the motion should be such that *equal areas* are swept out in *equal times*, the areal law. This hypothesis resulted in excellent agreement in predicting the longitudes of planets and, in particular, gave excellent agreement with the Earth's orbit about the Sun. Remarkably, there is no reference to this great discovery which we now know as *Kepler's second law of planetary motion* in his next great study, that of the orbit of Mars. The reason probably was that Mars did not appear to obey the areal law and we now understand that this was because Kepler had not yet described the orbit by an ellipse.

Kepler proceeded with the mammoth task of trying to fit ovoids and the areal law to the motion of Mars but could not obtain exact agreement, the minimum discrepancy amounting to about 4 arcmin. In parallel with his researches, he was writing his great treatise *A Commentary on the Motion of Mars* and he reached Chapter 51 before he realised that what was needed was a figure intermediate between ovoids and circles, the ellipse. He soon arrived at the crucial result that the planetary orbits are ellipses with the Sun in one focus. The treatise was renamed *The New Astronomy* with the subtitle *Based on Causes, or Celestial Physics* and was published in 1609, four years after his discovery of what we now know as *Kepler's first law of planetary motion.*

The third law of planetary motion appeared much later during work on his treatise *The Harmony of the World* in 1619. The various sections of this work are devoted to geometry, music, astrology and astronomy. In common with other philosophers of the period, he was profoundly impressed by the origins of musical harmonies in the just scale, i.e. the fact that the ratios of frequency of vibration to produce the intervals in the scale are $\frac{1}{2}$, $\frac{2}{3}$, $\frac{3}{4}$, $\frac{4}{5}$, $\frac{5}{6}$, $\frac{5}{3}$ and $\frac{5}{8}$. He had already shown as early as 1596 in his *Mysterium Cosmographorum* how planetary orbits could be related to symmetric solids. There are only five regular polyhedra and, in his own words, his model is as follows.

> The Earth's orbit is the measure of all things; circumscribe around it a dodecahedron and the circle containing this will be Mars; circumscribe around Mars a tetrahedron, and the circle containing this will be Jupiter; circumscribe around Jupiter a cube, and the circle containing this will be Saturn. Now inscribe within the Earth an icosahedron, and the circle contained in it will be Venus; inscribe within Venus an octahedron, and the circle contained in it will be Mercury. You now have the reason for the number of the planets.[4]

This model is shown in Figure 2.4. Remarkably, it could account for the relative radial distance of the planets from the Sun to within about 5%.

It is natural that he should seek relations between harmonic phenomena in music and in the heavens. He associated certain musical series with the planets (Figure 2.5) and discussed their astrological significance. The mystical element in *The Harmony of the World* was very much part of Kepler's intellectual make up and indeed latterly his income largely depended upon the production of horoscopes. By 1619, he was no longer satisfied with an accuracy of 5% in accounting for the radial distances of the planets from the Sun, which was the best that could be achieved using the model of nested polyhedra. Whilst trying to improve the accuracy of his model, he discovered the *third law of planetary motion* – that the period of any planet about the Sun is proportional to the

Figure 2.4. Kepler's model of nested polyhedra which he developed to account for the number of planets and their radial distances from the Sun. (From *Dictionary of Scientific Biography, Vol. VII*, 1973, p. 292, after Kepler's original drawing in his *Mysterium Cosmographorum*, 1596.)

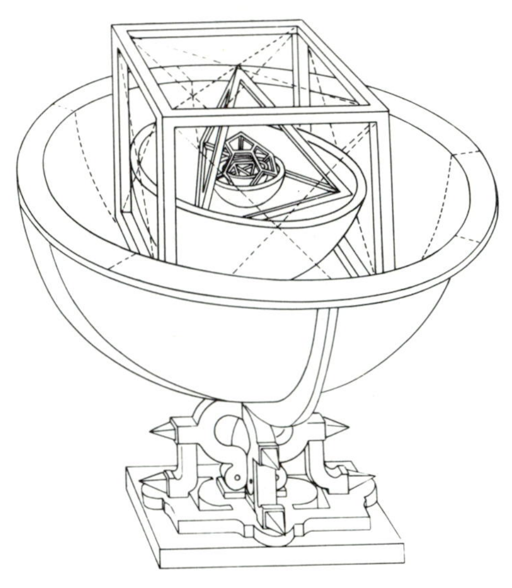

Figure 2.5. The 'music of the spheres' from Book V of Kepler's treatise *The Harmony of the World*, 1619, p. 207, Linz. (From the Crawford Collection, Royal Observatory, Edinburgh.)

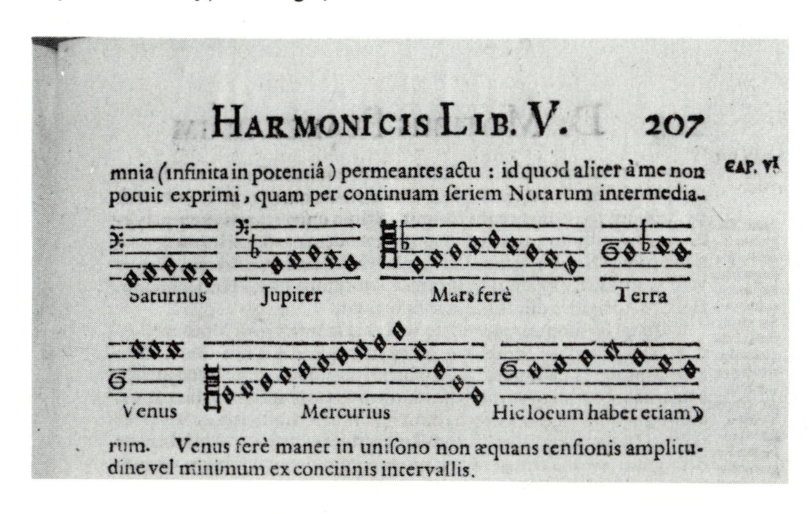

HARMONICIS LIB. V. 207

mnia (infinita in potentiâ) permeantes actu : id quod aliter à me non   CAP. VI
potuit exprimi, quam per continuam ſeriem Notarum intermedia-

rum. Venus ferè manet in uniſono non æquans tenſionis amplitudine vel minimum ex concinnis intervallis.

$\frac{3}{2}$ power of its average distance from the Sun. It is remarkable that this discovery of the greatest physical significance should lie buried in a welter of mystical harmonic speculation.

We should draw some important lessons from the story of Kepler's discovery of the three laws of planetary motion. First of all, it is apparent that Kepler's technical ability as a mathematician was of the highest order. All the calculations were carried out geometrically and it may be worth pondering how you would go about determining the orbits of planets if you were only supplied with Tycho's data and a ruler and compasses. I think that this is a tremendous achievement. Notice that Kepler possessed the mathematical tools needed to interpret Tycho's data. The second point of interest is the way in which he was guided by intuition, first in his discovering the areal law and then the ellipses which led to the first law. There is no logical way in which some of these leaps of the imagination come about. Rather, I imagine he was in the state of total absorption in the problem so that once the inspiration struck, he immediately recognised its importance and could follow through with the detailed analysis. Many of the greatest theoretical advances come through this state of total absorption in which one is so immersed in the problem and all its aspects that it only needs the right trigger to make everything fall into place. Most of us experience this in a modest way once or twice in a research career if we are lucky.

Another interesting aspect of Kepler's achievement was Galileo's reaction to his discoveries. It might be thought that the support of Kepler, an ardent Copernican, would have been invaluable to Galileo during his prosecution in 1633 for advocating the Copernican concept that the Earth and the planets orbit the Sun. Galileo was rather cautious about Kepler's support. Two years after Kepler's death, Galileo wrote 'I do not doubt but that the thoughts of Landsberg and some of Kepler's tend rather to the diminution of the doctrine of Copernicus than to its establishment as it seems to me that these (in the common phrase) have wished too much for it . . .' .[5] A glance at some of Kepler's purple prose will indicate the nature of Galileo's worries, as well as the fact that the solid achievements of mathematical physics were mixed up with heavenly harmonies and mysticism which could not have done Galileo's case any good. A contemporary analogy might be that it is difficult to be taken very seriously as a scientist if you mix up real science with quasi-scientific pursuits such as spoon-bending, parapsychology, unidentified flying objects, extrasensory perception, etc.

Another reason for Galileo's scepticism, which is equally interesting psychologically, is illustrated by his remark 'It seems to me that one may reasonably conclude that for the maintenance of perfect order among the parts of the Universe, it is necessary to say that movable bodies are movable only circularly'.[6] The fact that the orbits of the planets were ellipses rather than circles was intellectually repugnant to Galileo. Whilst we recognise now that this was a personal prejudice rather than a physical argument, we should not disguise the fact that we all make similar judgements in analysing physical problems. We will always seek the simplest, most elegant solution unless we are absolutely convinced it is inadequate. There is no question but that the mathematics of ellipses is much more complicated than that of circles.

At this stage, the profound importance of Kepler's laws was not appreciated. It was the genius of Newton which put the laws on a firm theoretical basis and in the process led to the discovery of the inverse square law of gravitation.

2.4 Newton and the law of gravitation

The last figure to appear in this story is Isaac Newton who was born in 1643, more than 20 years after Kepler's discovery of the third law of planetary motion. There is evidence from his student's notebooks of 1664 that he had learned of Kepler's first and third laws from the volume *Astronomia Carolina* by Streete which was published in 1661. In his notebooks of that year, he clarified the concept of inertia and derived for the first time the correct expression for centrifugal force,

$$f \propto v^2/r \tag{2.1}$$

Even at this early date, he propounded the idea of an inverse square law of gravitation which he derived from Kepler's third law and his own expression for centrifugal force. Kepler's law states that

$$T \propto r^{\frac{3}{2}} \tag{2.2}$$

where T is the period of the planet's orbit. For the case of circular orbits, we can deduce the force of attraction which holds the planets in their orbits. The velocity of the planet in its orbit is $v \propto r/T \propto r^{-\frac{1}{2}}$. Therefore the force which balances the centrifugal force (equation 2.1) must vary with r as

$$f \propto r^{-2} \tag{2.3}$$

This is a considerable triumph. A *single* force law was able to explain the orbits of *all* the planets. Likewise, because of the inverse square nature of the law, he could explain why the Moon orbited the Earth and is not dragged off towards the Sun. The law also explains why centrifugal force does not throw bodies off the Earth – they are restrained by the gravitational force. In a famous passage, he notes that the 'notion of gravitation' came to his mind as he 'sat in contemplative mood' and 'was occasioned by the fall of an apple'. Because of the inverse square law, the Moon attracts the apple with a force which is only $(\frac{1}{60})^2 = \frac{1}{3600}$ of that of the Earth, the factor $\frac{1}{60}$ being the ratio of the distance to the centre of the Earth to that of the Moon.

This work, however, was not published and a number of reasons have been advanced for this. First of all, Kepler had shown that the planetary orbits are ellipses and not circles. The mathematics of elliptical orbits was just too difficult at that time and it was only after he had discovered differential calculus that Newton was able to give the full answer in the 1680s. Second, he was unsure of the influence of the planets on each other's orbits and third, the simple theory could not account for the details of the Moon's orbit. Another possible concern was whether or not is is permissible to locate all the mass of the Earth at its centre when working out gravitational effects on the surface and beyond the Earth.

Newton returned to the problem in 1679, prompted by an interchange of letters with Hooke in that year. In this correspondence, he showed that a homogeneous sphere, or one composed of homogeneous concentric shells, gravitates as if all its mass is located at its centre, resolving one of his previous concerns. Hooke challenged Newton to work out the curve which would be followed by a particle in an inverse square law field of force. Over the next few years, 1680–4, Newton worked out the solution by the method of infinitesimals, a form of differential calculus, and found that the answer was an ellipse. Kepler's laws played a crucial part in this analysis. The second law (the areal law) is exactly equivalent to the statement that the force is a central force and the first law proved that the central force had to follow an inverse square law. These ideas are developed in the appendix. It is certain that Newton derived these results soon after the argument with Hooke but, because of the ill-will generated in the debate, he did not communicate the results of his calculations to Hooke or to anyone else.

In 1684, Halley journeyed to Cambridge to ask Newton precisely the same question which had been posed by Hooke. Newton's immediate response was that the orbit was an ellipse but he was unable to find the proof among his papers. Newton sent on the proof to Halley in November of that year. Halley then went back to Cambridge where he saw a manuscript by Newton entitled 'De Motu'. With some persuasion, Newton set about systematising the whole of his researches in mechanics, dynamics and gravitation. The treatise, 'De Motu', became the first part of Newton's great treatise *Philosophiae Naturalis Principia Mathematica*, referred to as the 'Principia' for short. This volume may rightly be considered the first genuine text on mathematical and theoretical physics as we know it today. It is also one of the greatest intellectual achievements of all time. The theory is entirely developed through mathematical relations and not through mechanistic interpretations of the physical origin of the forces. The text is not easy but what we now know as Newton's laws of motion are laid out in their definitive form for the first time. We will return to this topic in Chapter 5.

The great triumph is the fact that everything is worked out with mathematical strictness – nothing qualitative remains.

2.5 Reflections

Newton's achievements in accounting for Kepler's laws of planetary motion through a combination of his inverse square law of gravitation and his laws of motion must be considered to be amongst the greatest in theoretical physics. Newton's genius shines through in the way he arrived at his results, having in the process to discover the mathematics of differential calculus. We should, however, note the basic contributions of Tycho Brahe and Kepler without whose brilliant work Newton would have had no point of departure. Tycho provided the basic reliable experimental data which used the ultimate in contemporary technology to produce an order of magnitude improvement in the accuracy with which star positions and planetary motions were known. Kepler had the mathematical genius to be able to analyse these data and produce the three laws of planetary motion. As in so many of the great discoveries of physics and theoretical physics, the ultimate stimulus for new science is a *technological breakthrough* which provides an improvement in the precision with which an experiment can be done. This story illustrates clearly how important it is to have the right mathematical tools for the project. Finally, if by now anyone doubts it, there is no way of getting anywhere near this level of achievement without a great deal of very hard work, outstanding technical ability and, if one is lucky, the right inspiration at the right time.

APPENDIX TO CHAPTER 2

NOTES ON CONIC SECTIONS AND CENTRAL ORBITS

It is useful to recall some of the basic pieces of geometry and algebra associated with conic sections and central forces.

A2.1 Equations for conic sections

The basic geometric definition of conic sections is that they are the curves generated in a plane by the requirement that the ratio of the perpendicular distance of any point on the curve from a fixed straight line in the plane to the distance of that point on the curve from a fixed point be a constant. The fixed straight line is called the *directrix* and the fixed point the *focus*. From Figure A2.1, we see that this requirement can be written

$$\frac{AB}{BF} = \frac{AC + CB}{BF} = \text{constant}$$

i.e.

$$AC + r\cos\theta = r(\text{constant}) \tag{A2.1}$$

r and θ are polar coordinates with respect to F, the focus. Now AC and 'constant' are independent constants and so we can rewrite equation (A2.1).

Figure A2.1

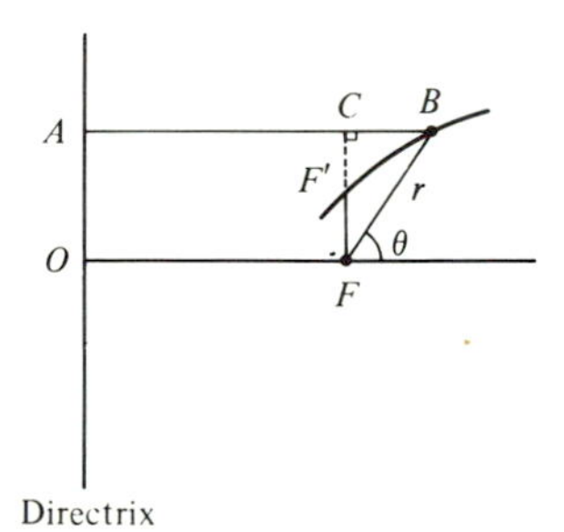

$$\frac{\lambda}{r} = 1 - e\cos\theta \tag{A2.2}$$

where λ and e are constants. We can find an immediate interpretation of λ. When $\theta = \pi/2$ and $3\pi/2$, $r = \lambda$, i.e. it is the distance FF' in Figure A2.1. Notice that the curves are symmetric with respect to the line $\theta = 0$. λ is known as the *semi-latus rectum*. The curves which are generated for various values of e are shown in Figure A2.2. If $e < 1$, we obtain an ellipse, if $e = 1$, a parabola and if $e > 1$, a hyperbola. Notice that two hyperbolae are generated in the case $e > 1$, the foci being referred to as the inner F_1 and outer F_2 foci for the hyperbola to the right-hand side of the directrix.

By simple algebra, the equation (A2.2) can be written in different ways. For example, suppose we choose a Cartesian coordinate system with origin O at the centre of the ellipse. In the polar coordinate system of equation (A2.2), the ellipse intersects the x axis at $\cos\theta = \pm 1$, i.e. at $x = -\lambda/(1+e)$ and $x = \lambda/(1-e)$. The semi-major axis of the ellipse therefore has length $a = \lambda/(1-e^2)$ and the new centre is at distance $x = e\lambda/(1-e^2)$ from F_1. In the new coordinate system, we therefore require that

$$x = r\cos\theta - e\lambda/(1-e^2)$$

$$y = r\sin\theta$$

With a bit of algebra, equation (A2.1) reduces to

$$\frac{x^2}{a^2} + \frac{y^2}{b^2} = 1 \tag{A2.3}$$

where $b = a(1-e^2)^{\frac{1}{2}}$. Equation (A2.3) shows that b is the semi-minor axis length. The meaning of e becomes apparent. If $e = 0$, the ellipse becomes a circle. It is therefore appropriate that e is called the *eccentricity*.

Figure A2.2

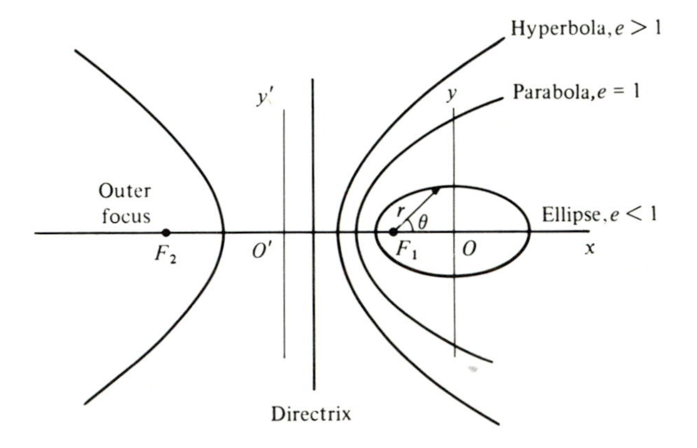

Exactly the same analysis can be carried out for hyperbolae, $e > 1$. The algebra looks the same but now the Cartesian coordinate system is referred to an origin at O' and $b^2 = a^2(1 - e^2)$ is a negative quantity. We therefore write

$$\frac{x^2}{a^2} - \frac{y^2}{b^2} = 1 \tag{A2.4}$$

where

$$b = a(e^2 - 1)^{\frac{1}{2}}$$

One of the simplest ways of relating these curves to the orbits of test particles in central fields of force is to rewrite the equations in what is known as *pedal form*. In this, the variable θ is replaced by the distance coordinate p which is the perpendicular distance from the tangent at a particular point on the curve to the focus. This is illustrated in Figure A2.3. From that diagram, it can be seen that $p = r \sin \phi$. Now we are interested in the tangent at the point B, so let us take the derivative of θ with respect to r. From equation (A2.2) we find that

$$\frac{\mathrm{d}\theta}{\mathrm{d}r} = -\frac{\lambda}{r^2 e \sin \theta} \tag{A2.5}$$

Now let us look at what happens when we vary θ and r, as shown in Figure (A2.4). It can be seen that

Figure A2.3

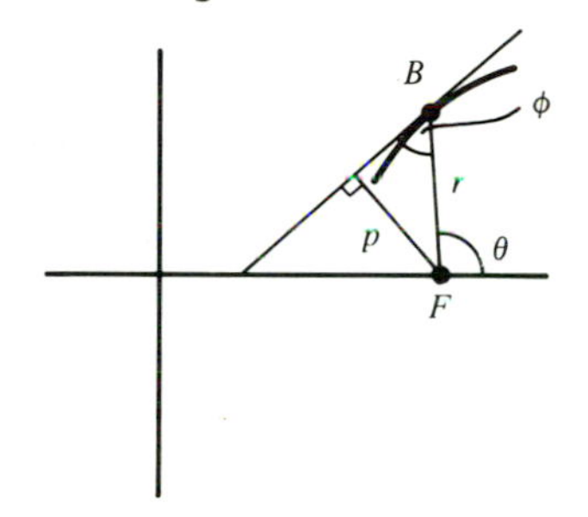

Figure A2.4

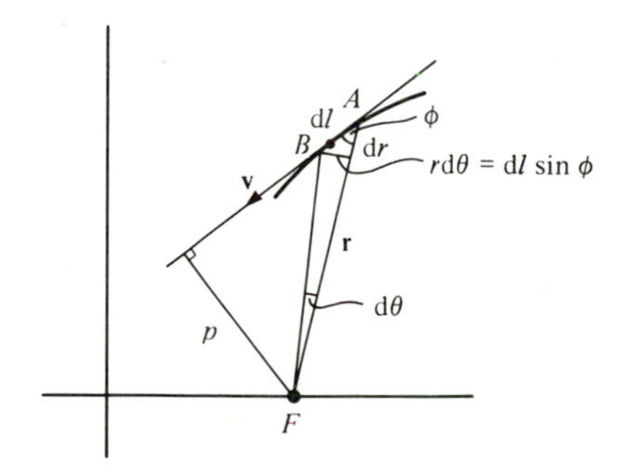

$$\tan \phi = r \frac{\mathrm{d}\theta}{\mathrm{d}r} \tag{A2.6}$$

We now have sufficient relations to eliminate θ from equation (A2.2) because

$$p = r \sin \theta \tag{A2.7}$$

$$\tan \phi = -\frac{\lambda}{re \sin \theta}$$

After a little algebra, we find that

$$\frac{\lambda}{p^2} = \frac{1}{A} + \frac{2}{r} \tag{A2.8}$$

where $A = \lambda/(e^2 - 1)$. This the *pedal* or *p–r equation* for conic sections. When A is positive, we obtain hyperbolae, when A is negative, ellipses and when A is infinite, parabolae. Notice that, in the case of the hyperbolae, equation (A2.8) refers to values of p relative to the inner focus. If the hyperbola is referred to the outer focus, the equation becomes

$$\frac{\lambda}{p^2} = \frac{1}{A} - \frac{2}{r}$$

We can now turn the whole thing round and say 'If we find equations of the form (A2.8) from a physical theory, the curves must be conic sections'.

A2.2 Kepler's laws and planetary motion

Let us write the three laws:

Kepler 1: Planetary orbits are ellipses with the Sun in one focus;

Kepler 2: Equal areas are swept out by the radius vector from the Sun to a planet in equal times;

Kepler 3: $T \propto r^{\frac{3}{2}}$, i.e. the period of the planetary orbit is proportional to the three halves power of the mean distance of the planet from the Sun.

Let us look at Kepler 2 first of all and consider the motion of a test particle under the influence of a *central field of force.* Newton's laws of motion lead directly to the *law of conservation of angular momentum.* Referring to Figure A2.4, the angular momentum of the particle is

$$\mathbf{L} = m(\mathbf{r} \times \mathbf{v}) = \text{constant} \tag{A2.9}$$

A central force is one which is always directed along the radius vector from the origin F. There is therefore no force acting outside the plane defined by the vectors $\mathbf{v}$ and $\mathbf{r}$, i.e. the motion is in the $\mathbf{v}$–$\mathbf{r}$ plane. The conservation equation can therefore be written

$$m\, r\, v \sin \phi = \text{constant} \tag{A2.10}$$

But $r \sin \phi = p$ and hence

$$pv = \text{constant} = h \tag{A2.11}$$

This indicates the usefulness of introducing the quantity p – the law of conservation of angular momentum reduces to a simple equation.

Finally, let us work out the area swept out per unit time. The area of the segment defined by FAB is $\frac{1}{2} r \, \mathrm{d}l \sin \phi$. Therefore, the area swept out per unit time is

$$\begin{aligned} \tfrac{1}{2} r \sin \phi \frac{\mathrm{d}l}{\mathrm{d}t} &= \tfrac{1}{2} r \sin \phi \, v \\ &= \tfrac{1}{2} p \, v = \tfrac{1}{2} h = \text{constant} \end{aligned} \tag{A2.12}$$

i.e. Kepler's second law is no more than a statement of the law of conservation of angular momentum in the presence of a central force, as was fully appreciated by Newton.

Going on to Kepler's first law, we can work out the solution in terms of the conservation of energy in a gravitational field. This law is a direct consequence of Newton's laws of motion. Newton was asked to work out the orbit of a test particle of mass m in an inverse square field of force and so let us consider only that case.

Let us write $\mathbf{F} = -\mathbf{i}_r GmM/r^2$. Setting $\mathbf{F} = -m \operatorname{grad} \phi$, we find that $\phi = -GM/r$. Therefore, the expression for conservation of energy in the gravitational field can be written

$$\tfrac{1}{2} m v^2 - \frac{GmM}{r} = C \tag{A2.13}$$

where C is a constant. But we have shown that, because of conservation of angular momentum, for any central field of force, we must have $pv = h =$ constant. Therefore,

$$\frac{h^2}{p^2} = \frac{2GM}{r} + \frac{2C}{m} \tag{A2.14}$$

or

$$\frac{(h^2/GM)}{p^2} = \frac{2}{r} + \frac{2C}{GMm} \tag{A2.15}$$

We recognise this equation as the pedal equation for conic sections, the exact form of the curve depending only on the sign and magnitude of the constant C. If C is negative, we find bound elliptical orbits; if C is positive, the orbits are hyperbolae. We find automatically that the centre of the force lies at the focus.

To find the period of the particle in its elliptical orbit, we note that the area of the ellipse is πab and the rate at which area is swept out is $\frac{1}{2}h$ (equation A2.12). Therefore the period is

$$T = \frac{\pi ab}{\frac{1}{2}h} \tag{A2.16}$$

Equation (A2.15) shows that the semi-latus rectum $\lambda = h^2/GM$ and from the analysis of Section A2.1, a, b and λ are related by

$$b = a(1-e^2)^{\frac{1}{2}}; \quad \lambda = a(1-e^2) \tag{A2.17}$$

Substituting into (A2.16) we obtain

$$T = \frac{2\pi}{(GM)^{\frac{1}{2}}} a^{\frac{3}{2}} \tag{A2.18}$$

Although a is the semi-major axis, for an ellipse this is proportional to the mean distance of the test particle from the focus and consequently we have derived Kepler's law for the general case of elliptical orbits.

A2.3 Rutherford scattering

The scattering of α particles by atoms was one of the classical experiments in atomic and nuclear physics carried out by Rutherford and his colleagues Geiger and Marsden in 1911. It established conclusively that the positive charge in atoms is contained within a point-like nucleus. The experiment involved measuring the distribution of α particles as a function of scattering angle, when α particles are fired at a thin gold sheet.

If we suppose that the positive charge is all centred on a nucleus, we can work out the deflection of an α particle due to the inverse square law of electrostatic repulsion between the particle and the nucleus using the formalism of orbits under an inverse square central force. Figure A2.5 shows the dynamics and geometry of the repulsion in both pedal and Cartesian form. The trajectory is a hyperbola with the nucleus located at the outer focus. If there were no repulsion, the α particle would have travelled along the diagonal AA' and passed by the nucleus at a perpendicular distance p_0 which is known as the *impact parameter*. The velocity of the α particle at infinity is v_0.

The details of the analysis follow very neatly from Figure A2.5. The trajectory of the α particle is given by equation (A2.4),

$$\frac{x^2}{a^2} - \frac{y^2}{b^2} = 1$$

from which we can find the asymptotes $x/y = \pm a/b$. Therefore, the scattering angle ϕ shown in Figure A2.5 is

$$\tan\frac{\phi}{2} = \frac{x}{y} = \frac{a}{b} = (e^2-1)^{-\frac{1}{2}} \tag{A2.19}$$

The pedal equation for the hyperbola with respect to its outer focus,

$$\frac{\lambda}{p^2} = \frac{1}{A} - \frac{2}{r}; \quad A = \lambda/(e^2 - 1)$$

should be compared with the equation of conservation of energy of the α particle in the field of the nucleus

$$\tfrac{1}{2}mv^2 + \frac{Zze^2}{4\pi\epsilon_0 r} = \tfrac{1}{2}mv_0^2 \qquad \text{(A2.20)}$$

Since $pv = h$ is a constant, we can rewrite equation (A2.20).

$$\left(\frac{4\pi\epsilon_0 mh^2}{Zze^2}\right)\frac{1}{p^2} = \frac{4\pi\epsilon_0 mv_0^2}{Zze^2} - \frac{2}{r} \qquad \text{(A2.21)}$$

We conclude that $A = Zze^2/4\pi\epsilon_0 mv_0^2$ and $\lambda = 4\pi\epsilon_0 mh^2/Zze^2$. When we recall that $p_0 v_0 = h$, we can combine the relation (A2.19) with the values of a and λ to find that

$$\cot\frac{\phi}{2} = \left(\frac{4\pi\epsilon_0 m}{Zze^2}\right) p_0 v_0^2 \qquad \text{(A2.22)}$$

Thus, the probability distribution of scattering angles ϕ is directly related to p_0. If we fire randomly a parallel beam of α particles at a single nucleus, the probability distribution of p_0 is

$$P(p_0)\mathrm{d}p_0 \propto 2\pi p_0\,\mathrm{d}p_0$$

i.e. just proportional to the area of an annulus of thickness $\mathrm{d}p_0$ at radius p_0. Therefore, the angular distribution of scattering angles ϕ is given by

Figure A2.5

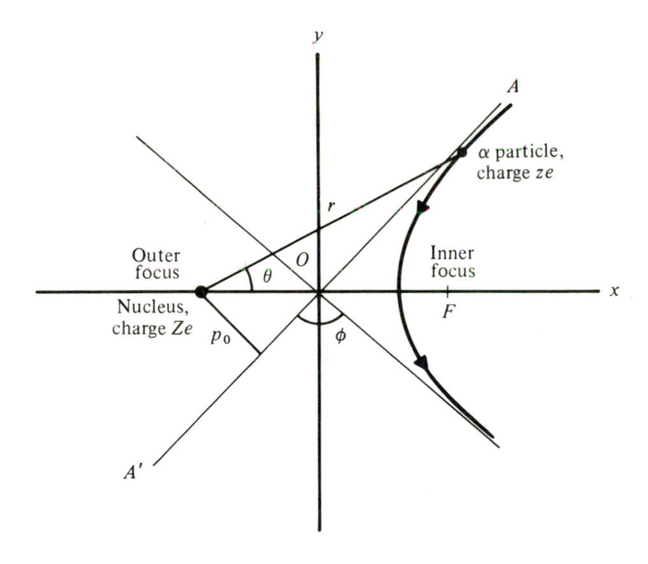

$$
\begin{aligned}
P(\phi)\,\mathrm{d}\phi &\propto p_0\,\mathrm{d}p_0 \\
&\propto \left(\frac{1}{v_0^2}\cot\frac{\phi}{2}\right)\left(\frac{1}{v_0^2}\operatorname{cosec}^2\frac{\phi}{2}\right)\mathrm{d}\phi \\
&= \frac{1}{v_0^4}\cot\frac{\phi}{2}\,\operatorname{cosec}^2\frac{\phi}{2}\,\mathrm{d}\phi \qquad \text{(A2.23)}
\end{aligned}
$$

This was the probability law which Rutherford and his colleagues found was followed by the α particles scattered by a very thin gold sheet. The law agreed with experiment over scattering angles between 5° and 150° over which range of angles the function $\cot(\phi/2)\operatorname{cosec}^2(\phi/2)$ varies by a factor of 40 000. From the known velocities of the α particles, and the fact that the law was obeyed up to large scattering angles, they deduced that the nucleus must be less than about 10^{-12} cm in radius, i.e. very much smaller than the size of atoms $\sim 10^{-8}$ cm.

Case Study 2

MAXWELL'S EQUATIONS

James Clerk Maxwell (1831–1879)
(From G. Holton & S.G. Brush, 1973, *Introduction to Concepts and Theories in Physical Science*, p. 364, Addison-Wesley Publishing Co.)

3

THE ORIGIN OF MAXWELL'S EQUATIONS AND THEIR EXPERIMENTAL VALIDATION

3.1 Electromagnetism before the time of Maxwell

By the end of the 18th century, many of the basic facts of electrostatics and magnetostatics had been established. In the 1770s and 1780s, Coulomb had established the inverse square laws of electrostatics and magnetostatics. In the SI notation which we will use throughout this exposition, the laws can be written

$$F = \frac{q_1 q_2}{4\pi\epsilon_0 r^2} \tag{3.1}$$

$$F = \frac{\mu_0 p_1 p_2}{4\pi r^2} \tag{3.2}$$

q_1 and q_2 are the electric charges of two points separated by distance r and p_1 and p_2 their magnetic pole strengths. The constants $1/4\pi\epsilon_0$ and $\mu_0/4\pi$ are written in this form simply because this is the modern convention and, although I would love to carry out the whole analysis in the original notation, such adherence to historical authenticity might obscure the argument. Note that in modern vector notation we can incorporate explicitly the directional dependence of the forces. For example

$$\mathbf{F} = \frac{q_1 q_2}{4\pi\epsilon_0 r^3}\,\mathbf{r} \quad \text{or} \quad \mathbf{F} = \frac{q_1 q_2}{4\pi\epsilon_0 r^2}\,\mathbf{i}_r \tag{3.3}$$

where $\mathbf{i}_r$ is the unit vector directed radially *away* from either charge in the direction of the other.

The mathematical foundations of the subject remained insecure until the early nineteenth century. In 1813, Poisson published his famous memoir in which he demonstrated that many of the problems of electrostatics and magnetostatics can be simplified by the introduction of the electrostatic (or magnetostatic) potential V (or V_{mag}) which is the solution of Poisson's equation

$$\frac{\partial^2 V}{\partial x^2} + \frac{\partial^2 V}{\partial y^2} + \frac{\partial^2 V}{\partial z^2} = -\frac{\rho_e}{\epsilon_0} \tag{3.4}$$

where the electric field strength $\mathbf{E}$ is given by

$$\mathbf{E} = -\operatorname{grad} V \tag{3.5}$$

and ρ_e is the electric charge density distribution.

The next experimental advance was made in 1820 when Oersted demonstrated that there is always a magnetic field associated with an electric current and this marks the beginning of the science of electromagnetism. As soon as his discovery was announced, the physicists Biot and Savart set out to discover the dependence of the strength of the magnetic field at distance $\mathbf{r}$ from a current element of length $\mathrm{d}\mathbf{l}$ in which a current I is flowing. In the same year, they came up with the answer, the *Biot–Savart law*, which in modern vector notation can be written

$$\mathrm{d}\mathbf{B} = \frac{\mu_0 I \mathrm{d}\mathbf{l} \times \mathbf{r}}{4\pi r^3} \tag{3.6}$$

Notice that the signs of the vectors are important in finding the correct direction of the field. The term $\mathrm{d}\mathbf{l}$ is the length of the current element in the direction of the current I and $\mathbf{r}$ is measured from the current element $\mathrm{d}\mathbf{l}$ to the point at distance r in the field. It is interesting to note that this is the generalised form of the law normally known in text books as *Ampère's law* which can be written

$$\int_C \mathbf{H} \cdot \mathrm{d}\mathbf{s} = I_{\text{enclosed}}$$

The story developed rapidly. In 1825, Ampère published his famous treatise in which the theory was developed and in which he showed how the magnetic field of a current loop could be represented by an equivalent magnetic shell. In the treatise, he also formulated the equation for the force between two current carrying elements, $\mathrm{d}\mathbf{l}_1$ and $\mathrm{d}\mathbf{l}_2$ with currents I_1 and I_2;

$$\mathrm{d}\mathbf{F}_2 = \frac{\mu_0 I_1 I_2 \, \mathrm{d}\mathbf{l}_1 \times (\mathrm{d}\mathbf{l}_2 \times \mathbf{r})}{4\pi r^3} \tag{3.7}$$

$\mathrm{d}\mathbf{F}_2$ is the force acting on the current element $\mathrm{d}\mathbf{l}_2$, $\mathbf{r}$ being measured from $\mathrm{d}\mathbf{l}_1$. Ampère also demonstrated the relation between this law and the Biot–Savart law. Notice that, so far, these developments are concerned with the forces between *stationary* charges, currents and magnets.

The essence of Maxwell's equations is, of course, that they deal with time varying phenomena as well. Many of the basic experimental results on time varying electric and magnetic fields were established by Faraday over the succeeding 20 years. Michael Faraday began life as a bookbinder's journeyman and learned his early science by reading the books he had to bind. These included the *Encyclopaedia Britannica* and his attention was particularly attracted by the section on electricity. By writing to Davy, he secured the post of assistant at the Royal Institution and began his investigations into electricity in the early 1820s. He is a classic example of a meticulous experimenter with essentially no mathematical training who was never able to express the results of his research in mathematical form. He had, however, an intuitive instinct amounting to genius for performing the correct experiments and in devising empirical models to account for the results.

As early as 1831, he had established the qualitative form of his law of induction – *the electromotive force induced in a current loop is directly related to the rate at which magnetic field lines are cut.* Faraday lay great emphasis upon the concept of *lines of force* and essentially all his work on electromagnetic induction is centred upon the idea that the magnetic field strength can be represented in magnitude and direction by magnetic lines of force (Figure 3.1). The idea certainly sprang from the observation of the patterns which iron filings take up around a magnet. It was a relatively simple extension from this concept to that of *tubes of force* which are defined by lines of force. We will see that these concepts which are no more than models for the forces associated with magnetic induction played a crucial role in Maxwell's mathematising of the processes of electromagnetic induction. I must confess that, when I first learned electromagnetism, lines of force were an obstacle to my understanding, largely because it was not explained clearly to me that they are only a *model* for what is going on. The things you actually measure in a laboratory experiment are vector forces at different points in the field and the fictitious lines of force are conceptual models to represent these forces.

Although the law of induction was enunciated at an early stage, it took Faraday many years to complete all the necessary experimental work to demonstrate the general validity of the law – namely, that it is the rate of change of the total magnetic flux linking the circuit, whatever its origin, which determines the

Figure 3.1. Illustrating the lines of force about a bar magnet.

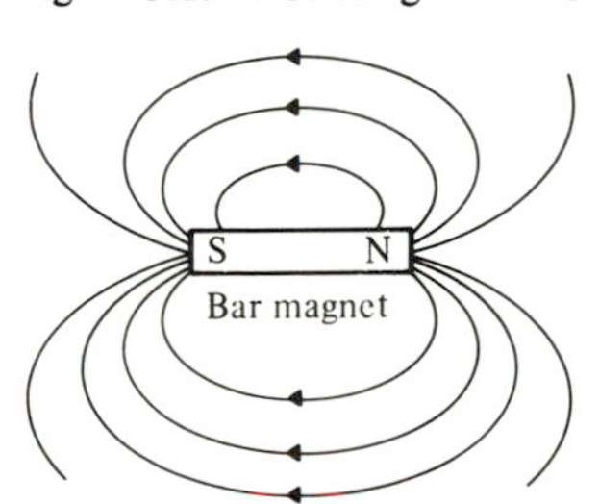

size of the electromotive force induced in the circuit. In 1834, Lenz enunciated his law which cleared up the problem of the direction of the induced electromotive force in the circuit – the electromotive force acts in such a direction as to oppose the change in magnetic flux. These laws were first put into mathematical form by Neumann who explicitly wrote down the proportionality of the induced electromotive force $\mathcal{E}$ to the rate of change of magnetic flux, Φ, in 1845.

$$\mathcal{E} = -\frac{d\Phi}{dt} \tag{3.8}$$

where Φ is the total magnetic flux through the circuit.

It is at this point in the story that we should introduce James Clerk Maxwell.

3.2 How Maxwell derived the complete set of equations for the electromagnetic field

Maxwell was born and educated in Edinburgh. In 1850, he went up to Cambridge where he studied mathematics and physics with considerable distinction. Allied to his formidable mathematical abilities was a physical imagination which could appreciate the empirical models of Faraday and give them mathematical substance. The very distinctive feature of his thinking was his ability to work by *analogy*. Indeed as early as 1856, he published the essence of his approach in an essay entitled 'Analogies in Nature'. The technique is best illustrated by the examples given below but in brief it may be considered a process of formalising partial resemblance. It consists of recognising mathematical similarities between quite distinct physical problems and seeing how far one can go in applying the successes of one theory to different circumstances. In Maxwell's case in relation to electromagnetism, he found formal analogies between mechanical systems and the phenomena of electrodynamics. I have no doubt at all that it was this kind of approach which caused feelings of 'distress' among most Frenchmen as we have already remarked.

Let us proceed further with the story before considering the technique any further. In the same year, 1856, he published the first of his great papers on electromagnetism. The first part of this enlarged upon the technique of analogy and drew particular attention to its application to incompressible fluid flow and magnetic lines of force. Let us recall the basic equation for incompressible fluid flow (see also the appendix to Chapter 5). Consider a volume v bounded by a surface S. Then, using vector notation, the mass flow per unit time through a surface element $d\mathbf{S}$ is $\rho\mathbf{u}\cdot d\mathbf{S}$ where $\mathbf{u}$ is the fluid velocity and ρ is the density distribution. Therefore, the total mass flux through the surface is $\int_S \rho\mathbf{u}\cdot d\mathbf{S}$. This is just the rate of loss of mass, i.e.

$$\frac{\mathrm{d}}{\mathrm{d}t}\int_v \rho \mathrm{d}v$$

Therefore,

$$-\frac{\mathrm{d}}{\mathrm{d}t}\int_v \rho \mathrm{d}v = \int_S \rho \mathbf{u} \cdot \mathrm{d}\mathbf{S}$$

Now applying the divergence theorem to the right-hand side and transposing we obtain

$$\int_v \left[\mathrm{div}\,(\rho \mathbf{u}) + \frac{\partial \rho}{\partial t}\right] \mathrm{d}v = 0$$

This result must be true for any volume element and hence

$$\mathrm{div}\,\rho \mathbf{u} = -\frac{\partial \rho}{\partial t} \tag{3.9}$$

If the fluid is incompressible, $\rho = \text{constant}$ and hence

$$\mathrm{div}\,\mathbf{u} = 0 \tag{3.10}$$

Now Maxwell was very impressed by the concepts of lines and tubes of force as expounded by Faraday and he drew an immediate analogy between the behaviour of magnetic field lines and the streamlines of incompressible fluid flow (Figure 3.2). The velocity **u** is entirely analogous to the magnetic flux density **B**. For example, if the tubes of force or streamlines diverge the strength of the field decreases, as does the fluid velocity. This suggests that the magnetic field should also be characterised by

$$\mathrm{div}\,\mathbf{B} = 0 \tag{3.11}$$

Notice the great subtlety with which he uses **B** rather than **H** in this argument; **u** is associated with a flux through a surface just as **B** is; **H** is associated with the force at a point in space. Indeed, in this paper, Maxwell recognised the important distinction between **B** and **H**, associating **B** with fluxes and **H** with forces.

Figure 3.2. Illustrating the analogy between magnetic field lines and the streamlines in the flow of an incompressible fluid.

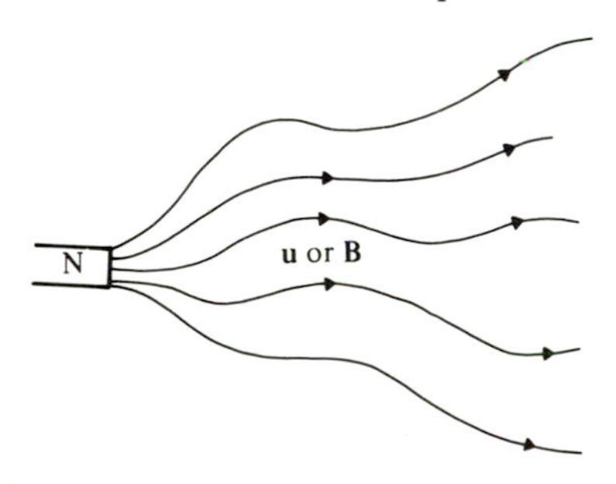

An interesting sidelight is that Maxwell did not actually use the expressions div, grad and curl in this paper but he did invent the names in a paper of 1870 and so we need not worry too much about using them here. In 1856, the vector operators were all written out in Cartesian form. One of the great achievements of this paper was that all the known relations between electromagnetic phenomena at that time were written down in vector form. In fact, he showed that equation (3.8)

$$\mathcal{E} = -\frac{\mathrm{d}\Phi}{\mathrm{d}t}$$

could be written

$$\int_C \mathbf{E} \cdot \mathrm{d}\mathbf{s} = -\frac{\mathrm{d}}{\mathrm{d}t}\int_S \mathbf{B} \cdot \mathrm{d}\mathbf{S} \tag{3.12}$$

The left-hand side is just the definition of electromotive force and the right-hand side contains the definition of magnetic flux through the surface S in terms of the magnetic flux density **B**. Now Stokes' theorem can be applied to equation (3.12)

$$\int_S \operatorname{curl} \mathbf{E} \cdot \mathrm{d}\mathbf{S} = -\frac{\mathrm{d}}{\mathrm{d}t}\int_S \mathbf{B} \cdot \mathrm{d}\mathbf{S}$$

Applying this to the elementary surface d**S**, it can be seen that

$$\operatorname{curl} \mathbf{E} = -\frac{\partial \mathbf{B}}{\partial t} \tag{3.13}$$

In the same way as for **B**, he could conclude that div **E** = 0 on the basis of the analogy with fluid flow, but he knew from Poisson's equation that this is only true in free space where there are no charges. In free space, he could deduce from that equation that

$$\operatorname{div} \mathbf{E} = \rho_e/\epsilon_0 \tag{3.14}$$

Finally, he rewrote the relation between the magnetic field produced by a current flowing in a wire (i.e. Ampère's law) in vector form.

$$\int_C \mathbf{H} \cdot \mathrm{d}\mathbf{s} = I_{\text{enclosed}}$$

where I_{enclosed} is the current flowing through the surface bounded by C; i.e.

$$\int_C \mathbf{H} \cdot \mathrm{d}\mathbf{s} = \int_S \mathbf{J} \cdot \mathrm{d}\mathbf{S}$$

where **J** is the current density. Applying Stokes' theorem,

$$\int_S \operatorname{curl} \mathbf{H} \cdot \mathrm{d}\mathbf{S} = \int_S \mathbf{J} \cdot \mathrm{d}\mathbf{S}$$

Applying this to an elementary area $d\mathbf{S}$, we find that

$$\text{curl}\,\mathbf{H} = \mathbf{J} \tag{3.15}$$

The final achievement of this paper was the formal introduction of a vector potential **A**. Such a vector had already been introduced by Neumann, Weber and Kirchoff in order to calculate induced currents.

$$\mathbf{B} = \text{curl}\,\mathbf{A} \tag{3.16}$$

This definition is clearly entirely consistent with equation (3.11) since div curl $\mathbf{A} = 0$. Maxwell went further and showed how the induced electric field in a circuit could be related to **A**. Incorporating the definition (3.16) into equation (3.13), we obtain

$$\text{curl}\,\mathbf{E} = -\frac{\partial}{\partial t}(\text{curl}\,\mathbf{A})$$

Interchanging the order of the time and spatial derivatives on the right-hand side, we obtain

$$\mathbf{E} = -\frac{\partial \mathbf{A}}{\partial t} \tag{3.17}$$

It is convenient to gather together this primitive and *incomplete* set of Maxwell's equations.

$$\text{curl}\,\mathbf{E} = -\frac{\partial \mathbf{B}}{\partial t} \tag{3.13}$$

$$\text{curl}\,\mathbf{H} = \mathbf{J} \tag{3.15}$$

$$\text{div}\,\epsilon_0\mathbf{E} = \rho_e \quad \text{in vacuum} \tag{3.14}$$

$$\text{div}\,\mathbf{B} = 0 \tag{3.11}$$

These new results gave formal coherence to the theory but Maxwell still lacked a physical model for the phenomena of electromagnetism. He developed his solution in 1861–2 and these new and quite remarkable results were published in a series of papers entitled 'On physical lines of force'. Since his previous work on the analogy between **u** and **B**, he had become convinced that magnetism was essentially rotational in nature. The aim was to devise a model for the medium that fills all space which could account for the stresses that Faraday had associated with magnetic lines of force.

He began with a model of a rotating vortex tube as an analogue for the magnetic field strength. The analogy is suggested by the following considerations. If left on their own, magnetic field lines expand apart and exactly the same occurs in the case of a vortex tube if the centrifugal force of rotation is not balanced. In addition, we can write down the energy in vortex motion as

$$\int_v \rho \mathbf{u}^2\, dv$$

where ρ is the density of the material and $\mathbf{u}$ its velocity. This is similar to the expression for the energy in a magnetic field, $\int_v (\mathbf{B}^2/2\mu_0)\, dv$. Thus, again $\mathbf{u}$ is analogous to $\mathbf{B}$. The greater the velocity of the tube, the stronger the magnetic field. In fact, he postulated that everywhere the local magnetic field strength should be proportional only to the angular velocity of the vortex.

Maxwell therefore began with a model in which the whole of space was filled up with vortex tubes (Figure 3.3(*a*)). There is, however, an immediate mechanical problem. Friction between neighbouring vortices will cause them to dissipate. Maxwell adopted the very practical solution of inserting 'ball-bearings' between the vortices so that they could all rotate in the same direction without friction (Figure 3.3(*b*)). Maxwell's original published picture for the vortices is shown in Figure 3.4. He then identified the 'ball-bearings' with electric particles which, if they are free to move, will carry an electric current. In conductors, these electric particles are free to move, whereas in insulators, *including free space*, they are fixed.

Figure 3.3. (*a*) Maxwell's original model of rotating vortices as a representation of a magnetic field. Friction at the points where the vortices touch would lead to dissipation of the rotational energy of the tubes. (*b*) Maxwell's model with 'ball-bearings' which prevent dissipation of the rotational energy of the vortices. If these particles are free to move they are identified with the particles which carry a current in a conductor.

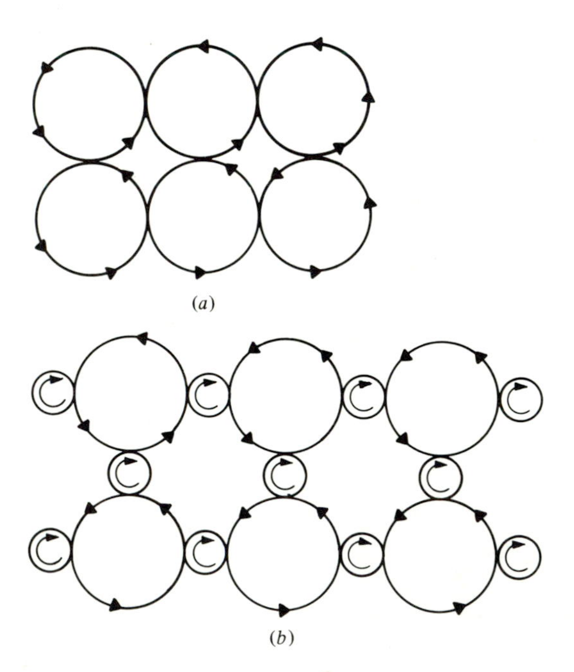

Remarkably, this model could explain all known phenomena of electromagnetism. For example, consider the magnetic field produced by a current in a wire. The electric current forces the vortices to rotate as an infinite series of vortex rings about the wire as illustrated in Figure 3.5. Correctly, the magnetic field lines form circular closed loops about the current. Another example is that of the interface between two regions in which the magnetic field strengths are different but parallel. In the region of the stronger field, the vortices rotate more rapidly and hence, in the interface, there must be a net force on the

Figure 3.4. Maxwell's own picture of the dynamic interaction of the vortices (represented by hexagons) and the current carrying particles (*Philosophical Magazine*, 1861, Series 4, Vol. 21, Plate V, Figure 2).

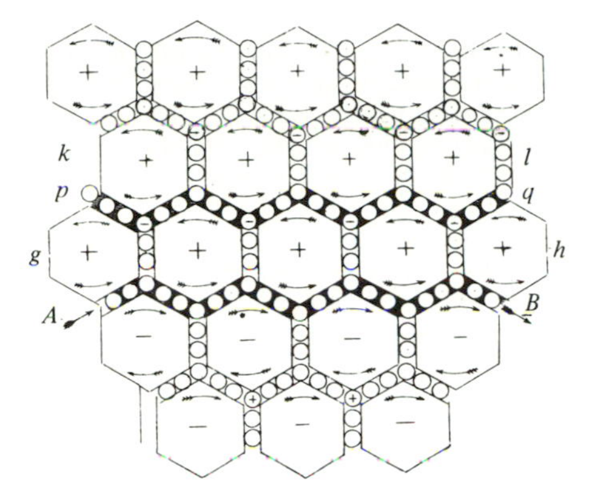

Figure 3.5. A representation of the magnetic field about a current carrying wire according to Maxwell's model. The vortices become circular tori concentric with the axis of the wire.

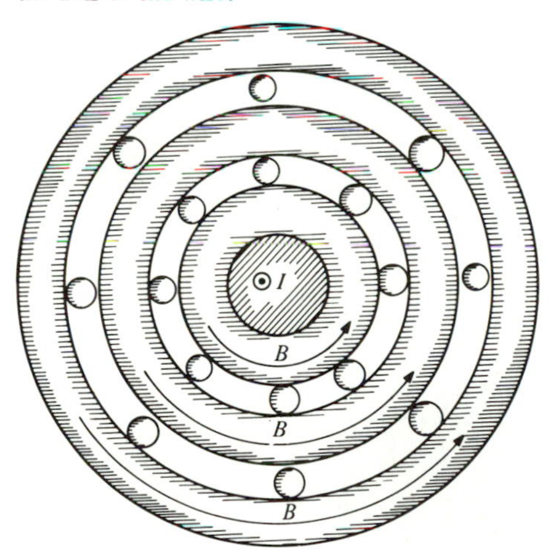

electric particles which drags them along the interface as shown in Figure 3.6 causing an electric current. Notice that the sense of the current in the interface agrees with what is found experimentally. As an example of induction, consider the effect of embedding a second wire in the magnetic field shown in Figure 3.5. If the current is steady, there will be no effect on the second wire. If, however, the current changes, an impulse is communicated through the intervening particles and vortices and a reverse current is induced in the second wire (Figure 3.7).

The last part of the paper contains the crucial step which led to the discovery of the complete set of Maxwell's equations. He now considered how insulators can store electrical energy. He made the further assumption that, in insulators, the medium is elastic so that the electric particles can be displaced from their equilibrium positions by the action of an electric field. Thus, he attributed the

Figure 3.6. Illustrating how flow along a current sheet J must cause a discontinuity in magnetic field strength according to Maxwell's model. The current carrying particles flow along the line indicated. Because of friction, the lower set of vortices are slowed down whilst the upper set are speeded up. The sense of the discontinuity in B is seen to be in the correct sense according to Maxwell's equations.

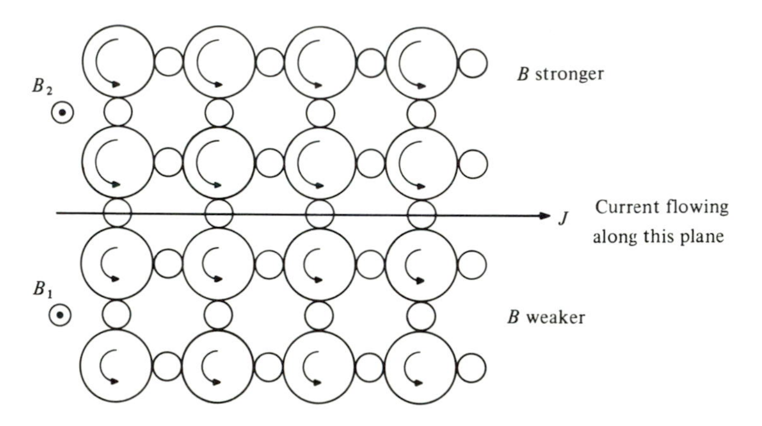

Figure 3.7. Illustrating the phenomenon of induction according to Maxwell's model. The induced current in conductor 2, ΔI_2, is in the opposite direction to ΔI_1 as can be seen from the sense of rotation of the current carrying particles.

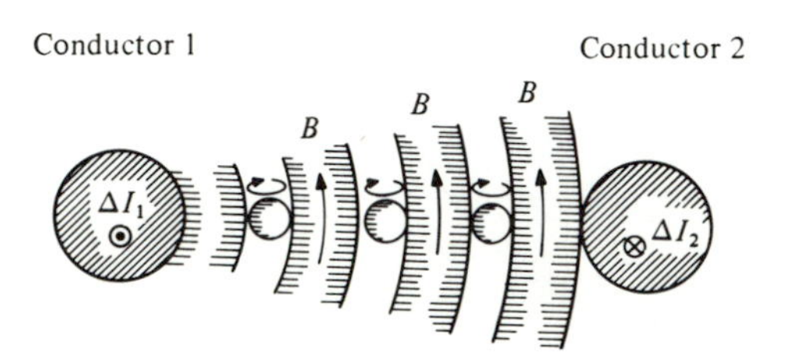

electrostatic energy in the medium to the elastic potential energy due to the displacement of the electric particles. This had two immediate and vital consequences. First, when the current in a wire is varying, there are small changes in the positions of the electric particles in the surrounding insulating medium or vacuum, i.e. the effect of the changing current is that small currents in the medium associated with the elastic motion of the particles are set up. In other words there is a current associated with the *displacement* of the electric particles. Second, by virtue of the medium being elastic, it is possible to calculate the velocity with which disturbances can be propagated through the insulator or vacuum.

The analysis is then straightforward. In an elastic medium, the displacement is assumed to be proportional to the electric field strength

$$\mathbf{r} = \alpha \mathbf{E} \tag{3.18}$$

When the strength of the field varies, the charges move causing a *displacement current*. If N_q is the number density of electric particles and q their charge, the displacement current density is

$$\mathbf{J}_\mathrm{d} = qN_q\dot{\mathbf{r}} = qN_q\alpha\dot{\mathbf{E}} = \beta\dot{\mathbf{E}} \tag{3.19}$$

This displacement current should be included in equation (3.15) which now reads

$$\text{curl}\,\mathbf{H} = \mathbf{J} + \mathbf{J}_\mathrm{d} = \mathbf{J} + \beta\dot{\mathbf{E}} \tag{3.20}$$

At this stage α and β are unknown constants to be determined from other electric properties of the medium. First of all, however, we can work out the velocity of propagation of a disturbance through the medium. Assuming that there are no currents, $\mathbf{J} = 0$, we have to solve the equations

$$\begin{aligned} \text{curl}\,\mathbf{H} &= \beta\dot{\mathbf{E}} \\ \text{curl}\,\mathbf{E} &= -\dot{\mathbf{B}} \end{aligned} \tag{3.21}$$

The properties of these waves can be found by standard procedures. The simplest technique is to assume wave solutions of the form $e^{i(\mathbf{k}\cdot\mathbf{r}-\omega t)}$ and then we can replace the vector operators by scalar and vector products as follows:

$$\text{curl} \rightarrow i\mathbf{k}\times$$

$$\partial/\partial t \rightarrow -i\omega$$

(see Appendix A3.6). Then equations (3.21) reduce to

$$\begin{aligned} i(\mathbf{k}\times\mathbf{H}) &= -i\omega\beta\mathbf{E} \\ i(\mathbf{k}\times\mathbf{E}) &= i\omega\mathbf{B} \end{aligned} \tag{3.22}$$

Now eliminating $\mathbf{E}$ from equations (3.22),

$$\mathbf{k}\times(\mathbf{k}\times\mathbf{H}) = -\omega^2\beta\mu\mu_0\mathbf{H} \tag{3.23}$$

Using the vector relation $\mathbf{A} \times (\mathbf{B} \times \mathbf{C}) = \mathbf{B}(\mathbf{A} \cdot \mathbf{C}) - \mathbf{C}(\mathbf{A} \cdot \mathbf{B})$, we find that

$$\mathbf{k}(\mathbf{k} \cdot \mathbf{H}) - \mathbf{H}(\mathbf{k} \cdot \mathbf{k}) = -\omega^2 \beta \mu \mu_0 \mathbf{H}$$

There is no solution for $\mathbf{k}$ parallel to $\mathbf{H}$, i.e. longitudinal waves, since the left-hand side is then zero. We obtain solutions for transverse waves if $\mathbf{k} \cdot \mathbf{H} = 0$. Thus, the solutions are for plane transverse waves with the $\mathbf{E}$ and $\mathbf{H}$ vectors perpendicular to each other and to the direction of propagation of the wave. The *dispersion relation* for the waves, i.e. the relation between k and ω, is then $k^2 = \omega^2 \beta \mu \mu_0$. Since the velocity of propagation of the wave is $c = \omega/k$, we immediately find that

$$c^2 = 1/\beta \mu \mu_0 \tag{3.24}$$

Maxwell knew how to evaluate the constant β. The energy density stored in the dielectric is just the work done per unit volume in displacing the electric particles a distance $\mathbf{r}$, i.e.

$$\text{Work done} = \int \mathbf{F} \cdot \mathrm{d}\mathbf{r} = \int N_q q \mathbf{E} \cdot \mathrm{d}\mathbf{r}$$

But

$$\mathbf{r} = \alpha \mathbf{E}$$

and hence

$$\mathrm{d}\mathbf{r} = \alpha \,\mathrm{d}\mathbf{E}$$

Therefore, the work done is just

$$\int_0^E N_q q \alpha E \,\mathrm{d}E = \tfrac{1}{2} \alpha N_q q E^2 = \tfrac{1}{2} \beta E^2$$

But we know that this is just the electrostatic energy density in the dielectric which is $\frac{1}{2}\mathbf{D} \cdot \mathbf{E} = \frac{1}{2}\epsilon\epsilon_0 E^2$. Therefore $\beta = \epsilon\epsilon_0$. Inserting this value into the above expression for the velocity of the waves, we find that

$$c = (\mu\mu_0 \epsilon\epsilon_0)^{-\frac{1}{2}} \tag{3.25}$$

Notice that even in a vacuum, $\mu = 1$, $\epsilon = 1$, the velocity of propagation of the waves is finite: $c = (\mu_0 \epsilon_0)^{-\frac{1}{2}}$. Maxwell then inserted the best available values for the electrostatic and magnetostatic constants and found, to his amazement, that c turned out to be the velocity of light; in his own words: 'we can scarcely avoid the inference that light consists in the transverse modulations of the same medium which is the cause of electric and magnetic phenomena'.

One cannot help but wonder at such 'pure gold' having come out of a specific mechanistic model for the electromagnetic field. Maxwell was quite straightforward about the value of the model.

> I do not bring it forward as a mode of connection existing in Nature . . . . It is however a mode of connection which is mechanically conceivable

and it serves to bring out the actual mechanical connections between known electromagnetic phenomena.[1]

In 1864, Maxwell developed the whole theory on a much more abstract basis without any special assumptions about the nature of the medium through which electromagnetic phenomena are propagated. This work was published in 1865 in a further classical paper entitled 'A dynamical theory of the electromagnetic field'. To quote Whittaker: 'In this, the architecture of his system was displayed, stripped of the scaffolding by aid of which it had been first erected'.[2]

In this paper, the equations appear in their final form, quite independent of the means by which they had been established. Maxwell's own view of the significance of this paper is revealed in what Everitt calls 'a rare moment of unveiled exuberance' in a letter to his cousin Charles Clay: 'I have also a paper afloat, containing an electromagnetic theory of light, which, till I am convinced to the contrary, I hold to be great guns'.[3]

In their final form, the equations read

$$\left.\begin{aligned} \text{curl}\,\mathbf{E} &= -\frac{\partial \mathbf{B}}{\partial t} \\ \text{curl}\,\mathbf{H} &= \mathbf{J} + \frac{\partial \mathbf{D}}{\partial t} \\ \text{div}\,\mathbf{D} &= \rho_e \\ \text{div}\,\mathbf{B} &= 0 \end{aligned}\right\} \tag{3.26}$$

Notice how the inclusion of what is still called the *displacement current* $\partial\mathbf{D}/\partial t$ resolves a problem with the equation of continuity in electromagnetism. Let us take the divergence of the second equation of (3.26). Then because div curl = 0,

$$\text{div}\,(\text{curl}\,\mathbf{H}) = \text{div}\,\mathbf{J} + \frac{\partial}{\partial t}(\text{div}\,\mathbf{D}) = 0$$

Since $\text{div}\,\mathbf{D} = \rho_e$, we obtain

$$\text{div}\,\mathbf{J} + \frac{\partial \rho_e}{\partial t} = 0$$

which is the correct continuity equation for conservation of charge in electrostatics (see Section 4.2). Without the displacement term, the continuity equation would be total nonsense and the primitive set of equations (3.11) to (3.15) would not be self consistent.

3.3 The subsequent history and Hertz's experiments

When one tells a story like this, there is an inevitability about how the hero will get the right answer. In fact, there was a great deal of theoretical activity at this time and there were several other viable theories of electromagnetism.

However, the very distinctive prediction of Maxwell's theory of electromagnetic waves and their identity with light was a clear triumph. It immediately gave real physical significance to the wave theory of light which could account for the phenomena of reflection, refraction, polarisation, etc. Despite the fact that we now recognise that light is one brand of electromagnetic radiation, this was far from obvious at the time. In addition, although Maxwell's paper of 1865 set out the theory in strictly mathematical form, the mechanistic analogies which were the basis of his previous paper were undoubtedly a hinderance to the general acceptance of the theory. Gradually, however, it became apparent that the beautiful symmetry and simplicity of the equations could account for all known electromagnetic phenomena. We indicate how this can be done in the next chapter. Indeed, the procedure by which Maxwell arrived at the displacement term is essentially identical to the way in which electromagnetic phenomena in simple dielectrics are treated in standard text books.

Maxwell died in 1879 before direct experimental evidence was obtained for electromagnetic waves. The matter was finally laid to rest ten years after his death in a classical series of experiments by Hertz. This was over 20 years after Maxwell first derived the equations and must be classed as one of the great predictions of theoretical physics. I strongly recommend perusal of Hertz's great paper 'On electric waves'[4] which sets out beautifully this remarkable set of experiments.

Hertz found that he could detect the effect of electromagnetic induction at considerable distances from his apparatus. His emitter and detector are shown in Figure 3.8. The tuning of the emitter is related to the size of the arms of the 'antenna'. The way in which it worked was that the sparks were produced in the emitter between the two small spheres and those signals in resonance with the arms of the transmitting aerial were emitted with the greatest intensity. The detector was of similar design and the technique of observation was to look for sparks produced in the receiver spark gap. What Hertz did was to bring the jaws of the spark gap as close together as possible so that it would be as sensitive as possible. After a great deal of trial and error, he found what proved to be waves of relatively short wavelength. He established the wavelength by placing a reflecting sheet at some distance from his spark gap emitter and finding the positions of minimum signal in the second spark gap detector placed in front of the screen.

With this arrangement he found standing waves as he moved the detector along the line between the emitter and the conducting sheet. The frequencies of the waves could be found from the frequency to which the receiving loop was tuned, $\omega = (LC)^{-\frac{1}{2}}$, where L and C are the inductance and capacitance of the emitting apparatus. This immediately gave him the velocity of the waves because the wavelength was twice the distance between the minima of the standing waves and $c = \nu\lambda$.

The velocity turned out to be the velocity of light in free space. He knew that the waves were electromagnetic in origin because they were produced by his spark gap emitter. Then, he began a series of experiments which demonstrated conclusively that these waves behaved in all respects exactly like light. His great paper on the subject has the headings: rectilinear propagation, polarisation, reflection, refraction. Some of the experiments were quite remarkable in their execution. To demonstrate refraction he constructed a prism weighing 12 cwt out of 'so-called hard pitch, a material like asphalt'. The experiments demonstrated convincingly that there exist electromagnetic waves of frequency about 1 GHz and wavelength 30 cm which behaved in all respects like light. Notice that he had to find a means of producing waves of frequency as high as about 1 GHz or else the wavelength of the radiation would have been too great to have fitted into his laboratory.

This was the final proof of the validity of Maxwell's equations.

Figure 3.8. Hertz's apparatus for the generation and detection of electromagnetic radiation. The emitter *a* produced electromagnetic radiation from discharges across the spherical conductors. The detector *b* consisted of a similar device with the jaws of the detector placed as close together as possible to achieve the maximum sensitivity. The emitter was placed at the focus of a cylindrical paraboloid reflector to produce a directed beam of radiation. (From H. Hertz, 1893, *Electric Waves*, pp. 183–4, MacMillan and Co., London.)

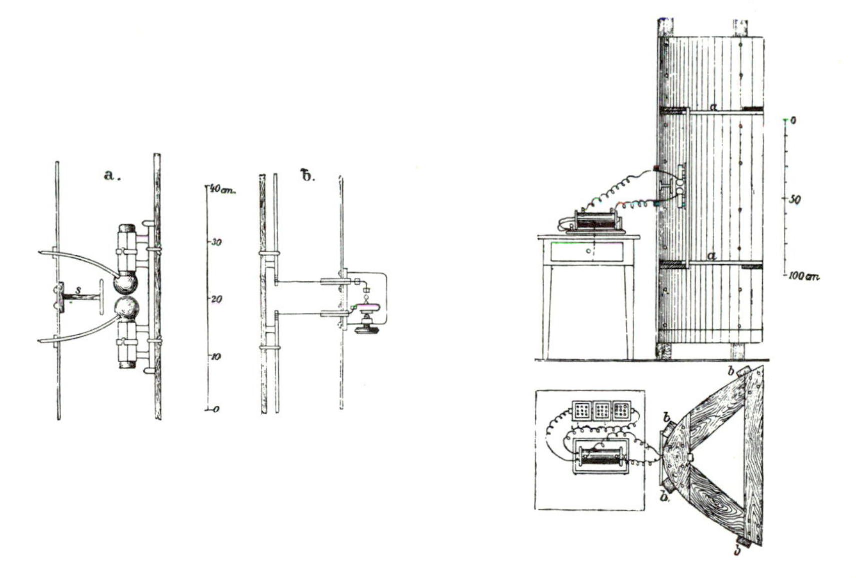

3.4 Reflections

This is as extreme an example of model building in theoretical physics as I know. One can detect a certain embarrassment among physicists about Maxwell's technique of model building by analogy, but how powerful it turned out to be. Not only in this example, but in all his great contributions to physics and theoretical physics, there is a very profound understanding of mathematics and its interpretation in terms of physical models. The technique of analogy in the hands of a genius produced new insights which enabled Maxwell to make this huge step forward in understanding the nature of electromagnetism. We can look back with hindsight and see how it all came together in as brilliant a series of contributions to science as anyone has ever made. At the time of his early death at the age of 49, he was exploring in detail the nature of the electromagnetic theory. His remarkable volume *A Treatise on Electricity and Magnetism* is a work in progress and one can see glimmerings of the revolution which was about to take place at the beginning of the next century with the theory of relativity.

APPENDIX TO CHAPTER 3

REVISION NOTES ON VECTOR FIELDS

These notes are intended to reinforce what you have already learned about vector fields and to summarise some useful results which will be used in Chapter 4.

A3.1 The divergence theorem and Stokes' theorem

The importance of these theorems is that we can either express the laws of physics in 'large-scale' form in terms of integrals over finite volumes of space or in 'small-scale' form in terms of the properties of the field in the vicinity of a particular point. The first form leads to integral equations and the second to differential equations. The fundamental theorems relating these two forms are as follows.

Divergence theorem

$$\underbrace{\int_S \mathbf{A}\cdot \mathrm{d}\mathbf{S}}_{\text{large-scale}} = \int_v \underbrace{\mathrm{div}\,\mathbf{A}}_{\text{small-scale}}\,\mathrm{d}v \tag{A3.1}$$

$\mathrm{d}\mathbf{S}$ is the element of surface area of the surface S which encloses volume v. The direction of the vector $\mathrm{d}\mathbf{S}$ is always taken to be directed *normally outwards* to the increment of surface area. The volume integral on the right-hand side is over the volume enclosed by the closed surface S.

Stokes' theorem

$$\underbrace{\int_C \mathbf{A}\cdot \mathrm{d}\mathbf{l}}_{\text{large-scale}} = \int_S \underbrace{\mathrm{curl}\,\mathbf{A}}_{\text{small-scale}}\cdot \mathrm{d}\mathbf{S} \tag{A3.2}$$

where C is a closed curve and the integral on the left is taken round the closed curve. S is *any* open surface bounded by the loop C, $\mathrm{d}\mathbf{S}$ being the element of surface area. The sign of $\mathrm{d}\mathbf{S}$ is decided by the right-hand screw convention.

Notice that if **A** is a vector field of force, $\int \mathbf{A}\cdot d\mathbf{l}$ is *minus* the work needed to take a particle once round the circuit.

A3.2 Results related to the divergence theorem

Problem 1

By taking $\mathbf{A} = f \operatorname{grad} g$, derive two forms of Green's theorem,

$$\int_v [f\nabla^2 g + (\nabla f)\cdot(\nabla g)]\,dv = \int_S f\nabla g\cdot d\mathbf{S} \qquad (A3.3)$$

$$\int_v [f\nabla^2 g - g\nabla^2 f]\,dv = \int_S (f\nabla g - g\nabla f)\cdot d\mathbf{S} \qquad (A3.4)$$

Substitute $\mathbf{A} = f \operatorname{grad} g$ in the divergence theorem, equation (A3.1):

$$\int_S f\nabla g\cdot d\mathbf{S} = \int_v \operatorname{div} f\nabla g\,dv$$

Then, because $\nabla\cdot(a\mathbf{b}) = a\nabla\cdot\mathbf{b} + \nabla a\cdot\mathbf{b}$,

$$\int f\nabla g\cdot d\mathbf{S} = \int_v (\nabla f\cdot\nabla g + f\nabla^2 g)\,dv \qquad (A3.3)$$

Then performing the same analysis for $g\nabla f$, we find that

$$\int g\nabla f\cdot d\mathbf{S} = \int_v (\nabla f\cdot\nabla g + g\nabla^2 f)\,dv \qquad (A3.5)$$

Subtracting (A3.5) from (A3.3), we obtain the desired result,

$$\int_v (f\nabla^2 g - g\nabla^2 f)\,dv = \int_S (f\nabla g - g\nabla f)\cdot d\mathbf{S} \qquad (A3.4)$$

This result can be particularly useful if one of the functions is a solution of Laplace's equation $\nabla^2 f = 0$.

Problem 2

Show that

$$\int_v \frac{\partial f}{\partial x_k}\,dv = \int_S f\,dS_k$$

In this case, take the vector **A** to be $f\mathbf{i}_k$ where f is a scalar function of position and $\mathbf{i}_k$ is the unit vector in a particular direction. Substituting in the divergence theorem, we find that

$$\int_S f\mathbf{i}_k\cdot d\mathbf{S} = \int_v \operatorname{div} f\mathbf{i}_k\,dv$$

The divergence on the right-hand side is just the gradient in the $\mathbf{i}_k$ direction, i.e. $\partial f/\partial x_k$. Therefore,

$$\int_S f \mathrm{d}S_k = \int_v \frac{\partial f}{\partial x_k} \mathrm{d}v \tag{A3.6}$$

Problem 3

Show that

$$\int_v (\nabla \times \mathbf{A})\, \mathrm{d}v = \int_S \mathrm{d}\mathbf{S} \times \mathbf{A}$$

We start by using the result (A3.6) applied to the scalar quantities A_x and A_y which are components of the vector **A**. Then

$$\int_v \frac{\partial A_y}{\partial x} \mathrm{d}v = \int_S A_y \, \mathrm{d}S_x \tag{A3.7}$$

$$\int_v \frac{\partial A_x}{\partial y} \mathrm{d}v = \int_S A_x \, \mathrm{d}S_y \tag{A3.8}$$

Subtracting (A3.8) from (A3.7),

$$\int_v \left(\frac{\partial A_y}{\partial x} - \frac{\partial A_x}{\partial y} \right) \mathrm{d}v = \int_S (A_y \, \mathrm{d}S_x - A_x \mathrm{d}S_y)$$

i.e.

$$\int_v \operatorname{curl} A_z \, \mathrm{d}v = -\int_S (\mathbf{A} \times \mathrm{d}\mathbf{S})_z$$

This must be true separately for all three directions, x, y and z and we can summarise these results in the single equation

$$\int_v \operatorname{curl} \mathbf{A} \, \mathrm{d}v = \int_S (\mathrm{d}\mathbf{S} \times \mathbf{A}) \tag{A3.9}$$

A3.3 Results related to Stokes' theorem

Problem

Show that an alternative way of writing Stokes' theorem is

$$\int_C f \mathrm{d}\mathbf{l} = \int_S \mathrm{d}\mathbf{S} \times \operatorname{grad} f$$

Let us write $\mathbf{A} = f\mathbf{i}_k$ where $\mathbf{i}_k$ is the unit vector in the k direction. Then Stokes' theorem,

$$\int_C \mathbf{A} \cdot \mathrm{d}\mathbf{l} = \int_S \operatorname{curl} \mathbf{A} \cdot \mathrm{d}\mathbf{S}$$

becomes

$$\mathbf{i}_k \cdot \int f \mathrm{d}\mathbf{l} = \int_S \operatorname{curl} f\mathbf{i}_k \cdot \mathrm{d}\mathbf{S}$$

We now need the expansion of curl **A** where **A** is the product of two functions, one of which is a vector.

$$\text{curl}\, f\mathbf{g} = f\,\text{curl}\,\mathbf{g} + (\text{grad}\, f) \times \mathbf{g} \qquad \text{(A3.10)}$$

In the present case, **g** is the unit vector $\mathbf{i}_k$ and hence curl $\mathbf{i}_k = 0$. Therefore,

$$\mathbf{i}_k \cdot \int_C f\,\mathrm{d}\mathbf{l} = \int_S [(\text{grad}\, f) \times \mathbf{i}_k] \cdot \mathrm{d}\mathbf{S}$$

$$= \mathbf{i}_k \cdot \int_S \mathrm{d}\mathbf{S} \times \text{grad}\, f$$

i.e.

$$\int_C f\,\mathrm{d}\mathbf{l} = \int_S \mathrm{d}\mathbf{S} \times \text{grad}\, f \qquad \text{(A3.11)}$$

A3.4 Vector fields with special properties

A vector field **A** for which curl $\mathbf{A} = 0$ is called *irrotational, conservative* or a *conservative field of force.* More generally, a vector field is conservative if it satisfies any one of the following conditions, all of which are equivalent.

(*a*) **A** can be expressed in the form $\mathbf{A} = -\text{grad}\,\phi$ where ϕ is a scalar function of position whose values depend only upon x, y and z.

(*b*) curl $\mathbf{A} = 0$.

(*c*) $\int_C \mathbf{A} \cdot \mathrm{d}\mathbf{l} = 0$.

(*d*) $\int_A^B \mathbf{A} \cdot \mathrm{d}\mathbf{l}$ is independent of the path from A to B.

The conditions (*b*) and (*c*) are plainly identical because of Stokes' theorem

$$\int_C \mathbf{A} \cdot \mathrm{d}\mathbf{l} = \int_S \text{curl}\,\mathbf{A} \cdot \mathrm{d}\mathbf{S} = 0$$

If $\mathbf{A} = -\text{grad}\,\phi$, then clearly curl $\mathbf{A} = 0$ because curl grad $= 0$. Finally, we can write

$$\int_A^B \mathbf{A} \cdot \mathrm{d}\mathbf{l} = -\int_A^B \text{grad}\,\phi \cdot \mathrm{d}\mathbf{l} = -(\phi_B - \phi_A)$$

and is independent of path between A and B. This last property is the reason for the name 'conservative field'. It does not matter at all what route one takes between A and B. If **A** is a field of force, ϕ is a potential and ϕ is the amount of work which needs to be expended to bring unit mass, charge, etc. to that point in the field.

A vector field **A** for which div $\mathbf{A} = \nabla \cdot \mathbf{A} = 0$ is called a *solenoidal field.* If $\mathbf{B} = \text{curl}\,\mathbf{A}$, div $\mathbf{B} = 0$. Conversely, if div $\mathbf{B} = 0$, **B** can be expressed as the curl of some vector **A** in many different ways. This is because we can add to **A** an arbitrary conservative vector field which will always vanish when **A** is curled. Thus, if $\mathbf{A}' = \mathbf{A} - \text{grad}\,\phi$,

$$\begin{aligned} \mathbf{B} &= \text{curl}\,\mathbf{A}' = \text{curl}\,\mathbf{A} - \text{curl grad}\,\phi \\ &= \text{curl}\,\mathbf{A} \end{aligned}$$

One of the most useful results in vector analysis is the identity

$$\nabla \times (\nabla \times \mathbf{A}) = \nabla(\nabla \cdot \mathbf{A}) - \nabla^2 \mathbf{A}$$

A3.5 Vector operators in curvilinear coordinates

It is useful to have a list of the vector operators grad, div, curl and ∇^2 in rectangular (or Cartesian), cylindrical and spherical polar coordinates. The standard books give the following.

grad

Cartesian
$$\text{grad}\,\Phi = \nabla\Phi = \mathbf{i}_x \frac{\partial \Phi}{\partial x} + \mathbf{i}_y \frac{\partial \Phi}{\partial y} + \mathbf{i}_z \frac{\partial \Phi}{\partial z}$$

Cylindrical polar
$$\text{grad}\,\Phi = \nabla\Phi = \mathbf{i}_r \frac{\partial \Phi}{\partial r} + \mathbf{i}_z \frac{\partial \Phi}{\partial z} + \mathbf{i}_\phi \frac{1}{r}\frac{\partial \Phi}{\partial \phi}$$

Spherical polar
$$\text{grad}\,\Phi = \nabla\Phi = \mathbf{i}_r \frac{\partial \Phi}{\partial r} + \mathbf{i}_\theta \frac{1}{r}\frac{\partial \Phi}{\partial \theta} + \mathbf{i}_\phi \frac{1}{r\sin\theta}\frac{\partial \Phi}{\partial \phi}$$

div

Cartesian
$$\text{div}\,\mathbf{A} = \nabla\cdot\mathbf{A} = \frac{\partial A_x}{\partial x} + \frac{\partial A_y}{\partial y} + \frac{\partial A_z}{\partial z}$$

Cylindrical polar
$$\text{div}\,\mathbf{A} = \nabla\cdot\mathbf{A} = \frac{1}{r}\left[\frac{\partial}{\partial r}(rA_r) + \frac{\partial}{\partial z}(rA_z) + \frac{\partial}{\partial \phi}(A_\phi)\right]$$

Spherical polar
$$\begin{aligned} \text{div}\,\mathbf{A} = \nabla\cdot\mathbf{A} = \frac{1}{r^2\sin\theta}\bigg[&\frac{\partial}{\partial r}(r^2\sin\theta\,A_r) + \frac{\partial}{\partial \theta}(r\sin\theta\,A_\theta) \\ &+ \frac{\partial}{\partial \phi}(rA_\phi)\bigg] \end{aligned}$$

curl

Cartesian
$$\begin{aligned} \text{curl}\,\mathbf{A} = \nabla\times\mathbf{A} = &\left(\frac{\partial A_y}{\partial z} - \frac{\partial A_z}{\partial y}\right)\mathbf{i}_x + \left(\frac{\partial A_z}{\partial x} - \frac{\partial A_x}{\partial z}\right)\mathbf{i}_y \\ &+ \left(\frac{\partial A_x}{\partial y} - \frac{\partial A_y}{\partial x}\right)\mathbf{i}_z \end{aligned}$$

Cylindrical polar
$$\begin{aligned} \text{curl}\,\mathbf{A} = \nabla\times\mathbf{A} = &\frac{1}{r}\left[\frac{\partial}{\partial z}(rA_\phi) - \frac{\partial}{\partial \phi}(A_z)\right]\mathbf{i}_r \\ &+ \frac{1}{r}\left[\frac{\partial}{\partial \phi}(A_r) - \frac{\partial}{\partial r}(rA_\phi)\right]\mathbf{i}_z \end{aligned}$$

$$+\left[\frac{\partial}{\partial r}(A_z)-\frac{\partial}{\partial z}(A_r)\right]\mathbf{i}_\phi$$

Spherical polar $\quad \operatorname{curl}\mathbf{A} = \nabla\times\mathbf{A} = \frac{1}{r^2\sin\theta}\left[\frac{\partial}{\partial\theta}(r\sin\theta A_\phi)-\frac{\partial}{\partial\phi}(rA_\theta)\right]\mathbf{i}_r$

$$+\frac{1}{r\sin\theta}\left[\frac{\partial}{\partial\phi}(A_r)-\frac{\partial}{\partial r}(r\sin\theta A_\phi)\right]\mathbf{i}_\theta$$

$$+\frac{1}{r}\left[\frac{\partial}{\partial r}(rA_\theta)-\frac{\partial}{\partial\theta}(A_r)\right]\mathbf{i}_\phi$$

Laplacian

Cartesian $\quad \nabla^2\Phi = \frac{\partial^2\Phi}{\partial x^2}+\frac{\partial^2\Phi}{\partial y^2}+\frac{\partial^2\Phi}{\partial z^2}$

Cylindrical polar $\quad \nabla^2\Phi = \frac{1}{r}\frac{\partial}{\partial r}\left(r\frac{\partial\Phi}{\partial r}\right)+\frac{1}{r^2}\frac{\partial^2\Phi}{\partial\phi^2}+\frac{\partial^2\Phi}{\partial z^2}$

Spherical polar $\quad \nabla^2\Phi = \frac{1}{r^2}\frac{\partial}{\partial r}\left(r^2\frac{\partial\Phi}{\partial r}\right)+\frac{1}{r^2\sin\theta}\frac{\partial}{\partial\theta}\left(\sin\theta\frac{\partial\Phi}{\partial\theta}\right)$

$$+\frac{1}{r^2\sin^2\theta}\frac{\partial^2\Phi}{\partial\phi^2}$$

$$=\frac{1}{r}\frac{\partial^2}{\partial r^2}(r\Phi)+\frac{1}{r^2\sin\theta}\frac{\partial}{\partial\theta}\left(\sin\theta\frac{\partial\Phi}{\partial\theta}\right)+\frac{1}{r^2\sin^2\theta}\frac{\partial^2\Phi}{\partial\phi^2}$$

A3.6 Vector operators in solutions of wave equations

In general three dimensional solutions of wave equations, such as Maxwell's equations, you will often find that it simplifies the working considerably to use the relations $\nabla\to \mathrm{i}\mathbf{k}$, $\partial/\partial t\to -\mathrm{i}\omega$, if the phase factor of the wave is written in the form $\exp\{\mathrm{i}(\mathbf{k}\cdot\mathbf{r}-\omega t)\}$. Let us prove the following relations.

$$\operatorname{grad} \mathrm{e}^{\mathrm{i}\mathbf{k}\cdot\mathbf{r}} = \mathrm{i}\mathbf{k}\,\mathrm{e}^{\mathrm{i}\mathbf{k}\cdot\mathbf{r}}$$

$$\operatorname{div}(\mathbf{A}\,\mathrm{e}^{\mathrm{i}\mathbf{k}\cdot\mathbf{r}}) = \mathrm{i}\mathbf{k}\cdot\mathbf{A}\,\mathrm{e}^{\mathrm{i}\mathbf{k}\cdot\mathbf{r}}$$

$$\operatorname{curl}(\mathbf{A}\,\mathrm{e}^{\mathrm{i}\mathbf{k}\cdot\mathbf{r}}) = \mathrm{i}[\mathbf{k}\times\mathbf{A}]\,\mathrm{e}^{\mathrm{i}\mathbf{k}\cdot\mathbf{r}}$$

In these equations, **A** is a constant vector. The exponential phase factor cancels through the wave equation resulting in a relation between **k** and ω which is called the *dispersion relation*. Notice that the vectors **k** and **A** may themselves be complex.

$$\operatorname{grad}\mathrm{e}^{\mathrm{i}\mathbf{k}\cdot\mathbf{r}} = \mathbf{i}_x\frac{\partial}{\partial x}\mathrm{e}^{\mathrm{i}\mathbf{k}\cdot\mathbf{r}}+\mathbf{i}_y\frac{\partial}{\partial y}\mathrm{e}^{\mathrm{i}\mathbf{k}\cdot\mathbf{r}}+\mathbf{i}_z\frac{\partial}{\partial z}\mathrm{e}^{\mathrm{i}\mathbf{k}\cdot\mathbf{r}}$$

$$= \mathrm{i}\mathbf{i}_xk_x\,\mathrm{e}^{\mathrm{i}\mathbf{k}\cdot\mathbf{r}}+\mathrm{i}\mathbf{i}_yk_y\,\mathrm{e}^{\mathrm{i}\mathbf{k}\cdot\mathbf{r}}+\mathrm{i}\mathbf{i}_zk_z\,\mathrm{e}^{\mathrm{i}\mathbf{k}\cdot\mathbf{r}}$$

$$= \mathrm{i}\mathbf{k}\,\mathrm{e}^{\mathrm{i}\mathbf{k}\cdot\mathbf{r}}$$

$$\begin{aligned}
\operatorname{div}(\mathbf{A}\,e^{i\mathbf{k}\cdot\mathbf{r}}) &= \frac{\partial}{\partial x}(A_x\,e^{i\mathbf{k}\cdot\mathbf{r}}) + \frac{\partial}{\partial y}(A_y\,e^{i\mathbf{k}\cdot\mathbf{r}}) + \frac{\partial}{\partial z}(A_z\,e^{i\mathbf{k}\cdot\mathbf{r}}) \\
&= ik_xA_x\,e^{i\mathbf{k}\cdot\mathbf{r}} + ik_yA_y\,e^{i\mathbf{k}\cdot\mathbf{r}} + ik_zA_z\,e^{i\mathbf{k}\cdot\mathbf{r}} \\
&= i\mathbf{k}\cdot\mathbf{A}\,e^{i\mathbf{k}\cdot\mathbf{r}} \\
\operatorname{curl}(\mathbf{A}\,e^{i\mathbf{k}\cdot\mathbf{r}}) &= \mathbf{i}_x(A_zik_y - A_yik_z)\,e^{i\mathbf{k}\cdot\mathbf{r}} + \mathbf{i}_y(A_xik_z - A_zik_x)\,e^{i\mathbf{k}\cdot\mathbf{r}} \\
&\quad + \mathbf{i}_z(A_yik_x - A_xik_y)\,e^{i\mathbf{k}\cdot\mathbf{r}} \\
&= i(\mathbf{k}\times\mathbf{A})\,e^{i\mathbf{k}\cdot\mathbf{r}}
\end{aligned}$$

Thus, for solutions of the wave equation involving travelling wave solutions, we can replace the operators grad, div and curl by the following vector products.

$$\begin{aligned}
\nabla\Phi &= \operatorname{grad}\Phi \rightarrow i\mathbf{k}\Phi \\
\nabla\cdot\mathbf{A} &= \operatorname{div}\mathbf{A} \rightarrow i\mathbf{k}\cdot\mathbf{A} \\
\nabla\times\mathbf{A} &= \operatorname{curl}\mathbf{A} \rightarrow i\mathbf{k}\times\mathbf{A}
\end{aligned}$$

i.e. we replace ∇ by the vector $i\mathbf{k}$.

Similarly, in all partial derivatives with respect to time

$$\frac{\partial}{\partial t}e^{-i\omega t} = -i\omega\,e^{-i\omega t}$$

and, hence

$$\frac{\partial}{\partial t} \rightarrow -i\omega$$

4

HOW TO REWRITE THE HISTORY OF ELECTROMAGNETISM

4.1 Introduction

Now that we have derived Maxwell's equations as he himself derived them, let us do everything backwards and start with a mathematical structure and see how often we have to look at the real world to keep our mathematical model on the right track.

We start with Maxwell's equations but regard them simply as a set of vector equations relating the vector fields **E**, **D**, **B**, **H** and **J**. Initially, these fields have *no physical significance.* We then make a minimum number of postulates from which we attempt to give them physical significance and derive from them all the laws of electromagnetism. This approach is taken by Stratton in his book *Electromagnetic Theory.*[1]

We are then in a position to apply the equations to new aspects of electromagnetic theory – the properties of electromagnetic waves, the emission of waves by accelerated charges, etc. – which provide tests of the equations which go far beyond the laws from which Maxwell's equations were deduced. If the theory goes wrong, the interlocking nature of many of the results, as illustrated below, indicates how the whole edifice would have to be changed.

A number of my colleagues objected strenuously to this approach to electromagnetism, principally on the grounds that no-one would ever have got the right answer by this approach. I am not prepared to speculate about that. What I know is that this procedure of starting with a mathematical structure which is then given physical meaning is found in many other aspects of fundamental physics, for example, the theory of linear operators and quantum mechanics, tensor calculus and the special and general theories of relativity. The mathematics gives a formal coherence to the physical theory and enables predictions to be made about the behaviour of real systems in unexplored regions of parameter space.

This study began as an examples class in mathematical physics and it is instructive to maintain that format. Much of the analysis which follows is mathematically simple – the emphasis is upon the clarity with which we are able to make the correspondence between the mathematics and the physics.

4.2 Maxwell's equations as a set of vector equations

We start with Maxwell's equations in the form

$$\text{curl}\,\mathbf{E} = -\frac{\partial \mathbf{B}}{\partial t} \tag{4.1}$$

$$\text{curl}\,\mathbf{H} = \mathbf{J} + \frac{\partial \mathbf{D}}{\partial t} \tag{4.2}$$

E, **D**, **B**, **H** and **J** are to be regarded as unspecified vector fields which are functions of space and time coordinates and which are intended to describe the electromagnetic field.

They are supplemented by a continuity equation for **J**,

$$\text{div}\,\mathbf{J} + \frac{\partial \rho}{\partial t} = 0 \tag{4.3}$$

We then make the first physical identification:

ρ = *the electric charge density* and *charge is conserved* (4.4)

Show from (4.3) that **J** must be identified as a *current density*, i.e. the rate of flow of charge through unit surface area.

Proof

We integrate equation (4.3) over a volume v bounded by a surface S, i.e.

$$\int_v \text{div}\,\mathbf{J}\,dv = -\frac{\partial}{\partial t}\int_v \rho\,dv$$

Now according to the divergence theorem,

$$\int_v \text{div}\,\mathbf{J}\,dv = \int_S \mathbf{J}\cdot d\mathbf{S} \tag{4.5}$$

is equal to the rate at which charge is lost from the volume,

$$-\frac{\partial}{\partial t}\int_v \rho\,dv$$

i.e.

$$\int_S \mathbf{J}\cdot d\mathbf{S} = -\frac{\partial}{\partial t}(\text{total enclosed charge}) \tag{4.6}$$

Thus, **J** must represent the rate of flow of charge per unit area through the surface S.

4.3 Gauss's theorem in electromagnetism

Show that the fields **B** and **D** must satisfy the relations

$$\text{div}\,\mathbf{B} = 0 \quad \text{and} \quad \text{div}\,\mathbf{D} = \rho \tag{4.7}$$

Derive the corresponding integral forms for these relations – these are different forms of Gauss's theorem.

Proof

Take the divergences of equations (4.1) and (4.2). Since the divergence of a curl is always zero, we obtain

$$\left.\begin{aligned} \text{div curl}\,\mathbf{E} &= -\frac{\partial}{\partial t}(\text{div}\,\mathbf{B}) = 0 \\ \text{div curl}\,\mathbf{H} &= \text{div}\,\mathbf{J} + \frac{\partial}{\partial t}(\text{div}\,\mathbf{D}) = 0 \end{aligned}\right\} \tag{4.8}$$

From (4.3), we obtain

$$\frac{\partial}{\partial t}(\text{div}\,\mathbf{D}) - \frac{\partial \rho}{\partial t} = \frac{\partial}{\partial t}(\text{div}\,\mathbf{D} - \rho) = 0 \tag{4.9}$$

Thus, the *partial derivatives* of div **B** and (div **D** $-\rho$) are both zero at all points in space. We must therefore have

$$\text{div}\,\mathbf{B} = \text{constant} \quad \text{and} \quad \text{div}\,\mathbf{D} - \rho = \text{constant}$$

At this point, we have to establish what these constants are. I have seen three approaches taken. (i) For *simplicity*, set both of the constants to zero and see if we obtain a self consistent story. (ii) At some time, we believe we could so arrange charges and currents in the Universe to reduce div **B** and div **D** $-\rho$ to zero. If we can do it for one moment, it must always be true. (iii) Look at the real world and see what the constants should be once we have a physical identification for the vector fields. I like argument (iii) best and then (i). We will find that we obtain a self consistent story if we take both constants to be zero, i.e.

$$\text{div}\,\mathbf{B} = 0 \tag{4.10}$$

$$\text{div}\,\mathbf{D} - \rho = 0 \tag{4.11}$$

Note that this is the sort of area where we have to abandon strict logic and adopt something which works.

Now, we can write these relations in integral form. Integrate both equations over a closed volume v and apply the divergence theorem

$$\int_v \text{div}\,\mathbf{B}\,\mathrm{d}v = 0$$

and hence

$$\int_S \mathbf{B}\cdot d\mathbf{S} = 0 \tag{4.12}$$

$$\int_v \text{div}\,\mathbf{D}\,dv = \int_v \rho\,dv$$

and hence

$$\int_S \mathbf{D}\cdot d\mathbf{S} = \int_v \rho\,dv \tag{4.13}$$

Notice that equation (4.13) tells us that the field **D** can originate on electric charges.

4.4 Time independent fields as conservative fields of force

Show that if the vector fields **E** and **B** are time independent, **E** must satisfy $\int_C \mathbf{E}\cdot d\mathbf{s} = 0$, i.e. it is a conservative field and therefore can be written in the form $\mathbf{E} = -\,\text{grad}\,\phi$ where ϕ is a scalar potential function. Prove that this is definitely not possible if the field **B** is time varying.

Proof

Since $\partial\mathbf{B}/\partial t = 0$, equation (4.1) tells us that

$$\text{curl}\,\mathbf{E} = 0$$

Therefore, taking the integral of **E** about a closed contour C, we must have

$$\int_C \mathbf{E}\cdot d\mathbf{s} = 0$$

This is one of the ways of defining a *conservative* field (see the appendix to Chapter 3). Since curl grad $\phi = 0$ where ϕ is a scalar function, **E** can be derived from the gradient of a scalar function,

$$\mathbf{E} = -\,\text{grad}\,\phi \tag{4.14}$$

If **B** is time varying, we find that

$$\text{curl}\,\mathbf{E} = -\frac{\partial\mathbf{B}}{\partial t} \neq 0$$

Thus, if we try to derive **E** entirely from the gradient of a scalar potential we find that

$$-\,\text{curl grad}\,\mathbf{E} = 0 = \frac{\partial\mathbf{B}}{\partial t} \neq 0$$

i.e. **E** cannot be wholly expressed as $-\,\text{grad}\,\phi$ if **B** is time varying.

4.5 Boundary conditions in electromagnetism

Across any sharp boundary between two media, the properties of the fields are expected to change discontinuously. There are three cases to consider:

(i) From Section 4.3, show that the component of the vector field **B** perpendicular to the surface between the media $\mathbf{B}\cdot\mathbf{n}$ is continuous at the boundary and likewise $\mathbf{D}\cdot\mathbf{n}$ is continuous if there are no surface charges on the boundary. If there are surface charges of surface charge density σ, show that $(\mathbf{D}_1 - \mathbf{D}_2)\cdot\mathbf{n} = \sigma$.

Proof

This is part of the standard analysis of boundary conditions. We erect a very short cylinder across the boundary (Figure 4.1(*a*)) and then apply the rules which we know are true for the fields **B** and **D** under all circumstances. The diagram shows **B** and **D** fields passing through the boundary.

First use Gauss's Law for **B** in integral form, (4.12),

$$\int_S \mathbf{B}\cdot d\mathbf{S} = 0 \tag{4.12}$$

Now we squash the cylinder until it is infinitesimally thin. Then the surface area round the edges $2\pi r\, dl$ goes to zero as dl tends to zero and the only components left are those through the upper and lower faces of the cylinder. If **n** is the unit vector normal to the surface, $\mathbf{B}\cdot\mathbf{n}$ must be the same on either side of the boundary. (Remember that $d\mathbf{S}$ must be taken pointing outwards (or inwards) through the closed surface.)

In exactly the same way,

Figure 4.1

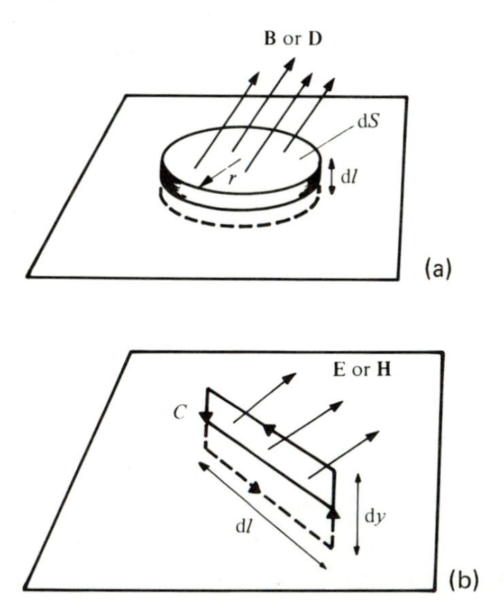

$$\int_S \mathbf{D}\cdot d\mathbf{S} = \int_v \rho dv$$

When we squash the cylinder, the left-hand side just becomes the difference in the values of $\mathbf{D}\cdot\mathbf{n}\, d\mathbf{S}$ on either side of the surface and $\int_v \rho dv$ becomes the surface charge σdS, i.e.

$$(\mathbf{D}_1 - \mathbf{D}_2)\cdot\mathbf{n} = \sigma \tag{4.15}$$

If there are no surface charges, $\sigma = 0$ and $\mathbf{D}\cdot\mathbf{n}$ is continuous.

(ii) If the fields **E** and **H** are static fields (as are also **D**, **B** and **J**), show that (*a*) the tangential component of **E** is continuous at the boundary and (*b*) the tangential component of **H** is continuous if there is no surface current density $\mathbf{J}_s$. If there is, show that

$$(\mathbf{H}_1 - \mathbf{H}_2)\times\mathbf{n} = \mathbf{J}_s$$

This is another standard analysis of boundary conditions.

Proof

(*a*) Create a little circuit in the surface as shown in Figure 4.1(*b*). We now integrate the fields **E** and **H** around this closed loop. All fields are static and therefore **E** is a conservative field since, as shown in Section 4.4, curl $\mathbf{E} = 0$ or $\int_C \mathbf{E}\cdot d\mathbf{s} = 0$.

Let $\mathbf{E}_1$ and $\mathbf{E}_2$ be the fields on either side of the surface. The loop can be squashed to infinitesimal thickness so that the ends make no contribution to the line integral. Then taking **n** normal to the surface again,

$$(\mathbf{E}_1 - \mathbf{E}_2)\times\mathbf{n} = 0$$

i.e. the tangential components of **E** are equal on either side of the interface.

(*b*) Now consider (4.2) for static fields.

$$\text{curl}\,\mathbf{H} = \mathbf{J} \tag{4.16}$$

As in Section 4.4, integrate these fields through the area of the little loop and apply Stokes' theorem.

$$\int_S \text{curl}\,\mathbf{H}\cdot d\mathbf{S} = \int_S \mathbf{J}\cdot d\mathbf{S}$$

$$\int_C \mathbf{H}\cdot d\mathbf{s} = \int_S \mathbf{J}\cdot d\mathbf{S}$$

Applying this result to the little current loop when it is squashed to zero thickness, $\int_S \mathbf{J}\cdot d\mathbf{S}$ just becomes the total surface current flowing normal to the loop, i.e. $J_S dl$, where dl is the length of the loop and J_S is the surface current density, i.e. the current per unit length of surface normal to the current.

As before, we can write the result $(\mathbf{H}_1 - \mathbf{H}_2) \times \mathbf{n}\, \mathrm{d}l = \mathbf{J}_s \mathrm{d}l$ or

$$(\mathbf{H}_1 - \mathbf{H}_2) \times \mathbf{n} = \mathbf{J}_s \tag{4.17}$$

If there are no surface currents, $\mathbf{J}_s = 0$, $\mathbf{H} \times \mathbf{n}$ is continuous, i.e. the tangential component of $\mathbf{H}$ is continuous.

Notice that, at this stage, it is not so obvious that the relations (4.14) and (4.15) are true in the presence of time varying fields because we have thrown away the time varying components. In fact, they are true as we demonstrate in (iii).

(iii) By consideration of equations (4.1) and (4.2), prove that the statements in (ii) are correct even in the presence of time varying fields.

Proof

This requires a little more attention to detail. Let us write down the integral of equation (4.1) over a surface S and then apply Stokes' theorem.

$$\int_S \operatorname{curl} \mathbf{E} \cdot \mathrm{d}\mathbf{S} = -\int_S \frac{\partial \mathbf{B}}{\partial t} \cdot \mathrm{d}\mathbf{S}$$

$$\int_C \mathbf{E} \cdot \mathrm{d}\mathbf{s} = -\int_S \frac{\partial \mathbf{B}}{\partial t} \cdot \mathrm{d}\mathbf{S}$$

Now apply this result to the little rectangle in Figure 4.1(*b*). The analysis proceeds as before except that there is the time varying component on the right-hand side.

$$(E_{2\parallel} - E_{1\parallel})\, \mathrm{d}l = -\frac{\partial B_\perp}{\partial t} \mathrm{d}l\, \mathrm{d}y + \text{small end contributions}$$

Now cancel out the dls on either side and let the breadth of the rectangle shrink to zero, $\mathrm{d}y \to 0$.

$$E_{2\parallel} - E_{1\parallel} = -\left(\frac{\partial B_\perp}{\partial t} \mathrm{d}y\right)_{\mathrm{d}y \to 0}$$

$\partial B_\perp / \partial t$ is a finite quantity in the case of time varying fields but in the limit in which dy tends to zero, the right-hand side makes no contribution, i.e. again

$$(\mathbf{E}_2 - \mathbf{E}_1) \times \mathbf{n} = 0$$

Exactly the same analysis can be carried out for time varying $\mathbf{D}$ fields, i.e.

$$\int_S \operatorname{curl} \mathbf{H} \cdot \mathrm{d}\mathbf{S} = \int_S \mathbf{J} \cdot \mathrm{d}\mathbf{S} + \int_S \frac{\partial \mathbf{D}}{\partial t} \cdot \mathrm{d}\mathbf{S}$$

$$\int_C \mathbf{H} \cdot \mathrm{d}\mathbf{S} = \int_S \mathbf{J} \cdot \mathrm{d}\mathbf{S} + \int_S \frac{\partial \mathbf{D}}{\partial t} \cdot \mathrm{d}\mathbf{S}$$

Squash the rectangle to zero thickness

$$\mathbf{n} \times (\mathbf{H}_2 - \mathbf{H}_1)\, \mathrm{d}l = \mathbf{J}_\perp \mathrm{d}l \mathrm{d}y + \left(\frac{\partial \mathbf{D}}{\partial t} \mathrm{d}l\, \mathrm{d}y\right) + \text{small end contributions}$$

i.e.

$$\mathbf{n} \times (\mathbf{H}_2 - \mathbf{H}_1) = \mathbf{J}_S + \left(\frac{\partial \mathbf{D}}{\partial t} \mathrm{d}y\right)_{\mathrm{d}y \to 0}$$

$$\mathbf{n} \times (\mathbf{H}_2 - \mathbf{H}_1) = \mathbf{J}_S$$

even in the presence of time varying **D** fields.

4.6 Ampère's law

If **H** is not time varying (we will eventually identify this sort of **H** with a magnetostatic field), show that **H** can be expressed in the form $-\text{grad } V_{\mathbf{mag}}$ when there are only permanent magnets and magnetisable materials but *no* currents. If there are steady currents show that $\int_C \mathbf{H} \cdot \mathrm{d}\mathbf{s} = I_{\mathbf{enclosed}}$. This equation is often used even for time varying fields. Why is this so?

Proof

From equation (4.2), curl $\mathbf{H} = 0$. **H** is therefore a conservative field and can be expressed as the gradient of a scalar field

$$\mathbf{H} = -\text{grad } V_{\mathbf{mag}}$$

If *steady* currents are present, curl $\mathbf{H} = \mathbf{J}$. Integrating over a surface S and applying Stokes' theorem.

$$\int_S \text{curl } \mathbf{H} \cdot \mathrm{d}\mathbf{S} = \int_S \mathbf{J} \cdot \mathrm{d}\mathbf{S}$$

$$\int_C \mathbf{H} \cdot \mathrm{d}\mathbf{s} = (\text{total enclosed current}) = I_{\mathbf{enclosed}} \tag{4.18}$$

This equation can be used in the presence of time varying **D** fields, provided that they are very slowly varying, i.e. we require that $\partial \mathbf{D}/\partial t \ll \mathbf{J}$. This is very often a good approximation but we have to check every time that it is a valid one when any of the fields or currents are varying.

4.7 Faraday's law

Derive Faraday's law, $\int_C \mathbf{E} \cdot \mathrm{d}\mathbf{s} = -\mathrm{d}\Phi/\mathrm{d}t$, where $\Phi = \int_S \mathbf{B} \cdot \mathrm{d}\mathbf{S}$ is defined to be the flux of the field **B** through the circuit C. In this formulation, **E** appears to be the *total* electric field, whereas the usual statement of Faraday's law refers only to the induced field in the circuit. Explain this discrepancy.

Proof

From equation (4.1), and using Stokes' theorem,

$$\int_S \text{curl } \mathbf{E} \cdot \mathrm{d}\mathbf{S} = -\frac{\partial}{\partial t}\left(\int_S \mathbf{B} \cdot \mathrm{d}\mathbf{S}\right)$$

i.e.

$$\int_C \mathbf{E}\cdot d\mathbf{s} = -\frac{\partial \Phi}{\partial t} \tag{4.19}$$

Normally, Faraday's law refers only to the induced part of the field **E**. There may also, however, be a component due to an electrostatic field, i.e.

$$\begin{aligned}
\mathbf{E} &= \mathbf{E}_{\text{induced}} + \mathbf{E}_{\text{electrostatic}} \\
&= \mathbf{E}_{\text{induced}} - \text{grad } V \\
\text{curl } \mathbf{E} &= \text{curl } \mathbf{E}_{\text{induced}} - \text{curl grad } V \\
&= -\frac{\partial \mathbf{B}}{\partial t} - 0
\end{aligned}$$

which implies that **E** refers to the total field including the electrostatic part but the latter does not survive curling.

4.8 The story so far

All the above analysis has been based upon the mathematical properties of the basic set of vector equations introduced in Section 4.2. Although we have used words like magnets, electrostatic field and so on, the properties of the equations are independent of these physical identifications. We now need to give the fields **E**, **D**, **H** and **B** some physical meaning.

We define $q\mathbf{E}$ to be the force acting on a stationary charge q. Let us see whether we have yet closed our set of equations. **J** is defined by equations (4.3) and (4.4) and **E** has just been defined. However, we have still to define **D**, **B** and **H** and we have only two independent equations left, (4.1) and (4.2). We have not yet closed the set of equations. In order to do so, we have to introduce a further set of definitions based upon experimental evidence. First of all, we consider the behaviour of electromagnetic fields *in vacuo* and make the further definitions, consistent with experiment, that

$$\left.\begin{aligned}
\mathbf{D} &= \epsilon_0 \mathbf{E} \\
\mathbf{B} &= \mu_0 \mathbf{H}
\end{aligned}\right\} \quad \text{in vacuo} \tag{4.20}$$

where ϵ_0 and μ_0 are constants. **D** is called the electric flux density, **E** the electric field strength, **H** the magnetic field strength and **B** the magnetic flux density.

Inside material media, these relations may not be correct and so we introduce vectors which describe the difference between the actual values and the vacuum definitions – these refer to the electric and magnetic *polarisation properties* of the material.

$$\mathbf{P} = \mathbf{D} - \epsilon_0 \mathbf{E} \tag{4.21}$$

$$\mathbf{M} = \frac{\mathbf{B}}{\mu_0} - \mathbf{H} \tag{4.22}$$

Granted these definitions, our objective is now to find physical meanings for the 'polarisation vectors', **P** and **M**. Let us proceed with the mathematical analysis of the equations and see if these definitions will lead to consistency with all the known laws of electricity and magnetism.

4.9 Derivation of Coulomb's law

In an electrostatic field with charges in a vacuum, show that $\nabla^2 V = -\rho/\epsilon_0$. From the definition of **E** in Section 4.8 and from Section 4.4, show that V is the work done per unit charge in bringing a charge to the point **r** from infinity. Show that the solution of the equation is

$$V(\mathbf{r}) = \int \frac{\rho(\mathbf{r}')}{4\pi\epsilon_0 |\mathbf{r} - \mathbf{r}'|} \mathrm{d}^3\mathbf{r}'$$

Hence derive Coulomb's law, $F = q_1 q_2/4\pi\epsilon_0 r^2$.

Proof

We have shown that div $\mathbf{D} = \rho$ and in a vacuum we have defined $\mathbf{D} = \epsilon_0 \mathbf{E}$. Therefore,

$$\operatorname{div}(\epsilon_0 \mathbf{E}) = \rho$$

and

$$\operatorname{div}(\epsilon_0 \operatorname{grad} V) = -\rho$$

i.e.

$$\nabla^2 V = -\frac{\rho}{\epsilon_0}$$

This is *Poisson's equation for electrostatics in a vacuum with a charge distribution* ρ.

We use the basic definition of work done as the work which has to be expended against the field **E** to bring the charge q from infinity to **r**, i.e.

$$\text{work done} = -\int_\infty^r \mathbf{F} \cdot \mathrm{d}\mathbf{r} = -q \int_\infty^r \mathbf{E} \cdot \mathrm{d}\mathbf{r} = q \int_\infty^r \operatorname{grad} V \cdot \mathrm{d}\mathbf{r} = qV \tag{4.23}$$

i.e. the electrostatic potential at a point measures the work done in an electrostatic field to bring a test particle from infinity to that point in the field. Notice that it is *not* the amount of energy needed to set up the electric field distribution as a whole. We will work that out in Section 4.12.

To solve Poisson's equation, it is simplest to work backwards. The reason is that *Laplace's equation* $\nabla^2 V = 0$ is a linear equation to which the principle of superposition applies, i.e. if V_1 and V_2 are two separate solutions of the equation, $a_1 V_1 + a_2 V_2$ is also a solution. Therefore, since we can imagine the charges q to be point charges in a vacuum, the field at any point is the superposition of vacuum solutions of Laplace's equation which have to be related to the sources of the field, i.e. the charges q. Let us therefore only consider a little bit of the charge distribution located at $\mathbf{r}'$:

$$\rho(\mathbf{r}')\,\mathrm{d}^3\mathbf{r}' = q(\mathbf{r}')$$

Then the proposed solution of Laplace's equation is

$$V(r) = \frac{\rho(\mathbf{r}')\,\mathrm{d}^3\mathbf{r}'}{4\pi\epsilon_0|\mathbf{r}-\mathbf{r}'|} = \frac{q}{4\pi\epsilon_0 r} \tag{4.24}$$

where r is the radial distance from the charge q at point $\mathbf{r}'$. Let us now test this solution away from $r = 0$. We recall that Laplace's equation in spherical polar coordinates is

$$\nabla^2 V = \frac{1}{r}\frac{\partial^2}{\partial r^2}(rV) + \frac{1}{r^2}\left[\frac{1}{\sin\theta}\frac{\partial}{\partial\theta}\left(\sin\theta\,\frac{\partial V}{\partial\theta}\right) + \frac{1}{\sin^2\theta}\frac{\partial^2 V}{\partial\phi^2}\right]$$

In the present case, we need only consider the radial coordinate because of the spherical symmetry of the problem, i.e. test $V(r)$ in $(1/r)(\partial^2/\partial r^2)(rV)$. Therefore,

$$\nabla^2 V = \frac{1}{r}\frac{\partial^2}{\partial r^2}(rV) = \frac{1}{r}\frac{\partial^2}{\partial r^2}\left(\frac{q}{4\pi\epsilon_0}\right) = 0$$

provided $r \neq 0$. Thus, the proposed solution is good away from the origin.

Now let us evaluate the charge Q enclosed by a sphere centred on the origin.

$$Q = \int_v \rho\,\mathrm{d}v = -\int_v \epsilon_0\nabla^2 V\,\mathrm{d}v$$

$V = q/4\pi\epsilon_0 r$ and hence the total enclosed charge is

$$Q = -\frac{q}{4\pi}\int_v \mathrm{div}\left(\mathrm{grad}\,\frac{1}{r}\right)\mathrm{d}v$$

Applying the divergence theorem,

$$Q = -\frac{q}{4\pi}\int_S \mathrm{grad}\left(\frac{1}{r}\right)\cdot\mathrm{d}\mathbf{S} = \frac{q}{4\pi}\int_S \frac{1}{r^2}r^2\,\mathrm{d}\Omega = \frac{q}{4\pi}\int_S \mathrm{d}\Omega = q$$

Thus, according to Poisson's equation, we obtain the correct charge at the origin from the proposed solution. We conclude that the solution

$$V(\mathbf{r}) = \frac{\rho(\mathbf{r}')\,\mathrm{d}^3\mathbf{r}'}{4\pi\epsilon_0|\mathbf{r}-\mathbf{r}'|}$$

satisfies Poisson's equation in electrostatics.

Returning again to a single particle q_1 at the origin, $\nabla^2 V = q_1/\epsilon_0$. The force on another particle q_2 is $q_2\mathbf{E}$, i.e.

$$\mathbf{F} = -q_2\,\mathrm{grad}\,V = -\frac{q_1 q_2}{4\pi\epsilon_0}\mathrm{grad}\left(\frac{1}{r}\right)$$

$$= \frac{q_1 q_2}{4\pi\epsilon_0}\frac{\mathbf{r}}{r^3} \tag{4.25}$$

This is the derivation of *Coulomb's inverse square law in electrostatics.* It may look as if we have taken rather a long time to get back to where most courses in electricity and magnetism begin. The point of interest is that we can start with Maxwell's equations and deduce Coulomb's law rigorously with a minimum number of assumptions.

4.10 Derivation of the Biot–Savart law

Steady or slowly varying currents are flowing in a vacuum. What is 'slowly varying'? Show that curl $\mathbf{H} = \mathbf{J}$.

It can be shown that the solution of this equation is

$$\mathbf{H}(\mathbf{r}) = \int \frac{\mathbf{J}(\mathbf{r}') \times (\mathbf{r} - \mathbf{r}')}{4\pi |\mathbf{r} - \mathbf{r}'|^3} \mathrm{d}^3\mathbf{r}' \tag{4.26}$$

Show that this is just the normal formula $\mathrm{d}H = I\mathrm{d}s \sin\theta / 4\pi r^2$ for the magnetic field from a current element $I\mathrm{d}s$.

Proof

We have already discussed the question of slowly varying fields in Section 4.6. 'Slowly varying' means that the displacement current $\partial\mathbf{D}/\partial t \ll \mathbf{J}$. Then directly from equation (4.2) we obtain curl $\mathbf{H} = \mathbf{J}$. Just as in Section 4.9 where we needed the solution of $\nabla^2 V = \rho/\epsilon_0$, so here we need the solution of curl $\mathbf{H} = \mathbf{J}$. The general solution is normally derived in the final-year course but we will not do that here. The solution we need is quoted above as (4.26). Let us remove the integration sign so that the relation becomes one between a current element and the field at a distance r from it.

$$\mathrm{d}\mathbf{H} = \frac{\mathbf{J}(\mathbf{r}') \times (\mathbf{r} - \mathbf{r}')}{4\pi |\mathbf{r} - \mathbf{r}'|^3} \mathrm{d}^3\mathbf{r}' \tag{4.27}$$

Now we identify $\mathbf{J}(\mathbf{r}')\mathrm{d}^3\mathbf{r}'$ with the current element $I\mathrm{d}\mathbf{s}$. Then we obtain

$$\mathrm{d}\mathbf{H} = \frac{I\mathrm{d}\mathbf{s} \times \mathbf{r}}{4\pi r^3} = \frac{I \sin\theta\, \mathrm{d}s}{4\pi r^2} \tag{4.28}$$

which is just the Biot–Savart law. With the equation in this form, by $\mathbf{r}$ we mean the vector from the current element to the point in the field.

4.11 The interpretation of Maxwell's equations in the presence of material media

Using the definitions in Section 4.8 and the results of Section 4.3, show that Maxwell's equations may be written in the form

$$\left.\begin{aligned}
\text{curl } \mathbf{E} &= -\frac{\partial \mathbf{B}}{\partial t} \\
\text{div } \mathbf{B} &= 0 \\
\text{div } \epsilon_0 \mathbf{E} &= (\rho - \text{div } \mathbf{P}) \\
\text{curl } (\mathbf{B}/\mu_0) &= \left(\mathbf{J} + \frac{\partial \mathbf{P}}{\partial t} + \text{curl } \mathbf{M}\right) + \frac{\partial \epsilon_0 \mathbf{E}}{\partial t}
\end{aligned}\right\} \tag{4.29}$$

Show that these equations may be interpreted as follows.

(i) 'In electrostatics the field **E** may be calculated correctly everywhere by replacing polarisable media by a vacuum together with a volume charge distribution $-\text{div } \mathbf{P}$ and a surface charge density $\mathbf{P} \cdot \mathbf{n}$'. Then write down the expression for the electrostatic potential V at any point in space in terms of the charges and show that **P** represents the dipole moment per unit volume within the medium. To do this, you may use the fact that the electrostatic potential due to an electric dipole of dipole moment **p** can be written

$$V = \frac{1}{4\pi\epsilon_0} \mathbf{p} \cdot \nabla \left(\frac{1}{r}\right) \tag{4.30}$$

(ii) In magnetostatics, the **B** field may be calculated by replacing polarisable bodies by a current distribution curl **M** with surface current densities of $-\mathbf{n} \times \mathbf{M}$.

Proof

I regard this as a particularly pleasant piece of analysis which gives remarkable insight into the mathematical structure of Maxwell's equations. The first part is straightforward and simply involves rewriting Maxwell's equations.

The first equation is simply the original version of Maxwell's equation (4.1) and the second the general result concerning magnetic fields derived in Section 4.2.

$$\text{curl } \mathbf{E} = -\frac{\partial \mathbf{B}}{\partial t} \tag{4.1}$$

$$\text{div } \mathbf{B} = 0 \tag{4.10}$$

In the third case, $\text{div } \mathbf{D} = \rho$ and since by definition $\mathbf{D} = \mathbf{P} + \epsilon_0 \mathbf{E}$,

$$\text{div } (\epsilon_0 \mathbf{E}) = \rho - \text{div } \mathbf{P} \tag{4.31}$$

Similarly, from equation (4.2) and the definition $\mathbf{H} = (\mathbf{B}/\mu_0) - \mathbf{M}$,

$$\text{curl } \mathbf{H} = \mathbf{J} + \frac{\partial \mathbf{D}}{\partial t}$$

$$\text{curl}\left(\frac{\mathbf{B}}{\mu_0} - \mathbf{M}\right) = \mathbf{J} + \frac{\partial \mathbf{P}}{\partial t} + \frac{\partial (\epsilon_0 \mathbf{E})}{\partial t}$$

$$\text{curl}\left(\frac{\mathbf{B}}{\mu_0}\right) = \left(\mathbf{J} + \frac{\partial \mathbf{P}}{\partial t} + \text{curl}\,\mathbf{M}\right) + \frac{\partial(\epsilon_0 \mathbf{E})}{\partial t} \tag{4.32}$$

Now let us look at questions (i) and (ii).

(i) Equation (4.31) tells us that, at any point in space, to calculate div $(\epsilon_0 \mathbf{E})$ we have to add on to ρ the quantity $-$ div **P** which we may think of as being an additional effective charge ρ^* to the charge ρ, i.e. $\rho^* = -\,\text{div}\,\mathbf{P}$.

We now have to look at the boundary conditions between the region containing the polarisable medium and the vacuum. From Section 4.5, we know that under all circumstances the component of **D** normal to the boundary is continuous.

$$(\mathbf{D}_2 - \mathbf{D}_1)\cdot\mathbf{n} = \sigma \tag{4.33}$$

where σ is the surface charge density. We will suppose that there are no free charges on the surface, $\sigma = 0$. Therefore,

$$(\mathbf{D}_2 - \mathbf{D}_1)\cdot\mathbf{n} = 0$$

$$[\epsilon_0 \mathbf{E}_2 - (\epsilon_0 \mathbf{E}_1 + \mathbf{P}_1)]\cdot\mathbf{n} = 0$$

$$[\epsilon_0 \mathbf{E}_2 - \epsilon_0 \mathbf{E}_1] = \mathbf{P}_1\cdot\mathbf{n} \tag{4.34}$$

i.e. the effective surface charge density σ^* is $\mathbf{P}_1\cdot\mathbf{n}$ per unit area.

Thus, in calculating the **E** field we get the correct answer if we replace the material medium by a charge distribution $\rho^* = -\,\text{div}\,\mathbf{P}$ and a surface charge distribution $\sigma^* = \mathbf{P}\cdot\mathbf{n}$ per unit area. Let us now interpret what this means. Suppose we have *real charge distributions* ρ^* and σ^*– what is the field at any point in space?

The total electric field due to these charges is obtained from the gradient of the electrostatic potential V where

$$V = \frac{1}{4\pi\epsilon_0}\int_v \frac{\rho^*}{r}\,dv + \frac{1}{4\pi\epsilon_0}\int_S \frac{\sigma^*}{r}\,dS$$

$$= -\frac{1}{4\pi\epsilon_0}\int_v \frac{\text{div}\,\mathbf{P}}{r}\,dv + \frac{1}{4\pi\epsilon_0}\int_S \frac{\mathbf{P}\cdot d\mathbf{S}}{r}$$

Applying the divergence theorem to the second term, we obtain

$$V = \frac{1}{4\pi\epsilon_0}\int_v \left[-\frac{\text{div}\,\mathbf{P}}{r} + \text{div}\left(\frac{\mathbf{P}}{r}\right)\right] dv$$

But

$$\text{div}\left(\frac{\mathbf{P}}{r}\right) = (\text{div}\,\mathbf{P})\frac{1}{r} + \mathbf{P}\cdot\text{grad}\,\frac{1}{r}$$

and hence

$$V = \frac{1}{4\pi\epsilon_0}\int_v \mathbf{P}\cdot\text{grad}\left(\frac{1}{r}\right) dv \tag{4.35}$$

This is the rather beautiful result we have been seeking. We recall that the potential at distance r from an electrostatic dipole is given by

$$V = \frac{1}{4\pi\epsilon_0}\mathbf{p}\cdot\text{grad}\left(\frac{1}{r}\right)$$

where $\mathbf{p}$ is the electric dipole moment of the dipole. Thus, we can interpret equation (4.35) as showing that the quantity $\mathbf{P}$ is simply the *dipole moment per unit volume* within the material.

If we want a physical picture for what this means, we can interpret the phenomenon of polarisation as the dipole moment per unit volume caused by putting the material in an electrostatic field $\mathbf{E}$. This is the origin of the term $\rho^* = -\,\text{div}\,\mathbf{P}$. However, the dipoles at the surfaces of the material will result in positive charges sticking out one side of the material and negative charges sticking out the other so that the whole system remains neutral. It is these charges which give rise to the surface charge distribution of $\sigma^* = \mathbf{P}\cdot\mathbf{n}$.

(ii) The analysis in the case of magnetostatics proceeds in exactly the same way. Equation (4.32) tells us that we have to include a current density distribution $\mathbf{J}^* = \text{curl}\,\mathbf{M}$ in the expression for $\mathbf{B}$. In the case of surfaces, we have from equation (4.17)

$$(\mathbf{H}_1 - \mathbf{H}_2)\times\mathbf{n} = \mathbf{J}_s$$

If there are no 'real' surface currents, this becomes

$$(\mathbf{H}_1 - \mathbf{H}_2)\times\mathbf{n} = 0$$

$$\left[\left(\frac{\mathbf{B}_1}{\mu_0} - \mathbf{M}_1\right) - \frac{\mathbf{B}_2}{\mu_0}\right]\times\mathbf{n} = 0$$

i.e.

$$\mathbf{n}\times\left[\frac{\mathbf{B}_2}{\mu_0} - \frac{\mathbf{B}_1}{\mu_0}\right] = -(\mathbf{M}_1\times\mathbf{n}) \tag{4.36}$$

i.e., in the notation of equation (4.17), there is a surface current distribution $\mathbf{J}_s^* = -(\mathbf{M}_1\times\mathbf{n})$. The ambitious reader may like to interpret the current distributions $\mathbf{J}^*$ and $\mathbf{J}_s^*$ in terms of a closed system of currents in a vacuum, similar to our analysis of the meanings of ρ^* and σ^* in (i).

4.12 The energy density of electromagnetic fields

From the definition of the electric field $\mathbf{E}$ in Section 4.8, show that the rate at which batteries have to work to push charges and currents against the electrostatic fields including emfs is (neglecting Ohmic heating)

$$\int_v \mathbf{J}\cdot(-\mathbf{E})\,dv$$

and hence the total energy of the system is

$$U = -\int_{\text{all space}} \mathrm{d}v \int_0^{\text{final fields}} \mathbf{J}\cdot\mathbf{E}\,\mathrm{d}t$$

By multiplying equation (4.2) by **E** and (4.1) by **H**, transform this equation into

$$U = \int_{\text{all space}} \mathrm{d}v \int_0^{\mathbf{D}} \mathbf{E}\cdot\mathrm{d}\mathbf{D} + \int_{\text{all space}} \mathrm{d}v \int_0^{\mathbf{B}} \mathbf{H}\cdot\mathrm{d}\mathbf{B}$$

Proof

This is one of the classic pieces of analysis in electromagnetic theory. We start with the work done by the electromagnetic field on a particle of charge q. In general, work is only done by the electric field component because, in the case of magnetic fields, no work is done on the particle, the force acting perpendicular to the displacement of the particle $\mathbf{F} = q(\mathbf{u}\times\mathbf{B})$. Therefore, as the particle moves from $\mathbf{r}$ to $\mathbf{r}+\mathrm{d}\mathbf{r}$ in unit time, the work done per second is $q\mathbf{E}\cdot\mathrm{d}\mathbf{r}$, i.e.

$$\text{work done} = q\mathbf{E}\cdot\mathbf{u}$$

Therefore, per unit volume, the work done is $qN\mathbf{E}\cdot\mathbf{u} = \mathbf{J}\cdot\mathbf{E}$ where N is the number density of charges q. Now integrate over all space and the total work done by the currents is

$$\int_v \mathbf{J}\cdot\mathbf{E}\,\mathrm{d}v$$

Now we have to get this work from somewhere and it must come from the batteries which are driving the currents, i.e. we must do work on the system which is $\int_v (-\mathbf{J})\cdot\mathbf{E}\,\mathrm{d}v$. Therefore, the total amount of energy which we put into the system is

$$U = -\int_{\text{all space}} \mathrm{d}v \int_0^t (\mathbf{J}\cdot\mathbf{E})\,\mathrm{d}t \tag{4.37}$$

This is the only tricky part of the analysis – the rest is straightforward. Now express $(\mathbf{J}\cdot\mathbf{E})$ in terms of **E**, **D**, **H** and **B**.

$$\mathbf{E}\cdot\mathbf{J} = \mathbf{E}\cdot\left(\text{curl}\,\mathbf{H} - \frac{\partial\mathbf{D}}{\partial t}\right) \tag{4.38}$$

Now take the scalar product of equation (4.1) with **H** and add it to equation (4.38).

$$\begin{aligned}\mathbf{E}\cdot\mathbf{J} &= \mathbf{E}\cdot\text{curl}\,\mathbf{H} - \mathbf{H}\cdot\text{curl}\,\mathbf{E} - \mathbf{E}\cdot\frac{\partial\mathbf{D}}{\partial t} - \mathbf{H}\cdot\frac{\partial\mathbf{B}}{\partial t}\\ &= \text{div}\,(\mathbf{E}\times\mathbf{H}) - \mathbf{E}\cdot\frac{\partial\mathbf{D}}{\partial t} - \mathbf{H}\cdot\frac{\partial\mathbf{B}}{\partial t}\end{aligned}$$

Therefore,

$$\int_0^t \mathbf{J}\cdot\mathbf{E}\,\mathrm{d}t = \int_0^t \text{div}\,(\mathbf{E}\times\mathbf{H})\,\mathrm{d}t - \int \mathbf{E}\cdot\mathrm{d}\mathbf{D} - \int \mathbf{H}\cdot\mathrm{d}\mathbf{B}$$

Let us apply the divergence theorem to the first term on the right-hand side.

$$\int_v \int_0^t \operatorname{div}(\mathbf{E}\times\mathbf{H})\,\mathrm{d}t\,\mathrm{d}v = \int_0^t \int_S (\mathbf{E}\times\mathbf{H})\cdot \mathrm{d}\mathbf{S}\,\mathrm{d}t$$

Clearly, this represents a flux of energy through the surface S which encloses the system. The quantity $\mathbf{E}\times\mathbf{H}$ which is known as the *Poynting vector* is the rate of flow of energy in the direction normal to both **E** and **H** per unit area. Evidently, the integral over the closed surface S represents a loss of energy through the surface, i.e. it represents a radiation loss. The other two terms clearly represent the energy stored in the **E** and **H** fields in the volume v. We can therefore write the expression for the energy in electric and magnetic fields:

$$U = \int\int_v \mathbf{E}\cdot\mathrm{d}\mathbf{D}\,\mathrm{d}v + \int\int_v \mathbf{H}\cdot\mathrm{d}\mathbf{B}\,\mathrm{d}v \tag{4.39}$$

This is the answer we were looking for.

Transform the first term of this equation into the form $\int \frac{1}{2}\rho V \mathrm{d}v$ if all polarisable media are linear, i.e. $\mathbf{D} \propto \mathbf{E}$.

Proof

Let us do this one backwards.

$$\tfrac{1}{2}\int_v \rho V \mathrm{d}v = \tfrac{1}{2}\int_v (\operatorname{div}\mathbf{D})V\mathrm{d}v$$

Now $\operatorname{div}(V\mathbf{D}) = \operatorname{grad} V\cdot\mathbf{D} + V\operatorname{div}\mathbf{D}$ and hence

$$\tfrac{1}{2}\int_v \rho V\mathrm{d}v = \tfrac{1}{2}\int_v \operatorname{div} V\mathbf{D}\,\mathrm{d}v - \tfrac{1}{2}\int_v \operatorname{grad} V\cdot\mathbf{D}\,\mathrm{d}v$$

$$= \tfrac{1}{2}\int_S V\mathbf{D}\cdot\mathrm{d}\mathbf{S} + \tfrac{1}{2}\int_v \mathbf{E}\cdot\mathbf{D}\,\mathrm{d}v \tag{4.40}$$

Now we only have to deal with the surface integral. We take the integral over a very large volume and then ask how V and **D** vary with radius r. If we have an isolated system, the lowest multipole electric field which could be present is that of an electric charge for which $D \propto E \propto r^{-2}$ and $V \propto r^{-1}$. If, for example, the system were neutral, the lowest possible multiple of the field at a large distance would be a dipole field for which $D \propto r^{-3}$, $V \propto r^{-2}$. Therefore, the best we can do is to consider the lowest multipole case.

$$\tfrac{1}{2}\int_S V\mathbf{D}\cdot\mathrm{d}\mathbf{S} \propto \tfrac{1}{2}\int \frac{1}{r}\times\frac{1}{r^2}r^2\,\mathrm{d}\Omega \propto \frac{1}{r} \to 0 \text{ as } r\to\infty$$

Thus, in the limit, the first term on the right-hand side of equation (4.40) vanishes and

$$\tfrac{1}{2}\int_v \rho V \,\mathrm{d}v = \tfrac{1}{2}\int_v \mathbf{E}\cdot\mathbf{D}\,\mathrm{d}v \tag{4.41}$$

The right-hand side is exactly the same as $\int\int \mathbf{E}\cdot\mathrm{d}\mathbf{D}\,\mathrm{d}v$ provided that $\mathbf{D} \propto \mathbf{E}$, i.e. the media are linear.

Similarly, show that in the absence of permanent magnets, the second term becomes $\Sigma\,\tfrac{1}{2}I_n\phi_n$ where I_n is the current in the nth circuit and ϕ_n is the magnetic flux threading that circuit, again assuming the media are linear, $\mathbf{B} \propto \mathbf{H}$.

Proof

It is simplest to introduce the vector potential **A** to solve this part of the problem. You will recall that this formed part of the story which led up to Maxwell's discovery of his equations (see Chapter 3). We define **A** as before by

$$\mathbf{B} = \mathrm{curl}\,\mathbf{A} \tag{4.42}$$

Now let us analyse the integral

$$\int_v \mathrm{d}v \int_0^{\mathbf{B}} \mathbf{H}\cdot\mathrm{d}\mathbf{B}$$

$\mathbf{B} = \mathrm{curl}\,\mathbf{A}$ and correspondingly $\mathrm{d}\mathbf{B} = \mathrm{curl}\,\mathrm{d}\mathbf{A}$. Now let us inspect $\nabla\cdot(\mathbf{H}\times\mathrm{d}\mathbf{A})$.

$$\begin{aligned}\nabla\cdot(\mathbf{H}\times\mathrm{d}\mathbf{A}) &= (\nabla\times\mathbf{H})\cdot\mathrm{d}\mathbf{A} - \mathbf{H}\cdot(\nabla\times\mathrm{d}\mathbf{A})\\ &= (\nabla\times\mathbf{H})\cdot\mathrm{d}\mathbf{A} - \mathbf{H}\cdot\mathrm{d}\mathbf{B}\end{aligned}$$

Substituting into the integral,

$$\begin{aligned}\int \mathrm{d}v \int_0^{\mathbf{B}} \mathbf{H}\cdot\mathrm{d}\mathbf{B} &= \int_v \mathrm{d}v \int_0^{\mathbf{A}} (\nabla\times\mathbf{H})\cdot\mathrm{d}\mathbf{A} - \int_v \mathrm{d}v \int_0^{\mathbf{A}} \nabla\cdot(\mathbf{H}\times\mathrm{d}\mathbf{A})\\ &= \int_v \mathrm{d}v \int_0^{\mathbf{A}} (\nabla\times\mathbf{H})\cdot\mathrm{d}\mathbf{A} - \int_S \int_0^{\mathbf{A}} (\mathbf{H}\times\mathrm{d}\mathbf{A})\cdot\mathrm{d}\mathbf{S}\end{aligned}$$

We use the same argument as before about evaluating the last surface integral at infinity. Since $A \propto r^{-1}, H \propto r^{-2}$ is the best one could possibly do, $\int_S \ldots \mathrm{d}\mathbf{S} \to 0$ as $r \to \infty$. Therefore,

$$\int \mathrm{d}v \int_0^{\mathbf{B}} \mathbf{H}\cdot\mathrm{d}\mathbf{B} = \int_v \mathrm{d}v \int_0^{\mathbf{A}} (\nabla\times\mathbf{H})\cdot\mathrm{d}\mathbf{A}$$

Now we assume there are no displacement currents present ($\partial\mathbf{D}/\partial t = 0$) and hence $\nabla\times\mathbf{H} = \mathbf{J}$, i.e.

$$\int_v \mathrm{d}v \int_0^{\mathbf{B}} \mathbf{H}\cdot\mathrm{d}\mathbf{B} = \int_v \mathrm{d}v \int_0^{\mathbf{A}} \mathbf{J}\cdot\mathrm{d}\mathbf{A}$$

Now consider a section of a flux tube of the current density **J**, cross section $\mathrm{d}\boldsymbol{\sigma}$ and length $\mathrm{d}l$ (Figure 4.2). The advantage of using a flux tube is that the current I is constant along the whole tube, i.e. $\mathbf{J}\cdot\mathrm{d}\boldsymbol{\sigma} = J\,\mathrm{d}\sigma = I = $ constant. Therefore,

$$\int_v \mathrm{d}v \int_0^{\mathbf{A}} \mathbf{J}\cdot \mathrm{d}\mathbf{A} = \int J\,\mathrm{d}\sigma\,\mathrm{d}\mathbf{A}\cdot \mathrm{d}\mathbf{l}$$

$$= \int_l \int_0^{\mathbf{A}} I\,\mathrm{d}\mathbf{A}\cdot \mathrm{d}\mathbf{l}$$

Now in taking the integral over d**A** we notice that we are, in fact, working out an energy through the product $I\mathrm{d}A$ and therefore the energy needed to attain the value **A** is one-half the product of I and **A** as occurs in electrostatics. We have disguised the dependence on the vector field **J** by including it in the constant current I. Therefore,

$$\int_v \mathrm{d}v \int_0^{\mathbf{A}} \mathbf{J}\cdot \mathrm{d}\mathbf{A} = \tfrac{1}{2}I \int_l \mathbf{A}\cdot \mathrm{d}\mathbf{l}$$

$$= \tfrac{1}{2}I \int_S \operatorname{curl} \mathbf{A}\cdot \mathrm{d}\mathbf{S}$$

$$= \tfrac{1}{2}I \int_S \mathbf{B}\cdot \mathrm{d}\mathbf{S}$$

$$= \tfrac{1}{2}I\Phi$$

Now we make up the whole of space by a superposition of current loops and flux linkages. Because of the linear relation between **B** and **H**, we can superimpose the total currents and flux linkages.

$$\tfrac{1}{2}\int \mathbf{H}\cdot \mathrm{d}\mathbf{B} = \tfrac{1}{2}\sum_n I_n\Phi_n.$$

Figure 4.2

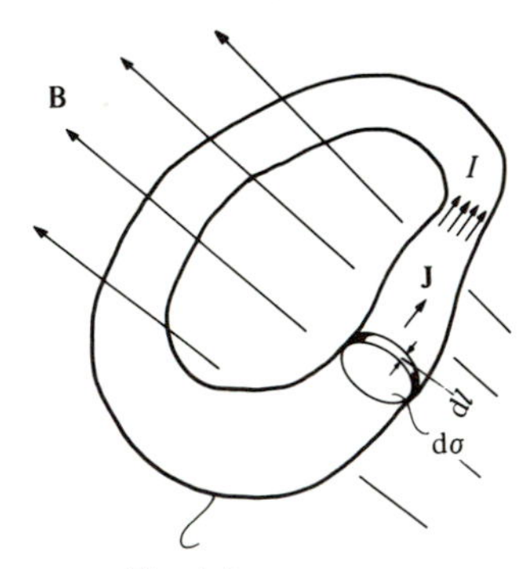

4.13 Concluding remarks

We are in great danger of writing a text book on vector fields in electromagnetism. It really is a remarkably elegant story. What comes through is the remarkable economy of Maxwell's equations in accounting for all the phenomena of classical electromagnetism. The formalism can be extended to treat much more complex systems involving anisotropic forces, for example, in the propagation of electromagnetic waves in magnetised plasmas, and in anisotropic material media. And all of this originated in Maxwell's mechanical analogy for the vacuum through which electromagnetic phenomena are propagated.

Case Study 3

MECHANICS AND DYNAMICS

Isaac Newton (1642–1727). (From Frontispiece, Memoirs of the Life, Writings and Discoveries of Isaac Newton, by Sir David Brewster, 2 vols., 1855, Constable, Edinburgh.)

5

APPROACHES TO MECHANICS AND DYNAMICS

5.1 Introduction

One of the key parts of any course on theoretical physics is the development of a wide range of more and more complex procedures for treating problems in classical mechanics and dynamics. In one way or another, these are all extensions of the basic principles enunciated by Newton in the *Principia* although some of them appear to bear only a distant resemblance to Newton's three laws of motion. As an example of the variety of ways in which the fundamentals of mechanics and dynamics can be expounded, here is a list of some of the different approaches which I found in a text book on theoretical physics:

Newton's laws of motion,
D'Alembert's principle,
The principle of virtual displacements,
Gauss's principle of least constraint,
Hertz's mechanics,
Hamilton's principle,
The principle of least action,
Generalised coordinates and Lagrange's equations,
The canonical equations of Hamilton,
Transformation theory of mechanics and the Hamilton–Jacobi equations.

Now this is not the place to go into what these different approaches are – this is more than adequately covered in standard texts such as Goldstein's *Classical Mechanics.*[1] Rather, I want to emphasise some of the basic features of these different approaches which provide us with important insights into different aspects of dynamical systems. Let me make a few points straightaway.

First of all, it is important to appreciate that all these different approaches are fully equivalent. A given problem in mechanics or dynamics can, in principle, be worked out by any of these techniques. However, the simplest statements of the laws of motion, as embodied in Newton's laws of motion, are not necessarily the simplest in application to any particular problem. Very often, other approaches provide a much more straightforward route to the answer.

A second important point is that some of these procedures result in a much deeper appreciation of some of the basic features of dynamical systems. Among these, I would classify as of special importance the concepts of *conservation laws* and of the *normal modes of oscillation* of a mechanical or dynamical system.

Finally, some of these procedures lead naturally to a formalism which can be taken over into quantum mechanics. We will use this last point as an illustration of how theoretical physicists work in practice.

In this case study, we will emphasise these three points and show how they lead to a deepening of our understanding of classical mechanics and its extension to quantum theory.

5.2 Newton's laws of motion

We have already described in Chapter 2 the history which led up to Newton's publication of his laws of motion in the *Principia* which was published by Samuel Pepys in 1687 (see Figure 5.1). The *Principia* consists of three books, preceded by two short sections called 'Definitions' and 'Axioms, or the Laws of Motion'. The definitions are eight in number and describe concepts such as mass, momentum or impulse, impressed force, inertia and centrifugal force. Then follow the three axioms which we now know as the three laws of motion. There has been considerable debate about the exact meanings of the definitions and axioms as they appear in the *Principia*, but in the analysis of Books I and III Newton uses these quantities unambiguously with their modern meaning. In fact, the three laws of motion appear as follows in a form remarkably similar to the form of words used in all standard texts.

Newton 1. 'Every body continues in its state of rest or of uniform motion in a straight line except in so far as it is compelled by forces to change that state.' That is, if

$$\mathbf{F} = 0, \quad \frac{\mathrm{d}\mathbf{v}}{\mathrm{d}t} = 0 \tag{5.1}$$

Newton 2. 'Change of motion (i.e. momentum) is proportional to the force and takes place in the direction of the straight line in which the force acts.'

In fact, Newton means that the rate of change of momentum is proportional to force as is exemplified by his use of this law throughout the *Principia.* In modern notation,

Figure 5.1. (*a*) The title page of Newton's *Principia*, 1687, London. (*b*) Newton's laws of motion as they appear in Newton's *Principia*.

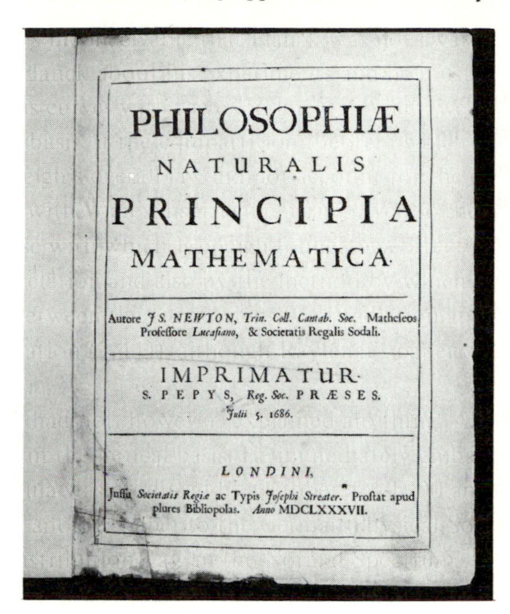

PHILOSOPHIÆ
NATURALIS
PRINCIPIA
MATHEMATICA.

Autore *J S. NEWTON*, *Trin. Coll. Cantab. Soc.* Matheseos Professore *Lucasiano*, & Societatis Regalis Sodali.

IMPRIMATUR.
S. PEPYS, *Reg. Soc.* PRÆSES.
Julii 5. 1686.

LONDINI,

Jussu *Societatis Regiæ* ac Typis *Josephi Streater*. Prostat apud plures Bibliopolas. *Anno* MDCLXXXVII.

(*a*)

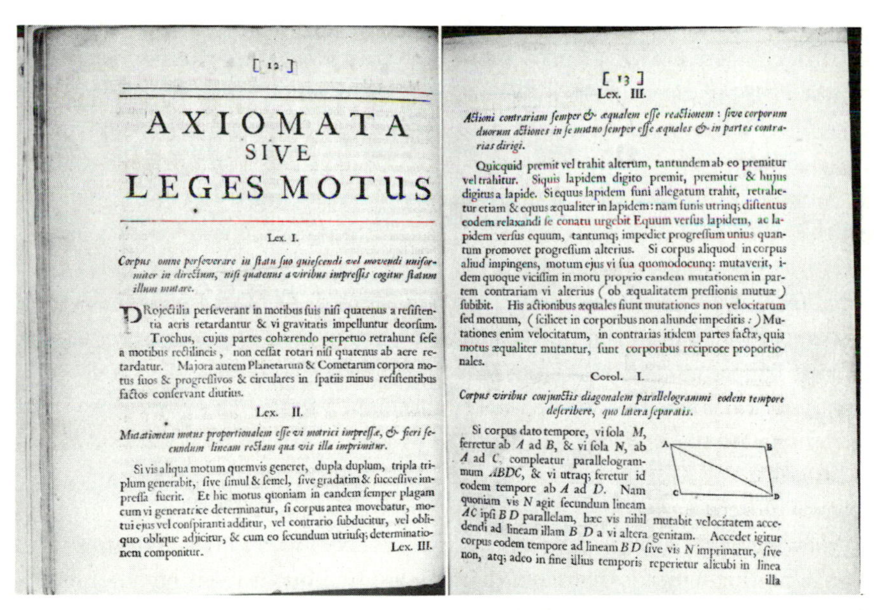

[12]

AXIOMATA
SIVE
LEGES MOTUS

Lex. I.

Corpus omne perseverare in statu suo quiescendi vel movendi uniformiter in directum, nisi quatenus a viribus impressis cogitur statum illum mutare.

PRojectilia perseverant in motibus suis nisi quatenus a resistentia aeris retardantur & vi gravitatis impelluntur deorsum. Trochus, cujus partes cohærendo perpetuo retrahunt sese a motibus rectilineis, non cessat rotari nisi quatenus ab aere retardatur. Majora autem Planetarum & Cometarum corpora motus suos & progressivos & circulares in spatiis minus resistentibus factos conservant diutius.

Lex. II.

Mutationem motus proportionalem esse vi motrici impressæ, & fieri secundum lineam rectam qua vis illa imprimitur.

Si vis aliqua motum quemvis generet, dupla duplum, tripla triplum generabit, sive simul & semel, sive gradatim & successive impressa fuerit. Et hic motus quoniam in eandem semper plagam cum vi generatrice determinatur, si corpus antea movebatur, motui ejus vel conspiranti additur, vel contrario subducitur, vel obliquo oblique adjicitur, & cum eo secundum utriusq; determinationem componitur.

Lex. III.

[13]

Lex. III.

Actioni contrariam semper & æqualem esse reactionem: sive corporum duorum actiones in se mutuo semper esse æquales & in partes contrarias dirigi.

Quicquid premit vel trahit alterum, tantundem ab eo premitur vel trahitur. Siquis lapidem digito premit, premitur & hujus digitus a lapide. Si equus lapidem funi allegatum trahit, retrahetur etiam & equus æqualiter in lapidem: nam funis utrinq; distentus eodem relaxandi se conatu urgebit Equum versus lapidem, ac lapidem versus equum, tantumq; impediet progressum unius quantum promovet progressum alterius. Si corpus aliquod in corpus aliud impingens, motum ejus vi sua quomodocunq: mutaverit, idem quoque vicissim in motu proprio eandem mutationem in partem contrariam vi alterius (ob æqualitatem pressionis mutuæ) subibit. His actionibus æquales fiunt mutationes non velocitatum sed motuum, (scilicet in corporibus non aliunde impeditis:) Mutationes enim velocitatum, in contrarias itidem partes factæ, quia motus æqualiter mutantur, sunt corporibus reciproce proportionales.

Corol. I.

Corpus viribus conjunctis diagonalem parallelogrammi eodem tempore describere, quo latera separatis.

Si corpus dato tempore, vi sola *M*, ferretur ab *A* ad *B*, & vi sola *N*, ab *A* ad *C*, compleatur parallelogrammum *ABDC*, & vi utraq; feretur id eodem tempore ab *A* ad *D*. Nam quoniam vis *N* agit secundum lineam *AC* ipsi *BD* parallelam, hæc vis nihil mutabit velocitatem accedendi ad lineam illam *B D* a vi altera genitam. Accedet igitur corpus eodem tempore ad lineam *B D* sive vis *N* imprimatur, sive non, atq; adeo in fine illius temporis reperietur alicubi in linea illa

(*b*)

$$\mathbf{F} = \frac{\mathrm{d}}{\mathrm{d}t}(\mathbf{p}) \quad \mathbf{p} = m\mathbf{v} \tag{5.2}$$

where $\mathbf{p}$ is momentum.

Newton 3. 'To every action there is always an equal and contrary reaction; or, the mutual actions of any two bodies are always equal and oppositely directed along the same straight line.'

As you are aware there are considerable logical problems about the definitions and the three laws of motion. Newton apparently regarded a number of the concepts as self evident, e.g. mass and force. Nowadays, it is preferable to regard the three laws of motion themselves as providing the definitions of the quantities involved and to consider the laws to reflect our experimental experience.

As an example, we can use Newton's second law in order to define *mass.* We devise a means by which we can apply a force of fixed magnitude to different bodies. By performing experiments, we find that different bodies suffer different accelerations under the action of this force. The masses of the bodies can be defined as quantities proportional to the inverse of their accelerations. By using Newton's second and third laws we can show that we can define a relative scale of mass. Suppose we have two point-like masses A and B interacting with each other. Then $\mathbf{F} = -\mathbf{F}$. If we measure their accelerations relative to each other, we will find that they are in a certain ratio a_{AB}/a_{BA} which we can call M_{AB}. If we then compare C with A and B separately, we will measure another ratio of accelerations, M_{BC} and M_{AC}. We observe experimentally that $M_{AB} = M_{AC}/M_{BC}$. Therefore, we can now define a mass scale by choosing one of them as a standard.

Now all of this may seem very trivial to you but it emphasises one point which I made in the introduction. Even in something which is apparently as intuitive as Newton's laws, there are fundamental assumptions and results of experiment upon which the whole structure is based. The fact is that the system works very well indeed and the equations are a very good approximation to what actually happens in the real world. However, we emphasise that there is no strictly logical way of setting up the foundations *ab initio.* This gives us no operational problem so long as we adhere to Dirac's dictum 'that we want the equations which describe nature . . . and have to reconcile ourselves to an absence of strict logic.'

We will not go into the development of conservation laws from Newton's laws of motion which is treated in the standard text books. We simply recall that the laws which can be derived from Newton's laws in a straightforward manner are

(i) *the law of conservation of momentum,*
(ii) *the law of conservation of angular momentum,*
(iii) *the law of conservation of energy,* once suitable potential energy functions have been defined.

We will derive these later from the Euler–Lagrange equation (Section 5.6).

One point worth making is the *invariance* of Newton's laws of motion with respect to transformation between frames of reference in uniform relative motion. This subject is the key to the development of special relativity which is discussed in Chapter 13. Figure 13.1 shows frames of reference S and S' in uniform relative motion with relative velocity V. According to Newtonian theory, the coordinates in the frames of reference S and S' are related by

$$\left.\begin{aligned} x' &= x - Vt \\ y' &= y \\ z' &= z \\ t' &= t \end{aligned}\right\} \tag{5.3}$$

Taking the second derivatives of the first of the relations (5.3) with respect to time, we see that $\ddot{x}' = \ddot{x}$, i.e. the acceleration and consequently the forces are the same in any inertial frame of reference. Notice that time is absolute. These transformations are implicit in the works of Galileo who stated that there should be no difference in the laws of nature whether you are stationary or in uniform rectilinear motion. For this reason, the equations (5.3) are often referred to as *Galilean transformations.* It is intriguing that his motivation for advancing this argument was to demonstrate that there is nothing inconsistent with our view of the Universe if the Earth happens to be in motion rather than at rest at the centre of the Universe, i.e. it was really part of a pro-Copernican argument.

5.3 Principles of 'least action'

Some of the most powerful alternative approaches to mechanics and dynamics involve finding the function which results in a minimum value of some suitably defined function. In the formal development of mechanics, these procedures are stated axiomatically and the formalism for treating a wide range of problems elaborated. As in so many aspects of teaching basic theoretical physics, I find Feynman's approach in Chapter 19 of Volume 2 of his *Lectures on Physics*,[2] entitled 'The Principle of Least Action', quite brilliant in exposing the basic ideas behind these approaches to dynamics. We will follow his exposition but rather more briefly since we already have most of the mathematical apparatus at our fingertips (or should have!).

Let us consider the simple case of the dynamics of a particle in a conservative field of force, i.e. one which can be derived from the derivative of a scalar potential, $\mathbf{F} = -\,\text{grad}\; V$. Examples of such fields of force include electrostatic forces $\mathbf{F} = -q\,\text{grad}\,\Phi_e$ and gravitational forces $\mathbf{F} = -m\,\text{grad}\,\Phi_g$, where Φ_e and Φ_g are the electrostatic and gravitational potentials respectively. The potential energy of the particle is $q\Phi_e$ or $m\Phi_g$, i.e. the V introduced above is the potential energy of the particle.

Now let us introduce a set of axioms which enables us to work out the path of the particle subject to this force field.

First, we define the quantity

$$\mathscr{L} = (\tfrac{1}{2}mv^2 - V) \tag{5.4}$$

which is the difference between the kinetic and potential energy of the particle at any point in the field. Then to derive the trajectory of the particle between two points in the field in a fixed time interval t_1 to t_2, we must find that path which minimises the function

$$S = \int_{t_1}^{t_2} (\tfrac{1}{2}mv^2 - V)\,\mathrm{d}t \tag{5.5}$$

$$= \int_{t_1}^{t_2} \left(\tfrac{1}{2}m\left(\frac{\mathrm{d}\mathbf{r}}{\mathrm{d}t}\right)^2 - V\right)\mathrm{d}t \tag{5.6}$$

We should regard these statements as exactly equivalent to Newton's laws of motion or, rather, what he accurately called his 'axioms'.

$\mathscr{L}$ is called the Lagrangian. S does not have a simple official name but Feynman calls it the 'action' so that minimising S can be considered as a principle of 'least action'. The problem with this definition is that the word 'action' is used for something rather different from S in the formal development of mechanics. We will simply call it the function S.

We can see that this definition is certainly consistent with Newton's first law of motion. If there are no forces present, $V = 0$ and hence we must minimise $S = \int_{t_1}^{t_2} v^2\,\mathrm{d}t$. The minimum value of S must correspond to a constant velocity v between t_1 and t_2. If the particle travels between the two points by accelerating and decelerating in such a way that t_1 and t_2 are the same, the integral must be greater than that for constant velocity between t_1 and t_2 because it is v^2 which appears in the integral and a basic rule of analysis tells us that $\langle v^2\rangle \geqslant \langle\bar{v}\rangle^2$. Thus, in this special case, Newton's first law of motion is found to be the solution corresponding to minimising S, i.e., in the absence of forces, $\mathbf{v} =$ constant.

To proceed further, we need the techniques of the *calculus of variations.* We will deal with some simple aspects of this in a moment but let us first analyse the case of motion in a scalar potential field in more detail. I find Feynman's analysis very attractive. Suppose we have a function of a single variable, say $f(x)$. When we find the minimum value of the function, we find the point on the curve where $\mathrm{d}f(x)/\mathrm{d}x$ is zero (Figure 5.2). If we approximate the variation of the function about the minimum at $x = 0$ by a power series,

$$f(x) = a_0 + a_1 x + a_2 x^2 + a_3 x^3 + \ldots$$

it is evident that the function can only have $\mathrm{d}f/\mathrm{d}x = 0$ at $x = 0$ if $a_1 = 0$. Thus, the function approximates to a parabola since the first non-zero coefficient in front of the expansion of $f(x)$ about the minimum must be in the term x^2, i.e.

$$f(x) = a_0 + a_2 x^2 + a_3 x^3 + \ldots$$

In other words, for a small displacement x from the minimum, the change in the function $f(x)$ is only second order in x.

Exactly the same principle is used to work out the path of the particle in the procedure for minimising S. If the true path of the particle is $\mathbf{x}_0(t)$, then another path between t_1 and t_2 is given by

$$\mathbf{x}(t) = \mathbf{x}_0(t) + \boldsymbol{\eta}(t) \tag{5.7}$$

where $\boldsymbol{\eta}(t)$ describes the deviation of $\mathbf{x}(t)$ from the minimum path $\mathbf{x}_0(t)$. Just as we can define the minimum of a function as the point at which there is no first order dependence of the function upon x, so we can define the minimum of the function $\mathbf{x}(t)$ as that function for which the dependence on $\boldsymbol{\eta}(t)$ is at least *second order*, i.e. there should be no term in $\boldsymbol{\eta}(t)$. Let us now do this.

We substitute equation (5.7) into (5.6) and then

$$S = \int_{t_1}^{t_2} \left[\frac{m}{2}\left(\frac{\mathrm{d}\mathbf{x}_0}{\mathrm{d}t} + \frac{\mathrm{d}\boldsymbol{\eta}}{\mathrm{d}t}\right)^2 - V(\mathbf{x}_0 + \boldsymbol{\eta})\right] \mathrm{d}t$$

$$= \int_{t_1}^{t_2} \left\{\frac{m}{2}\left[\left(\frac{\mathrm{d}\mathbf{x}_0}{\mathrm{d}t}\right)^2 + 2\,\frac{\mathrm{d}\mathbf{x}_0}{\mathrm{d}t}\cdot\frac{\mathrm{d}\boldsymbol{\eta}}{\mathrm{d}t} + \left(\frac{\mathrm{d}\boldsymbol{\eta}}{\mathrm{d}t}\right)^2\right] - V(\mathbf{x}_0 + \boldsymbol{\eta})\right\} \mathrm{d}t \tag{5.8}$$

Now we are only interested in eliminating quantities to first order in $\mathrm{d}\boldsymbol{\eta}$ and hence we can drop the term $(\mathrm{d}\boldsymbol{\eta}/\mathrm{d}t)^2$. Furthermore, we can expand $V(\mathbf{x}_0 + \boldsymbol{\eta})$ to first order in $\boldsymbol{\eta}$ by a Taylor expansion,

$$V(\mathbf{x}_0 + \boldsymbol{\eta}) = V(\mathbf{x}_0) + \nabla V \cdot \boldsymbol{\eta} \tag{5.9}$$

Figure 5.2

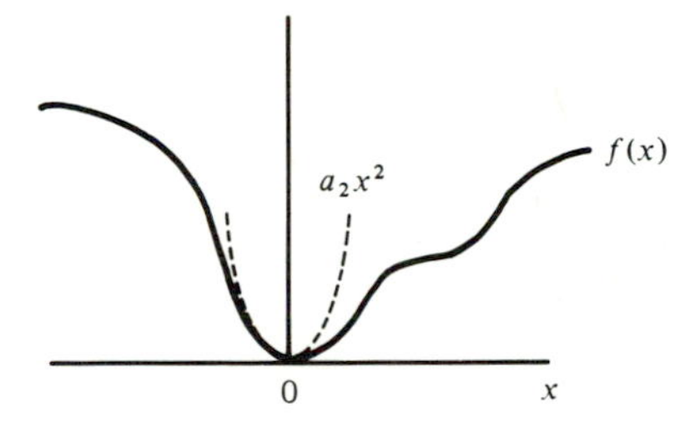

Substituting the relation (5.9) into (5.8) and preserving only quantitites to first order in $\boldsymbol{\eta}$, we find that

$$S = \int_{t_1}^{t_2} \left[\frac{m}{2} \left(\frac{\mathrm{d}\mathbf{x}_0}{\mathrm{d}t} \right)^2 - V(\mathbf{x}_0) + m \frac{\mathrm{d}\mathbf{x}_0}{\mathrm{d}t} \cdot \frac{\mathrm{d}\boldsymbol{\eta}}{\mathrm{d}t} - \nabla V \cdot \boldsymbol{\eta} \right] \mathrm{d}t \tag{5.10}$$

Now the first two terms inside the integral are what we calculate for the minimum path and are a constant. We are interested in ensuring that the last two terms have no dependence upon $\boldsymbol{\eta}$, the condition for a minimum. Let us therefore concentrate upon the last two terms, i.e.

$$S = \int_{t_1}^{t_2} \left(m \frac{\mathrm{d}\mathbf{x}_0}{\mathrm{d}t} \cdot \frac{\mathrm{d}\boldsymbol{\eta}}{\mathrm{d}t} - \boldsymbol{\eta} \cdot \nabla V \right) \mathrm{d}t \tag{5.11}$$

We integrate the first term by parts so that $\boldsymbol{\eta}$ alone appears in the integrand, i.e.

$$S = m \left[\boldsymbol{\eta} \cdot \frac{\mathrm{d}\mathbf{x}_0}{\mathrm{d}t} \right]_{t_1}^{t_2} - \int_{t_1}^{t_2} \left(\frac{\mathrm{d}}{\mathrm{d}t} \left(m \frac{\mathrm{d}\mathbf{x}_0}{\mathrm{d}t} \right) \cdot \boldsymbol{\eta} + \boldsymbol{\eta} \cdot \nabla V \right) \mathrm{d}t \tag{5.12}$$

We know that the function $\boldsymbol{\eta}$ must be zero at t_1 and t_2 because that is where the path must begin and end. Therefore, the first term in equation (5.12) is zero and we can write

$$S = - \int_{t_1}^{t_2} \left\{ \boldsymbol{\eta} \cdot \left[\frac{\mathrm{d}}{\mathrm{d}t} \left(m \frac{\mathrm{d}\mathbf{x}_0}{\mathrm{d}t} \right) + \nabla V \right] \right\} \mathrm{d}t = 0 \tag{5.13}$$

This must be true for arbitrary perturbations about $\mathbf{x}_0(t)$ and hence the term in square brackets must be zero, i.e.

$$\frac{\mathrm{d}}{\mathrm{d}t} \left(m \frac{\mathrm{d}\mathbf{x}_0}{\mathrm{d}t} \right) = - \nabla V \tag{5.14}$$

We recognise that we have recovered Newton's second law of motion since $\mathbf{F} = - \nabla V$, i.e.

$$\mathbf{F} = \frac{\mathrm{d}}{\mathrm{d}t} \left(m \frac{\mathrm{d}\mathbf{x}_0}{\mathrm{d}t} \right) = \frac{\mathrm{d}}{\mathrm{d}t} (\mathbf{p}) \tag{5.15}$$

Thus, our alternative formulation of the laws of motion in terms of an action principle is exactly equivalent to Newton's statement of the laws of motion. There is plainly a great deal to be done in working out all the ramifications of this procedure.

It will have to suffice to say that we can generalise these procedures to take account of conservative *and* non-conservative forces such as those which depend upon velocity, e.g. friction and the force on a charged particle in a magnetic field. The key point is that we have a prescription which involves us in writing down directly the kinetic and potential energy (T and V respectively) of the system, then forming the Lagrangian $\mathscr{L}$ and finding the minimum value of the function S. The big advantage of this procedure is that it is often a straightforward matter to write down these energies in some suitable set of coordinates. The next step is therefore clear. We need rules which tell us how to find the minimum value of S in any set of coordinates which is convenient for the problem in hand. The rules turn out to be the *Euler–Lagrange equations*.

One final point about action principles in physics. In general, we do not have a definite prescription for finding the function equivalent to the Lagrangian. To quote Feynman, 'The question of what the action (S) should be for a particular case must be determined by some kind of trial and error. It is just the same problem as determining the laws of motion in the first place. You just have to fiddle around with the equations that you know and see if you can get them into the form of a principle of least action.' An interesting and important example is that of a relativistic particle moving in an electromagnetic field. The appropriate Lagrangian is

$$\mathscr{L} = -m_0c^2\,(1 - v^2/c^2)^{\frac{1}{2}} - q\phi + \mathbf{v}\cdot\mathbf{A} \tag{5.16}$$

where ϕ is the electrostatic potential and $\mathbf{A}$ the vector potential. You may like to derive the appropriate equations of motion for a relativistic particle in electric and magnetic fields in their more conventional guises.

5.4 The Euler–Lagrange equation

For simplicity, we will only consider forces which can be derived from a scalar potential V. As an example, suppose we consider a problem involving N particles interacting via a scalar potential function V. The positions of the N particles are given by the vector $[\mathbf{r}_1, \mathbf{r}_2, \mathbf{r}_3 \ldots \mathbf{r}_N]$. In fact, since we need three numbers to describe each position, e.g. $x_i y_i z_i$, this vector has $3N$ coordinates. It may well prove more convenient to work in terms of a different set of coordinates which we can write as $[q_1, q_2, q_3 \ldots q_{3N}]$. There will be a set of relations between the two sets of coordinates which we can write as

$$\left.\begin{aligned} q_i &= q_i\,(\mathbf{r}_1, \mathbf{r}_2, \mathbf{r}_3 \ldots \mathbf{r}_N) \\ \text{and}\qquad r_i &= r_i\,(q_1, q_2, q_3 \ldots q_{3N}) \end{aligned}\right\} \tag{5.17}$$

Notice that this is no more than a change of variables.

The aim of the procedure now is to write down equations for the dynamics of the particles, i.e. an equation for each independent coordinate, in terms of the coordinates q_i rather than $\mathbf{r}_i$. We are guided by the analysis of the previous section on action principles to form the quantities T and V, the kinetic and potential energies respectively, in terms of the new set of coordinates and then to find the stationary value of the 'action' S, i.e.

$$\mathscr{L} = (T - V) \qquad \delta\int_{t_1}^{t_2} (T - V)\,\mathrm{d}t = 0 \tag{5.18}$$

This formulation is called *Hamilton's principle* and $\mathscr{L}$ as before is the *Lagrangian*. The key point is that Hamilton's principle does not say anything about the coordinate system in which we are working. It is the extension to more general coordinate systems of the arguments which we developed in Section 5.3 for Cartesian coordinates.

Now the kinetic energy of the system is

$$T = \sum_i m_i \dot{\mathbf{r}}_i^2 \tag{5.19}$$

In terms of our new coordinate system, we can write without loss of generality

$$r_i = r_i(q_1, q_2, \ldots q_{3N}, t)$$
$$\dot{r}_i = \dot{r}_i(\dot{q}_1, \dot{q}_2, \dot{q}_3 \ldots \dot{q}_{3N}, q_1, q_2 \ldots q_{3N}, t)$$

Notice that we have now included explicitly the time dependence of r_i and $\dot{r}_i$. Therefore, we can rewrite the kinetic energy as some function of the coordinates $\dot{q}_i, q_i$ and t, i.e. $T(\dot{q}_i, q_i, t)$, where we understand that all the values of i from 1 to $3N$ are included. Similarly, we can write the expression for the potential energy entirely in terms of the coordinates q_i and t, i.e. $V(q_i, t)$. We require a procedure for finding the stationary value of

$$S = \int_{t_1}^{t_2} [T(\dot{q}_i, q_i, t) - V(q_i, t)]\, \mathrm{d}t \tag{5.20}$$

We now repeat our analysis of Section 5.3 in which we found the condition for S to be independent of first order perturbations about the minimum path. As before, we let $q_0(t)$ be the minimum solution and write the expression for another function $q(t)$ in the form

$$q(t) = q_0(t) + \eta(t) \tag{5.21}$$

We rewrite S as

$$S = \int_{t_1}^{t_2} \mathscr{L}(\dot{q}_i, q_i, t)\, \mathrm{d}t \tag{5.22}$$

Now we insert the trial solution (5.21) into (5.22). Then

$$S = \int_{t_1}^{t_2} \mathscr{L}(\dot{q}_0(t) + \dot{\eta}(t),\, q_0(t) + \eta(t),\, t)\, \mathrm{d}t$$

Performing a Taylor expansion to first order in $\dot{\eta}(t)$ and $\eta(t)$,

$$S = \int_{t_1}^{t_2} \mathscr{L}(\dot{q}_0(t), q_0(t), t)\, \mathrm{d}t + \int_{t_1}^{t_2} \left(\frac{\partial \mathscr{L}}{\partial \dot{q}_i}\, \dot{\eta}(t) + \frac{\partial \mathscr{L}}{\partial q_i}\, \eta(t)\right) \mathrm{d}t$$

As before, we integrate the term in $\dot{\eta}(t)$ by parts and then

$$S = S_0 + \left[\frac{\partial \mathscr{L}}{\partial \dot{q}_i}\, \eta(t)\right]_{t_1}^{t_2} - \int_{t_1}^{t_2} \left[\frac{\mathrm{d}}{\mathrm{d}t}\left(\frac{\partial \mathscr{L}}{\partial \dot{q}_i}\right) \eta(t) - \frac{\partial \mathscr{L}}{\partial q_i}\, \eta(t)\right] \mathrm{d}t \tag{5.23}$$

Again because $\eta(t)$ must always be zero at end points, the first term in square brackets disappears and the result can be written

$$S = S_0 - \int_{t_1}^{t_2} \eta(t) \left[\frac{\mathrm{d}}{\mathrm{d}t} \left(\frac{\partial \mathscr{L}}{\partial \dot{q}_i} \right) - \frac{\partial \mathscr{L}}{\partial q_i} \right] \mathrm{d}t \tag{5.24}$$

We require the integral to be zero for all first order perturbations about the minimum solution. Therefore, we find the condition

$$\frac{\partial \mathscr{L}}{\partial q_i} - \frac{\mathrm{d}}{\mathrm{d}t} \left(\frac{\partial \mathscr{L}}{\partial \dot{q}_i} \right) = 0 \tag{5.25}$$

Equation (5.25) represents $3N$ equations for the time evolution of the $3N$ coordinates. This fundamental equation is known as the *Euler–Lagrange equation.* From our previous studies, it is apparent that this is no more than Newton's second law of motion written in the q_i coordinate system.

5.5 Small oscillations and normal modes

We do not want to proceed to the formal development of mechanics using the Euler–Lagrange equations but only to give a simple example to show the power of the technique in analysing the dynamical behaviour of systems which undergo *small oscillations.* This leads naturally to the concept of the *normal modes* of vibration of the system. Let us solve a simple example first of all to illustrate the general procedure.

A typical problem is that of a cylinder which is suspended by strings of equal length at either end. The strings are attached to the circumference of the cylinder as shown in Figure 5.3(*a*) so that it can wobble and swing at the same time. We will consider first the swinging motion in which the string and the flat end of the cylinder remain in a single plane. An end-on view of the displaced cylinder is shown in Figure 5.3(*b*). For the sake of definiteness we take the length of the string to be $3a$ and the radius of the cylinder $2a$.

First of all, we must choose a suitable set of coordinates which will define completely the position of the cylinder when it suffers a small perturbation. It can be seen that the coordinates θ and ϕ fulfil this role and are the natural system to use for what we expect to be pendulum-like motion. We now write down the Lagrangian for the system in terms of the coordinates $\dot{\theta}, \dot{\phi}, \theta, \phi$.

The *kinetic energy T* consists of two parts, one associated with the *translational motion* of the cylinder and the other with its *rotation*. In the analysis, we consider only small displacements. Therefore, the displacement of the centre of mass of the cylinder from the vertical position is

$$x = 3a \sin\theta + 2a \sin\phi = 3a\theta + 2a\phi \tag{5.26}$$

and consequently the translation motion is $\dot{x} = 3a\dot{\theta} + 2a\dot{\phi}$. The linear kinetic energy is therefore $\frac{1}{2}ma^2(3\dot{\theta} + 2\dot{\phi})^2$. The rotational motion of the cylinder is

$\frac{1}{2}I\dot{\phi}^2$ where I is the moment of inertia of the cylinder about the point of attachment of the string. In this case $I = 4a^2m$ where m is the mass of the cylinder. Thus, the total kinetic energy of the cylinder is

$$\begin{aligned} T &= \tfrac{1}{2}ma^2(3\dot{\theta} + 2\dot{\phi})^2 + 2a^2m\dot{\phi}^2 \\ &= \tfrac{1}{2}ma^2(9\dot{\theta}^2 + 12\dot{\theta}\dot{\phi} + 8\dot{\phi}^2) \end{aligned} \tag{5.27}$$

The *potential energy* is entirely associated with the displacement of the centre of mass of the cylinder above its equilibrium position.

$$\begin{aligned} y &= 3a(1 - \cos\theta) + 2a(1 - \cos\phi) \\ &= \frac{3a\theta^2}{2} + a\phi^2 \end{aligned} \tag{5.28}$$

for small values of θ and ϕ. Consequently, the potential energy is

$$V = \frac{mg}{2}(3a\theta^2 + 2a\phi^2) \tag{5.29}$$

The Lagrangian is therefore

$$\mathscr{L} = T - V = \tfrac{1}{2}ma^2(9\dot{\theta}^2 + 12\dot{\theta}\dot{\phi} + 8\dot{\phi}^2) - \frac{mg}{2}(3a\theta^2 + 2a\phi^2) \tag{5.30}$$

Notice that we have ended up with a form which is quadratic in $\dot{\theta}$, $\dot{\phi}$, θ and ϕ. The reason for this is that the kinetic energy is a quadratic function of velocity. Also, the potential energy is evaluated relative to the equilibrium position, $\theta = \phi = 0$, and consequently the lowest order expansion about the minimum is second order in the displacement from the minimum (see Section 5.3).

Figure 5.3

(a)

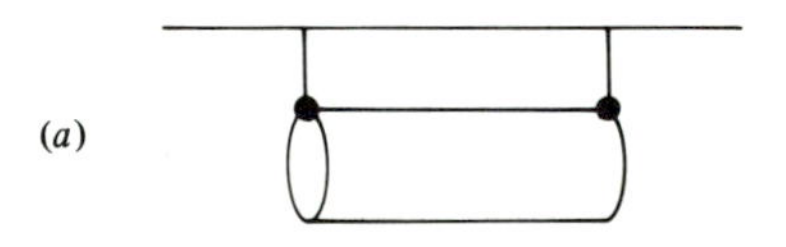

(b)

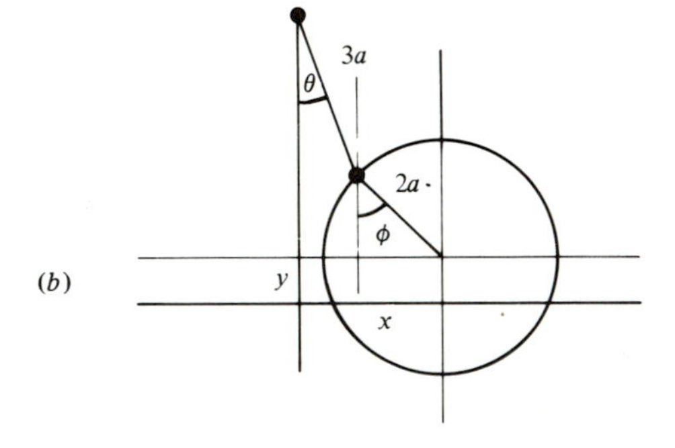

We now use the Euler–Lagrange equation,

$$\frac{\partial \mathscr{L}}{\partial q_i} - \frac{\mathrm{d}}{\mathrm{d}t}\left(\frac{\partial \mathscr{L}}{\partial \dot{q}_i}\right) = 0 \tag{5.31}$$

Let us take the $\theta, \dot{\theta}$ pair of coordinates first. Substituting (5.30) into (5.31) with $\dot{q}_i = \dot{\theta}$ and $q_i = \theta$, we find that

$$\tfrac{1}{2}ma^2 \frac{\mathrm{d}}{\mathrm{d}t}(18\dot{\theta} + 12\dot{\phi}) = -\tfrac{1}{2}mga\,6\theta \tag{5.32}$$

Similarly, for the coordinate pair $\dot{q}_i = \dot{\phi}$, $q_i = \phi$, we find that

$$\tfrac{1}{2}ma^2 \frac{\mathrm{d}}{\mathrm{d}t}(12\dot{\theta} + 16\dot{\phi}) = -\tfrac{1}{2}mga\,4\phi \tag{5.33}$$

We end up with two differential equations:

$$\left.\begin{aligned} 9\ddot{\theta} + 6\ddot{\phi} &= -\frac{3g}{a}\theta \\ 6\ddot{\theta} + 8\ddot{\phi} &= -\frac{2g}{a}\phi \end{aligned}\right\} \tag{5.34}$$

The key to finding the normal modes of oscillation of the system is that, in a normal mode, all components oscillate at the same frequency, i.e. we seek oscillatory solutions of the form

$$\left.\begin{aligned} \ddot{\theta} &= -\omega^2\theta \\ \ddot{\phi} &= -\omega^2\phi \end{aligned}\right\} \tag{5.35}$$

We now insert these trial solutions into equation (5.34). If we set $\lambda = a\omega^2/g$,

$$\left.\begin{aligned} (9\lambda - 3)\theta + 6\lambda\phi &= 0 \\ 6\lambda\theta + (8\lambda - 2)\phi &= 0 \end{aligned}\right\} \tag{5.36}$$

The condition that equations (5.36) be satisfied for all θ and ϕ is that the determinant of the coefficients in the relation (5.36) should be zero. Therefore,

$$\begin{vmatrix} (9\lambda - 3) & 6\lambda \\ 6\lambda & (8\lambda - 2) \end{vmatrix} = 0 \tag{5.37}$$

Multiplying out this determinant,

$$6\lambda^2 - 7\lambda + 1 = 0 \tag{5.38}$$

i.e. the solutions are $\lambda = \frac{1}{6}$ and $\lambda = 1$. Therefore, the angular frequencies of oscillation are $\omega^2 = \lambda(g/a)$, $\omega_1 = (g/a)^{\frac{1}{2}}$ and $\omega_2 = (g/6a)^{\frac{1}{2}}$, i.e. the ratio of frequencies of oscillation of the normal modes is $6^{\frac{1}{2}}:1$.

We can now find the physical nature of these modes by inserting the solutions $\lambda = \frac{1}{6}$ and $\lambda = 1$ into equations (5.36). The results are

$$\left.\begin{aligned} \lambda &= 1, \quad \omega_1 = (g/a)^{\frac{1}{2}}, \quad \phi = -\theta \\ \lambda &= \tfrac{1}{6}, \quad \omega_2 = (g/6a)^{\frac{1}{2}}, \quad \phi = \tfrac{3}{2}\theta \end{aligned}\right\} \tag{5.39}$$

These modes of oscillation are illustrated in Figure 5.4. According to our analysis, if we set the system oscillating in either of the modes (*a*) or (*b*) shown in Figure 5.4, it will continue to do so for all time at frequencies ω_1 and ω_2 respectively. We also note that we can represent any initial set of conditions of the system by a superposition of the modes 1 and 2 if we choose suitable amplitudes of the normal modes. Suppose we represent the normal modes by the vectors

$$a(\mathbf{i}_\theta + \tfrac{3}{2}\,\mathbf{i}_\theta)$$
$$b(\mathbf{i}_\theta - \mathbf{i}_\phi)$$

Then if we start with an arbitrary configuration of θ and ϕ, we can find suitable amplitudes a and b from the relations

$$\theta = a + b$$
$$\phi = \tfrac{3}{2}a - b$$

By a similar procedure, we can fit suitable initial values for $\dot{\theta}$ and $\dot{\phi}$. The beauty of this procedure is that we can find the behaviour of the system at *any* subsequent time by adding together the independent behaviour of the two normal modes of oscillation, i.e.

$$x = a\mathrm{e}^{\mathrm{i}\omega_2 t}(\mathbf{i}_\theta + \tfrac{3}{2}\,\mathbf{i}_\phi) + b\mathrm{e}^{\mathrm{i}\omega_1 t}(\mathbf{i}_\theta - \mathbf{i}_\phi)$$

This is the fundamental importance of *normal modes.* Any configuration of the system can be represented by a suitable superposition of normal modes and this enables us to predict the subsequent dynamical behaviour of the system.

Of particular importance are *complete sets of orthonormal functions.* The functions themselves are independent and normalised to unity so that any

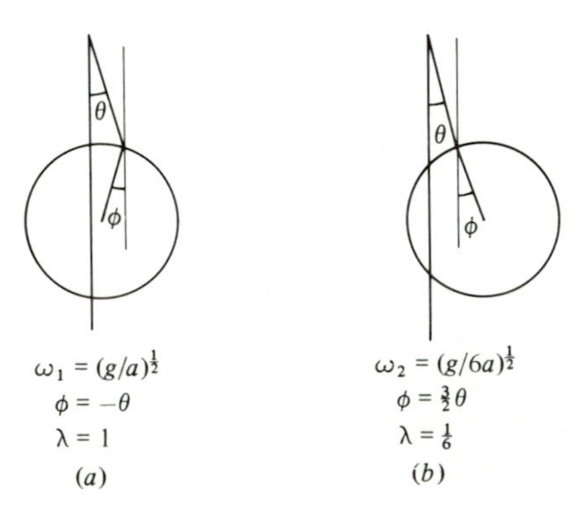

Figure 5.4. Illustrating the two normal modes associated with the swinging of a cylinder with its ends remaining in one plane.

function can be represented by a superposition of them. The simplest example is probably the Fourier series,

$$f(x) = \frac{a_0}{2} + \sum_{n=1}^{\infty} a_n \cos\frac{2\pi nx}{L} + \sum_{n=1}^{\infty} b_n \sin\frac{2\pi nx}{L}$$

which can describe precisely any function defined in the interval $0 < x < L$. The separate terms with coefficients a_0, a_n, b_n can be thought of as the normal modes of oscillation with fixed end points at 0 and L. There exist hosts of such orthonormal functions which find application in a wide range of aspects of physics and theoretical physics. For example, in spherical polar coordinates, the Legendre polynomials and associated Legendre polynomials provide a convenient complete set of orthogonal functions defined on a sphere.

In reality, the modes will not be completely independent physically. In real situations, there will exist higher order processes by which energy can be exchanged between modes. These are the mechanisms by which energy will eventually be shared equally between the modes. Furthermore, if they are not maintained, the modes will eventually decay to zero by dissipative processes. The time evolution of the system can be accurately determined by following the decay of each mode with time. In the example given above, the time evolution of the system would be

$$x = a\,\mathrm{e}^{\mathrm{i}\omega_2 t}\,\mathrm{e}^{-\gamma_2 t}(\mathbf{i}_\theta + \tfrac{3}{2}\mathbf{i}_\phi) + b\,\mathrm{e}^{\mathrm{i}\omega_1 t}\,\mathrm{e}^{-\gamma_1 t}(\mathbf{i}_\theta - \mathbf{i}_\phi)$$

One final point about the normal modes. It is clear that dynamical systems can become very complicated. Suppose, for example, we allowed the cylinder to move arbitrarily rather than it being obliged to oscillate with the end of the cylinder and the string in one plane. The motion would become more complex but we can guess what the normal modes would have to be. There are two further degrees of freedom which are introduced in the general case.

These are indicated schematically in Figure 5.5. We guess that the other

Figure 5.5. Other normal modes associated with the general motion of a cylinder.

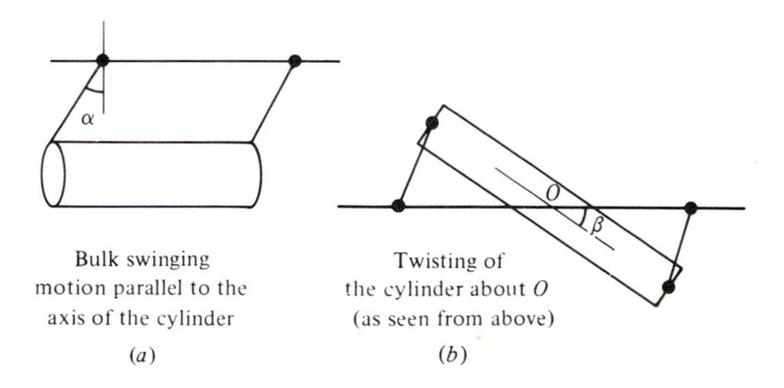

normal modes are associated with swinging motion as in Figure 5.5(a) and with oscillations of the cylinder about O (Figure 5.5(b)). Indeed, in most physical problems, one can get a long way by guessing what the forms of the modes must be by inspection. The key point is that all the parts of the system must oscillate with the same frequency ω. We will have a great deal more to say about normal modes in our case study on the origins of the concept of quanta.

5.6 Conservation laws and symmetry

We recall that, in Newtonian mechanics, conservation laws are derived from the first integral of the equations of motion. In exactly the same way, we can derive a set of conservation laws from the first integral of the Euler–Lagrange equations. Approaching the problem from the point of view of these equations brings out clearly the close relation between symmetry and conservation laws. In fact, a great deal simply depends upon the form of the Lagrangian itself.

5.6.1 Lagrangian not a function of q_i

In this case, $\partial\mathcal{L}/\partial q_i = 0$ and hence the Euler–Lagrange equation reads

$$\frac{\mathrm{d}}{\mathrm{d}t}\left(\frac{\partial\mathcal{L}}{\partial\dot{q}_i}\right) = 0$$

$$\frac{\partial\mathcal{L}}{\partial\dot{q}_i} = \text{constant}$$

An example of such a Lagrangian is the motion of a particle in the absence of a field of force. Then $\mathcal{L}=\frac{1}{2}m\dot{\mathbf{r}}^2$ and

$$\frac{\partial\mathcal{L}}{\partial\dot{q}_i} = \frac{\partial\mathcal{L}}{\partial\dot{\mathbf{r}}} = m\dot{\mathbf{r}} = \text{constant} \tag{5.40}$$

i.e. Newton's first law of motion.

This calculation is suggestive of how we can define a generalised momentum p_i. For an arbitrary coordinate system, we can define a *conjugate momentum* as

$$p_i \equiv \frac{\partial\mathcal{L}}{\partial\dot{q}_i} \tag{5.41}$$

Now this is not necessarily anything like a normal momentum but it does have the property that, if $\mathcal{L}$ does not depend on q_i, it is a constant of the motion.

5.6.2 Lagrangian independent of time

This is obviously related to energy conservation. Let us perform the straightforward mathematical analysis. Dropping terms higher than $\dot{q}_i$,

$$\frac{\mathrm{d}\mathcal{L}}{\mathrm{d}t} = \frac{\partial\mathcal{L}}{\partial t} + \sum_i \dot{q}_i \frac{\partial\mathcal{L}}{\partial q_i} \tag{5.42}$$

From the Euler–Lagrange equation, we know that

$$\frac{\mathrm{d}}{\mathrm{d}t}\left(\frac{\partial \mathscr{L}}{\partial \dot{q}_i}\right) = \frac{\partial \mathscr{L}}{\partial q_i}$$

Therefore,

$$\frac{\mathrm{d}\mathscr{L}}{\mathrm{d}t} - \frac{\mathrm{d}}{\mathrm{d}t}\left(\sum_i \dot{q}_i \frac{\partial \mathscr{L}}{\partial \dot{q}_i}\right) = \frac{\partial \mathscr{L}}{\partial t}$$

$$\frac{\mathrm{d}}{\mathrm{d}t}\left(\mathscr{L} - \sum_i \dot{q}_i \frac{\partial \mathscr{L}}{\partial \dot{q}_i}\right) = \frac{\partial \mathscr{L}}{\partial t} \tag{5.43}$$

But the Lagrangian does not have any explicit dependence on time and hence $\partial \mathscr{L}/\partial t = 0$. Therefore,

$$\sum_i \dot{q}_i \frac{\partial \mathscr{L}}{\partial \dot{q}_i} - \mathscr{L} = \text{constant} \tag{5.44}$$

This expression is exactly the same as the *law of conservation of energy* in Newtonian mechanics. The quantity which is conserved is known as the *Hamiltonian H*, where

$$H = \sum_i \dot{q}_i \frac{\partial \mathscr{L}}{\partial \dot{q}_i} - \mathscr{L} \tag{5.45}$$

We can rewrite this in terms of the conjugate momentum $p_i = \partial \mathscr{L}/\partial \dot{q}_i$:

$$H = \sum_i p_i \dot{q}_i - \mathscr{L} \tag{5.46}$$

Notice that, in the case of Cartesian coordinates, the Hamiltonian becomes

$$\begin{aligned} H &= \sum_i (m_i \dot{\mathbf{r}}_i) \cdot \dot{\mathbf{r}}_i - \mathscr{L} \\ &= 2T - (T - V) \\ &= T + V \end{aligned} \tag{5.47}$$

which explicitly shows the relation to energy conservation.

Notice that, formally, the conservation law arises from the homogeneity of the Lagrangian with respect to time.

5.6.3 *Lagrangian independent of the absolute position of the particles*

By this statement, we mean that $\mathscr{L}$ depends only on $\mathbf{r}_1 - \mathbf{r}_2$ and not on $\mathbf{r}_1$. $\mathscr{L}$ remains unchanged on shifting all the particles a small distance ϵ, i.e.

$$\mathscr{L} + \sum_i \frac{\partial \mathscr{L}}{\partial q_i} \delta q_i = \mathscr{L} + \sum_i \epsilon \frac{\partial \mathscr{L}}{\partial q_i} = \mathscr{L}$$

Invariance requires that

$$\sum_i \frac{\partial \mathscr{L}}{\partial q_i} = 0$$

But, from the Euler–Lagrange equation,

$$\frac{\mathrm{d}}{\mathrm{d}t}\left(\sum_i \frac{\partial \mathscr{L}}{\partial \dot{q}_i}\right) = \sum_i \frac{\partial \mathscr{L}}{\partial q_i} = 0$$

$$\frac{\mathrm{d}}{\mathrm{d}t}\sum_i p_i = 0 \tag{5.48}$$

This is just the law of *conservation of linear momentum* derived from the requirement that the Lagrangian be homogeneous with respect to spatial translations.

5.6.4 *Lagrangian independent of the orientation of the whole system in space*

By this statement we mean that $\mathscr{L}$ is invariant under rotations. If the system is rotated through a small angle $\delta\boldsymbol{\theta}$ (Figure 5.6), the position and velocity of a particle $\mathbf{r}_i$ and $\mathbf{v}_i$ change by

$$\delta \mathbf{r}_i = \mathrm{d}\boldsymbol{\theta} \times \mathbf{r}_i$$

$$\delta \mathbf{v}_i = \mathrm{d}\boldsymbol{\theta} \times \mathbf{v}_i$$

We require that $\delta\mathscr{L} = 0$ under this rotation and hence

$$\delta\mathscr{L} = \sum_i \left(\frac{\partial \mathscr{L}}{\partial \mathbf{r}_i} \cdot \delta \mathbf{r}_i + \frac{\partial \mathscr{L}}{\partial \mathbf{v}_i} \cdot \delta \mathbf{v}_i\right) = 0$$

Using the Euler–Lagrange relations,

$$\sum_i \left[\frac{\mathrm{d}}{\mathrm{d}t}\left(\frac{\partial \mathscr{L}}{\partial \dot{\mathbf{r}}_i}\right) \cdot (\mathrm{d}\boldsymbol{\theta} \times \mathbf{r}_i) + \frac{\partial \mathscr{L}}{\partial \mathbf{v}_i} \cdot (\mathrm{d}\boldsymbol{\theta} \times \mathbf{v}_i)\right] = 0$$

Reordering the vector products, this becomes

Figure 5.6

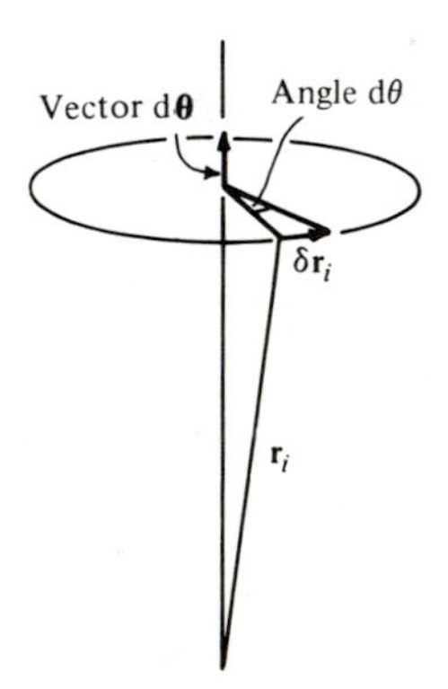

$$\sum_i \left\{ \mathrm{d}\boldsymbol{\theta} \cdot \left[\mathbf{r}_i \times \frac{\mathrm{d}}{\mathrm{d}t}\left(\frac{\partial \mathscr{L}}{\partial \dot{\mathbf{r}}_i}\right)\right] + \mathrm{d}\boldsymbol{\theta} \cdot \left(\mathbf{v}_i \times \frac{\partial \mathscr{L}}{\partial \mathbf{v}_i}\right)\right\} = 0$$

$$\sum_i \mathrm{d}\boldsymbol{\theta} \cdot \left[\frac{\mathrm{d}}{\mathrm{d}t}\left(\mathbf{r}_i \times \frac{\partial \mathscr{L}}{\partial \mathbf{v}_i}\right)\right] = 0$$

$$\mathrm{d}\boldsymbol{\theta} \cdot \sum_i \left[\frac{\mathrm{d}}{\mathrm{d}t}\left(\mathbf{r}_i \times \frac{\partial \mathscr{L}}{\partial \mathbf{v}_i}\right)\right] = 0 \qquad (5.49)$$

Therefore, if the Lagrangian is to be independent of orientation,

$$\sum_i \left(\mathbf{r}_i \times \frac{\partial \mathscr{L}}{\partial \mathbf{v}_i}\right) = \text{constant} \qquad (5.50)$$

i.e.

$$\sum_i (\mathbf{r}_i \times \mathbf{p}_i) = \text{constant} \qquad (5.51)$$

This is simply the *law of conservation of angular momentum* and results from the requirement that the Lagrangian be invariant under rotations.

5.7 Hamilton's equations and Poisson brackets – how Dirac discovered their application in quantum mechanics

The next extension is to express the equations of motion in terms of p_i and q_i rather than in terms of $\dot{q}_i$ and q_i, i.e. we use the canonical momentum p_i as defined by $\partial\mathscr{L}/\partial\dot{q}_i$. Remember that this does not necessarily correspond to what we mean by momentum in Newtonian mechanics. The equation relating the Hamiltonian to the Lagrangian may be written

$$H = \sum_i p_i \dot{q}_i - \mathscr{L}(q, \dot{q}) \qquad (5.52)$$

It looks as though H depends upon p_i, $\dot{q}_i$ and q_i but, in fact, we can rearrange the equation so that H is a function of only p_i and q_i. Let us take the total differential of H in the usual way, assuming $\mathscr{L}$ is time independent. Then

$$\mathrm{d}H = \sum_i p_i\,\mathrm{d}\dot{q}_i + \sum_i \dot{q}_i\,\mathrm{d}p_i - \sum_i \frac{\partial \mathscr{L}}{\partial \dot{q}_i}\,\mathrm{d}\dot{q}_i - \sum_i \frac{\partial \mathscr{L}}{\partial q_i}\,\mathrm{d}q_i \qquad (5.53)$$

Since $p_i = \partial\mathscr{L}/\partial\dot{q}_i$, the first and third terms on the right-hand side are equal and hence cancel. Therefore,

$$\mathrm{d}H = \sum_i \dot{q}_i\,\mathrm{d}p_i - \sum_i \frac{\partial \mathscr{L}}{\partial q_i}\,\mathrm{d}q_i \qquad (5.54)$$

This differential depends only on the increments $\mathrm{d}p_i$ and $\mathrm{d}q_i$ and hence we can compare $\mathrm{d}H$ with its formal expansion in terms of p_i and q_i:

$$\mathrm{d}H = \sum_i \frac{\partial H}{\partial p_i}\,\mathrm{d}p_i + \sum_i \frac{\partial H}{\partial q_i}\,\mathrm{d}q_i$$

It follows immediately that

$$\frac{\partial H}{\partial q_i} = -\frac{\partial \mathscr{L}}{\partial q_i}; \quad \frac{\partial H}{\partial p_i} = \dot{q}_i$$

and since

$$\frac{\partial \mathscr{L}}{\partial q_i} = \frac{\mathrm{d}}{\mathrm{d}t}\left(\frac{\partial \mathscr{L}}{\partial \dot{q}_i}\right)$$

from the Euler–Lagrange equation,

$$\frac{\partial H}{\partial q_i} = -\dot{p}_i$$

We thus reduce the equations of motion to the pair of relations

$$\left.\begin{aligned} \dot{q}_i &= \frac{\partial H}{\partial p_i} \\ \dot{p}_i &= -\frac{\partial H}{\partial q_i} \end{aligned}\right\} \tag{5.55}$$

This pair of equations is known as *Hamilton's equations.* They are a pair of first order differential equations for each of the $3N$ coordinates. We are now treating the p_is and the q_is on the same footing.

The last comment I want to make about these techniques is on the concept of *Poisson brackets.* In conjunction with Hamilton's equations of motion, they reduce the formalism to a compact form. Poisson brackets are defined by

$$[g, h] = \sum_{i=1}^{n}\left(\frac{\partial g}{\partial p_i}\frac{\partial h}{\partial q_i} - \frac{\partial g}{\partial q_i}\frac{\partial h}{\partial p_i}\right) \tag{5.56}$$

This is called the Poisson bracket for g and h. Clearly, we may write, in general,

$$\dot{g} = \sum_{i=1}^{n}\left(\frac{\partial g}{\partial q_i}\dot{q}_i + \frac{\partial g}{\partial p_i}\dot{p}_i\right) \tag{5.57}$$

for the variation of any physical quantity g and hence, using Hamilton's equations, we can write

$$\dot{g} = [H, g] \tag{5.58}$$

Therefore, Hamilton's equations can be written

$$\dot{q}_i = [H, q_i]; \quad \dot{p}_i = [H, p_i]$$

These brackets have a number of very useful properties. If we identify g with q_i and h with q_j, we find that

$$[q_i, q_j] = 0$$

Similarly,

$$[p_j, p_k] = 0$$

and if $j \neq k$,

$$[p_j, q_k] = 0$$

But if $g = p_k$ and $h = q_k$,

$$[p_k, q_k] = 1, \quad [q_k, p_k] = -1$$

Quantitites whose Poisson brackets are zero are said to *commute.* Those whose Poisson brackets are equal to 1 are said to be canonically conjugate. From the relation established in equation (5.58), we see that any quantity which commutes with the Hamiltonian does not change with time. In particular, H itself is constant in time because it commutes with itself. Yet again we have got back to the conservation of energy.

You will find that these quantities play an important role in quantum mechanics (see e.g. Dirac *The Principles of Quantum Mechanics*[3]). There is a very nice story about how Dirac came to realise their importance in his memoirs. In October 1925, he was worrying about the fact that, according to his formulation of quantum mechanics, the dynamical variables did not commute, i.e. if you have two variables u and v, uv is not the same as vu. Dirac apparently had an almost religious rule about taking walks on Sunday afternoons. He writes:

> It was during one of the Sunday walks in October 1925 when I was thinking very much about this $uv-vu$, in spite of my intention to relax, that I thought about Poisson brackets . . . . I did not remember very well what a Poisson bracket was. I did not remember the precise formula for a Poisson bracket and only had some vague recollections. But there were exciting possibilities there and I thought that I might be getting on to some big new idea.
>
> Of course, I could not [find out what a Poisson bracket was] right out in the country. I just had to hurry home and see what I could then find about Poisson brackets. I looked through my notes and there was no reference there anywhere to Poisson brackets. The text books which I had at home were all too elementary to mention them. There was nothing I could do, because it was Sunday evening then and the libraries were all closed. I just had to wait impatiently through that night without knowing whether this idea was any good or not but still I think that my confidence grew during the course of the night. The next morning, I hurried along to one of the libraries as soon as it was open and then I looked up Poisson brackets in Whittaker's 'Analytic Dynamics' and I found that they were just what I needed. They provided the perfect analogy with the commutator.[4]

This is a good example of how professionals work. They do not actually know all the mathematics there is to know but they keep their eyes and ears open to everything they hear, no matter how remote it might seem from their immediate interests. Then, someday, these little bits of information end up being important. They may not remember it exactly but they know where to find the necessary knowledge. This is how all of us try to approach physics in the context of research.

5.8 A warning

My approach to classical mechanics has intentionally been non-rigorous so that I can bring out certain specific features of different approaches to mechanics and dynamics. The subject often appears somewhat abstruse and mathematical and I have intentionally concentrated upon the simple parts which relate most directly to our intuitive understanding of Newton's laws. I would emphasise that the subject can be made rigorous and such an approach will be found in books such as Goldstein's *Classical Mechanics*[1] or Landau and Lifshitz's text book *Mechanics.*[5] My aim has been to show that there are good physical reasons for developing these more complex approaches to mechanics and dynamics.

APPENDIX TO CHAPTER 5

THE MOTION OF FLUIDS

A5.1 Introduction

The aim of this appendix is to revise the application of Newton's laws of motion to fluids. Fluid dynamics is a vast and complex subject for reasons which will become apparent in a moment. The intention of these notes is to emphasise some of the basic differences we encounter in dealing with fluids as opposed to systems of particles or solid bodies.

A5.2 The equation of continuity

First of all, we have to derive the basic equation which tells us that the fluid does not 'get lost'. By a procedure similar to that used for electromagnetism (Section 4.2), we consider the mass flux from a volume v bounded by a surface S. If $\mathbf{dS}$ is an element of surface area, the direction of the vector being taken normally outwards, the mass flow through $\mathbf{dS}$ is $\rho\mathbf{u}\cdot\mathbf{dS}$ (see Figure A5.1). Integrating over the surface of the volume, the rate of outflow of mass must equal the rate of loss of mass within v, i.e.

Figure A5.1

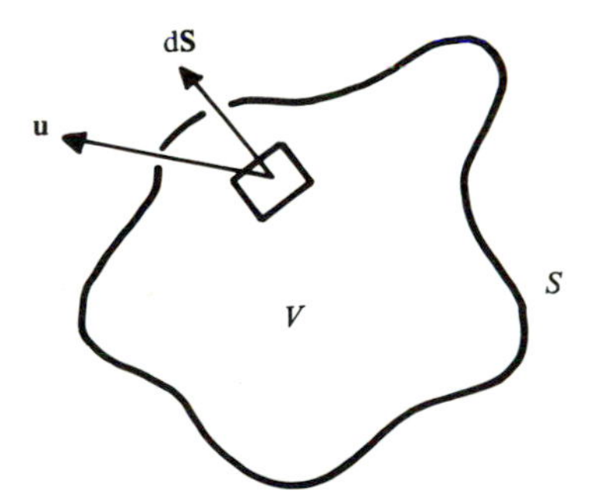

$$\int_S \rho \mathbf{u} \cdot d\mathbf{S} = -\frac{d}{dt} \int_v \rho dv \tag{A5.1}$$

Using the divergence theorem,

$$\int_v \operatorname{div}(\rho \mathbf{u})\, dv = -\frac{d}{dt} \int_v \rho dv$$

i.e.

$$\int_v \left(\operatorname{div} \rho \mathbf{u} + \frac{d\rho}{dt} \right) dv = 0$$

This result must be true for any region at any point in space and hence

$$\operatorname{div} \rho \mathbf{u} + \frac{\partial \rho}{\partial t} = 0 \tag{A5.2}$$

This is the *equation of continuity*. Notice that we have changed from total to partial derivatives between the last two equations. The reason is that, in microscopic form, we use partial derivatives to denote variations in the properties of the fluid *at a fixed point in space*. This is distinct from the derivatives of quantities which follow a particular element of the fluid, which are denoted by total derivatives, e.g. $d\rho/dt$. The reason for this is as follows.

We define $d\rho/dt$ to be the rate of change of density when we follow an element of the fluid. This is shown schematically in Figure A5.2 in which we follow the motion of that element over the time interval δt. If the velocity of the fluid at (x, y, z) is $\mathbf{u} = (u_x, u_y, u_z)$, we see that $d\rho/dt$ is given by

$$\frac{d\rho}{dt} = \underset{\delta t \to 0}{\text{limit}} \frac{1}{\delta t} [\rho(x + u_x \delta t, y + u_y \delta t, z + u_z \delta t, t + \delta t) - \rho(x, y, z, t)] \tag{A5.3}$$

Now we perform a Taylor expansion of the first term in square brackets.

$$\frac{d\rho}{dt} = \underset{\delta t \to 0}{\text{limit}} \frac{1}{\delta t} \left[\rho(x, y, z, t) + \frac{\partial \rho}{\partial x} u_x \delta t + \frac{\partial \rho}{\partial y} u_y \delta t + \frac{\partial \rho}{\partial z} u_z \delta t + \frac{\partial \rho}{\partial t} \delta t - \rho(x, y, z, t) \right]$$

$$= \frac{\partial \rho}{\partial t} + u_x \frac{\partial \rho}{\partial x} + u_y \frac{\partial \rho}{\partial y} + u_z \frac{\partial \rho}{\partial z}$$

Figure A5.2

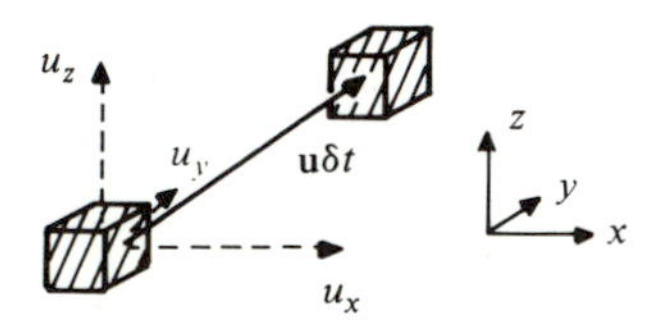

$$\frac{\mathrm{d}\rho}{\mathrm{d}t} = \frac{\partial \rho}{\partial t} + \mathbf{u}\cdot\text{grad}\ \rho \tag{A5.4}$$

Notice that equation (A5.4) is no more than the chain rule and this explains our definitions of $\mathrm{d}/\mathrm{d}t$ and $\partial/\partial t$. Notice also that equation (A5.4) is a relation between differential operators:

$$\frac{\mathrm{d}}{\mathrm{d}t} = \left(\frac{\partial}{\partial t} + \mathbf{u}\cdot\text{grad}\right) \tag{A5.5}$$

and this recurs throughout fluid dynamics. One can work in either of these frames of reference. If we work in the coordinate system which follows the fluid elements, coordinates are called *Lagrangian coordinates.* If we work in a fixed external reference frame, they are called *Eulerian coordinates.* Generally, one uses the Eulerian specification of coordinates although there are occasions when it is simpler to use a Lagrangian approach.

The relation (A5.4) enables us to rewrite the equation of continuity (A5.2) as follows.

$$\text{div}\ \rho\mathbf{u} + \frac{\mathrm{d}\rho}{\mathrm{d}t} = \mathbf{u}\cdot\text{grad}\ \rho$$

Expanding div $\rho\mathbf{u}$,

$$\mathbf{u}\cdot\text{grad}\ \rho + \rho\ \text{div}\ \mathbf{u} + \frac{\mathrm{d}\rho}{\mathrm{d}t} = \mathbf{u}\cdot\text{grad}\ \rho$$

$$\frac{\mathrm{d}\rho}{\mathrm{d}t} = -\rho\ \text{div}\ \mathbf{u} \tag{A5.6}$$

If the fluid is incompressible, $\rho = \text{constant}$ and the flow is described by div $\mathbf{u} = 0$.

A5.3 The equations of motion for an incompressible fluid in the absence of viscosity

To derive the equations of motion, we consider the forces acting upon unit element of volume of the fluid. Clearly Newton's laws of motion apply to the fluid in Lagrangian coordinates. Then, if we neglect viscous forces and assume that the fluid is incompressible,

$$\rho\frac{\mathrm{d}\mathbf{u}}{\mathrm{d}t} = -\text{grad}\ p - \rho\ \text{grad}\ \phi \tag{A5.7}$$

where p is pressure and ϕ the gravitational potential. We now rewrite this equation in terms of Eulerian coordinates, i.e. in terms of the partial differentials rather than total differentials. The analysis proceeds exactly as in Section A5.2 but now we are considering the variation of the vector quantity $\mathbf{u}$ rather than the scalar quantity ρ. This provides only a minor complication because we can consider the three components of the vector **u**, i.e. (u_x, u_y, u_z), to be scalar functions for which the relation (A5.3) applies. For example,

$$\frac{\mathrm{d}}{\mathrm{d}t} u_x = \frac{\partial u_x}{\partial t} + \mathbf{u} \cdot \operatorname{grad} u_x$$

There are similar equations for u_y and u_z and therefore, adding all three together vectorially,

$$\mathbf{i}_x \frac{\mathrm{d}u_x}{\mathrm{d}t} + \mathbf{i}_y \frac{\mathrm{d}u_y}{\mathrm{d}t} + \mathbf{i}_z \frac{\mathrm{d}u_z}{\mathrm{d}t} = \mathbf{i}_x \frac{\partial u_x}{\partial t} + \mathbf{i}_y \frac{\partial u_y}{\partial t} + \mathbf{i}_z \frac{\partial u_z}{\partial t}$$

$$+ \mathbf{i}_x (\mathbf{u} \cdot \operatorname{grad} u_x) + \mathbf{i}_y (\mathbf{u} \cdot \operatorname{grad} u_y) + \mathbf{i}_z (\mathbf{u} \cdot \operatorname{grad} u_z)$$

$$\frac{\mathrm{d}\mathbf{u}}{\mathrm{d}t} = \frac{\partial \mathbf{u}}{\partial t} + (\mathbf{u} \cdot \operatorname{grad}) \mathbf{u} \tag{A5.8}$$

Notice that the operator $(\mathbf{u} \cdot \operatorname{grad})$ means that we perform the operation $[u_x (\partial/\partial x) + u_y (\partial/\partial y) + u_z (\partial/\partial z)]$ on all three of the components of the vector **u** to find the magnitude of $(\mathbf{u} \cdot \operatorname{grad})$ in each direction. The equation of motion therefore becomes

$$\frac{\partial \mathbf{u}}{\partial t} + (\mathbf{u} \cdot \operatorname{grad}) \mathbf{u} = -\frac{1}{\rho} \operatorname{grad} p - \operatorname{grad} \phi \tag{A5.9}$$

This equation indicates clearly where the problems of fluid mechanics come from. The second term on the left introduces a nasty non-linearity in the velocity **u** and this causes all sorts of complications when trying to find exact solutions in fluid dynamical problems. It is obvious that the subject can rapidly become one of great mathematical complexity.

One interesting way of rewriting equation (A5.9) is to introduce a new vector $\boldsymbol{\omega}$ which is called the *vorticity* of the flow and is defined by $\boldsymbol{\omega} = \operatorname{curl} \mathbf{u} = \nabla \times \mathbf{u}$. You may like to show that we can write

$$\mathbf{u} \times \boldsymbol{\omega} = \mathbf{u} \times (\operatorname{curl} \mathbf{u}) = \tfrac{1}{2} \operatorname{grad} \mathbf{u}^2 - (\mathbf{u} \cdot \operatorname{grad}) \mathbf{u} \tag{A5.10}$$

Therefore, the equation of motion becomes

$$\frac{\partial \mathbf{u}}{\partial t} - (\mathbf{u} \times \boldsymbol{\omega}) = -\operatorname{grad} \left(\tfrac{1}{2} u^2 + \frac{p}{\rho} + \phi \right) \tag{A5.11}$$

This form is useful for obtaining particular solutions of some fluid dynamical problems. For example, if we are concerned only with steady motion, $\partial \mathbf{u}/\partial t = 0$ and then

$$\mathbf{u} \times \boldsymbol{\omega} = \operatorname{grad} \left(\tfrac{1}{2} u^2 + \frac{p}{\rho} + \phi \right) \tag{A5.12}$$

If the flow is irrotational, i.e. curl $u = 0$ and consequently $\mathbf{u}$ can be derived from the gradient of a scalar potential (see Appendix A3.4), the right-hand side must be zero and thus

$$\tfrac{1}{2}u^2 + \frac{p}{\rho} + \phi = \text{constant} \tag{A5.13}$$

Another way in which the conservation law can be applied is to introduce the concept of *streamlines* which are lines in the fluid whose tangent is everywhere parallel to $\mathbf{u}$ instantaneously. For example, Figure A5.3 shows the streamlines associated with the flow of an incompressible fluid past a solid sphere. If we follow the flow along a streamline, the quantity $\mathbf{u} \times \boldsymbol{\omega}$ must be perpendicular to $\mathbf{u}$ and hence if we take $\text{grad}(\frac{1}{2}u^2 + p/\rho + \phi)$ along the streamline, $\mathbf{u} \times \boldsymbol{\omega} = 0$ and again

$$\tfrac{1}{2}u^2 + \frac{p}{\rho} + \phi = \text{constant}$$

provided we follow a particular streamline. Notice that this result is true in any case, even if $\boldsymbol{\omega} \neq 0$. We recognise that we have derived *Bernoulli's theorem.*

A5.4 The equation of motion of an incompressible fluid including viscous forces

We have come almost as far as we can without writing down the stress tensor for the fluid and the equations of motion in tensor form. The proper derivation of the equations of motion when viscous forces are included needs the full tensor treatment and we will do no more than rationalise the form the equations must have. Let us consider steady unidirectional flow of an incompressible fluid in the positive x direction and work out the viscous force on an element of the fluid of volume $\delta V = \mathrm{d}x\mathrm{d}y\mathrm{d}z$ which has dimensions $\mathrm{d}x$ and

Figure A5.3. A representation of the flow of an incompressible, non-viscous fluid about a sphere by streamlines.

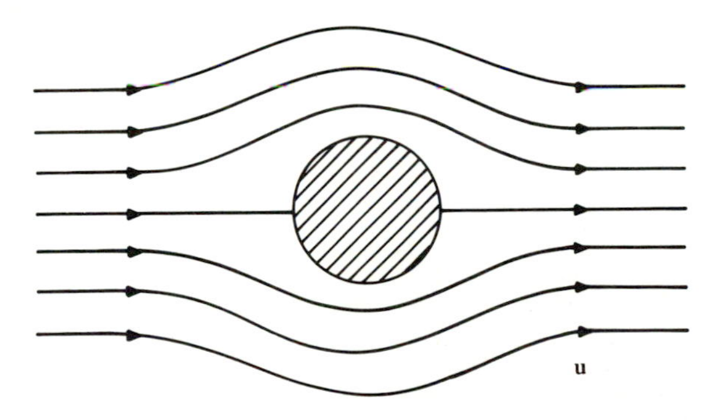

cross-sectional area dy dz (Figure A5.4). Consider first the viscous forces acting on the top and bottom faces of the volume at y and $y + dy$. The viscous force on the bottom surface is $\mu\, dx\, dz\, \partial u_x(y)/\partial y$ where μ is the viscosity of the fluid. The force on the top surface is $\mu\, dx\, dz\, \partial u_x(y + dy)/\partial y$. The net force on the element of fluid is the difference between these forces. Performing a Taylor expansion,

$$\mu\, dx\, dz \frac{\partial u_x(y + dy)}{\partial y} = \mu\, dx\, dz \left(\frac{\partial u_x(y)}{\partial y} + \frac{\partial^2 u_x(y)}{\partial y^2}\, dy\right)$$

Therefore, the net force on the upper and lower faces is

$$\mu\, (dx\, dz) \frac{\partial^2 u_x}{\partial y^2}\, dy$$

The same calculation can be performed for the forces acting on the faces defined by dx dy. Then the net force is

$$\mu\, (dx\, dy) \frac{\partial^2 u_x}{\partial z^2}\, dz$$

The equation of motion of the element of fluid is therefore

$$\rho\, \delta V \frac{du_x}{dt} = \mu \left(\frac{\partial^2 u_x}{\partial y^2} + \frac{\partial^2 u_x}{\partial z^2}\right) dx dy dz$$

i.e.

$$\rho \frac{du_x}{dt} = \mu \left(\frac{\partial^2 u_x}{\partial y^2} + \frac{\partial^2 u_x}{\partial z^2}\right)$$

Since the flow is unidirectional in the x direction, $\partial u_x/\partial x = 0$ and hence

$$\rho \frac{du_x}{dt} = \mu \nabla^2 u_x$$

The complete analysis shows that the equation of motion of an incompressible fluid is

$$\frac{\partial \mathbf{u}}{\partial t} + (\mathbf{u} \cdot \text{grad})\, \mathbf{u} = -\frac{1}{\rho}\, \text{grad}\, p - \text{grad}\, \phi + \frac{\mu}{\rho}\, \nabla^2 \mathbf{u} \qquad \text{(A5.14)}$$

Figure A5.4. Illustrating the viscous forces acting on an element of fluid in unidirectional flow.

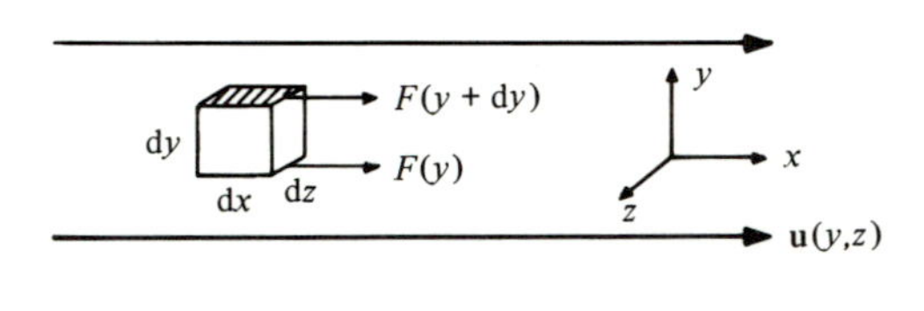

This equation is known as the *Navier–Stokes equation.* We can see that the viscous forces appear in the form suggested by our simple analysis of steady unidirectional flow. For a full treatment, the reader should consult, for example, Batchelor's text *An Introduction to Fluid Dynamics*[6] or Landau and Lifshitz's text book *Fluid Mechanics.*[7]

Case Study 4

THERMODYNAMICS AND STATISTICAL MECHANICS

Sadi Carnot (1796–1832)[1]

Rudolf Clausius (1822–1888)[2]

James Clerk Maxwell (1831–1879)[3]

Ludwig Boltzmann (1844–1902)[4]

[1] From *Scientific American*, **245**, 103, 1981.
[2] From *Introduction to Concepts and Theories in Physical Science*, G. Holton & S.G. Brush, p. 347, Addison-Wesley, 1973.
[3] From the frontispiece of *Clerk Maxwell 1831–1931*, Cambridge University Press, 1931.
[4] From *The Boltzmann Equation. Theory and Applications*, ed. E.G.D. Cohen & W. Thirring, Frontispiece, Springer-Verlag, 1973.

6

SIMPLE THERMODYNAMICS

6.1 The unique status of thermodynamics

Thermodynamics is the science of how the properties of matter and systems change with temperature. The subject may be viewed on the *microscopic scale* in which case we study the interactions of atoms and molecules as the temperature changes. In that case, we must build a specific model for those physical phenomena. It is, however, when we take the opposite view and look only on the *macroscopic scale* that the unique status of what we will call *classical thermodynamics* becomes evident. In that approach, we only look at the behaviour of matter and radiation in bulk and quite specifically deny that they have any internal structure at all. In other words, the science of classical thermodynamics is solely concerned with relations between macroscopic observable quantities.

Now this may sound rather dull but, in fact, it is quite the opposite. In many physical problems, one may not know in detail the correct microscopic physics and yet the thermodynamic approach can provide answers about the macroscopic behaviour of the system which are independent of the unknown detailed physics. Another way of looking at the subject is to think of classical thermodynamics as providing the boundary conditions which any microscopic model must satisfy. The thermodynamic arguments have absolute validity independent of the model adopted to explain any particular phenomenon.

It is remarkable that these profound statements can be made on the basis of the *two laws of thermodynamics*. The laws themselves are remarkable in that they are no more than reasonable hypotheses formulated as a result of practical experience. They prove, however, to be hypotheses of the greatest power. They have been applied to myriads of phenomena and time and again have proved to be correct. They have been applied to matter in extreme physical conditions such as matter in bulk at nuclear densities inside neutron stars and in the early stages of the hot big bang model of the Universe ($\rho \sim 10^{15}$ g cm^{-3}) and at ultralow temperatures in laboratory experiments. It is important to emphasise that there is no way in which one proves the laws of thermodynamics – they are simply expressions of our common experience of the thermal properties of matter and radiation.

I have emphasised the unique status of classical thermodynamics in the above paragraphs. We should distinguish between this 'thermodynamic' approach to the problems of physics and the approach of model building in which we seek to interpret the nature of the laws in terms of microscopic processes. The present chapter will remain strictly *model-free* but in the next chapter we will study two models. These will be the *kinetic theory of gases* as enunciated by Clausius and Maxwell and *statistical mechanics* which was discovered by Boltzmann. These theories are explanatory in a sense that thermodynamics is not. Both of them are very successful and yet there remains problems which become too complicated to be handled by them. It is in these highly complicated problems that the thermodynamical approach comes into its own.

The object of this chapter is twofold. First of all, the history of the discovery of the laws of thermodynamics illuminates certain conceptual problems which confronted the 19th century pioneers. The resolution of these problems helps to clarify the definitions of concepts such as heat, energy, work, etc. Second, when we come to the nature of the second law and the concept of entropy, I find that one has to be particularly careful about defining exactly what one means by the various terms. The reason is that, for much of the argument, we deal with the *ideal* behaviour of perfect machines or systems and then contrast this with what happens in the actual imperfect world. I believe that many of the uncertainties students have about the second law result from a poor exposition of the basic definitions.

6.2 The origin of the first law of thermodynamics

In anticipation of the development of later sections, we recall that the first law is a statement about the conservation of energy. A simple formulation is

Energy is conserved when heat is taken into account.

The second law tells us about the way in which thermodynamic systems evolve. The statement by Clausius is

No process is possible whose sole result is the transfer of heat from a colder to a hotter body.

To most students, the first law is the easier of the two and yet historically it was the first law which proved to be the more difficult to establish. The development of these concepts is admirably described in the book *Energy, Force and Matter. The Conceptual Development of Nineteenth Century Physics* by

P.M. Harman.[1] The difficulties with the first law can be traced to the problem of understanding exactly what heat is. In the 18th century, the dominant theory was that heat was a form of 'imponderable fluid', i.e. a massless fluid, which was called *caloric.* If one body is at a higher temperature than another and they are brought into thermal contact, caloric is said to flow from the hotter to the cooler body until they come into equilibrium at the same temperature. There were problems with this theory – for example, when a warm body is brought into contact with ice, the caloric flows from the warm body into the ice but, although ice is converted into water, the temperature of the ice–water mixture remains the same. It had to be supposed that caloric could combine with ice to form water. An alternative view developed during the 18th century according to which heat is associated with the motions or vibrations of the microscopic particles which make up matter. This theory, known as the *kinetic or dynamic theory*, associated heat with the kinetic energy of the motions of the microscopic constituents of matter.

The two theories came into conflict at the end of the 18th century. Among evidence against the caloric theory were the experiments of Count Rumford who demonstrated in 1798 that heat could be produced by friction. This soldier of fortune illustrated the phenomenon by attempting to bore cannons with a blunt drill. In this experiment, there is no obvious source of the caloric which evidently can be produced in indefinite amounts.

Among the contributions which were to prove to be of prime importance in the future mathematical development was Fourier's treatise *Analytical Theory of Heat*[2] published in 1822. In this treatise, Fourier worked out the mathematical theory of heat transfer in the form of differential equations which did not require the construction of any particular model of the physical nature of heat. Fourier's methods were firmly based in the French tradition of rational mechanics and gave mathematical expression to the *effects* of heat without enquiring into its causes. This process of giving mathematical substance to physical effects was to have profound influence upon the future generation of Scottish physicists, in particular, William Thomson (later Lord Kelvin) who received a thorough training in the French school of mathematical physics.

The idea of conservation laws, as applied to what we would now call energy, for mechanical systems had been discussed in 18th century treatises, in particular, in the work of Leibniz. He argued that 'living force' or '*vis viva*', which we now call the kinetic energy of particles, is conserved in mechanical processes. By the 1820s, the relation of the kinetic energy to the work done was clarified. Then, in the 1840s, a number of scientists independently came to the correct conclusion about the interconvertibility of heat and work. In 1842, Mayer proposed that heat and work are interchangeable and derived a value for the mechanical equivalent of heat from the adiabatic expansion of gases. The most important of these contributions was, however, by Joule who performed an outstanding series of experiments which was fundamental to the mathematical elaboration of the laws of conservation of energy.

James Prescott Joule was born into a family which had grown wealthy through the foundation of a brewery by his grandfather. Joule's pioneering experiments were carried out in laboratories which he installed at his own expense in his own home or in the brewery. His genius was as a meticulous experimenter and, indeed, the reason for singling out his name from the other pioneers of the 1840s is that he put the science of thermodynamics on a firm experimental basis. Perhaps the most important aspect of his work was his ability to measure accurately very small temperature changes in his experiments.

The first set of key experiments concerned the 'Production of heat by voltaic electricity'.[3] In the earliest experiments, he established that the quantity of heat produced is proportional to RI^2 where R is the resistance and I the current. By 1843, he was able to derive from these electrical experiments a value for the mechanical equivalent of heat. In a postscript to his paper, we find a remark that in other experiments he had shown 'that heat is evolved by the passage of water through narrow tubes'. This hints at the origin of his most famous experiments, the paddle wheel experiments, which are illustrated in a paper published in 1850 (Figure 6.1). The work done by the weights in driving the paddle wheel goes into heat through the frictional force between the water and the vanes of the paddle wheel. By taking the greatest care to estimate all heat losses, he found the value for the mechanical equivalent of heat to be 4.13 Joules calorie^{-1} in the modern notation. The modern value is 4.187 Joules calorie^{-1}.

It was the early results of the paddle wheel experiments which greatly excited William Thomson in 1847 when he was only 22. Thomson immediately appreciated the fundamental significance of Joule's experiments and used them as the foundation for the subject which we now call thermodynamics, the name itself being invented by Thomson. These basic results became widely known in continental Europe and, by 1850, Helmholtz and Clausius formulated what is now known as the law of conservation of energy or the first law of thermodynamics. Helmholtz, in particular, was the first to express the conservation laws in mathematical form in such a way as to incorporate mechanical and electrical phenomena, heat and work.

6.3 The first law of thermodynamics

I emphasise that there is no intention here of providing a systematic introduction to this huge subject. My concern is to describe the fundamentals in as clear a manner as possible. My approach to both laws has been strongly influenced by Pippard's *The Elements of Classical Thermodynamics*.[4]

6.3.1 The zeroth law and the definition of empirical temperature

In many of the arguments which lead to the formulation of the laws of thermal physics, we make many statements of the form 'It is a fact of experience that . . . '. From these axioms, we derive a mathematical structure.

I am going to consider the thermal properties of *fluids* for the simple reason that, if we change the shape of the containing vessel without changing the volume, no work is done. We also note that at this stage we have no definition of temperature of any kind. We now make our first dogmatic statement:

'It is a fact of experience that the properties of a fluid are entirely determined by only two properties, the pressure p and the volume V of the vessel.' This assumes that we do not subject the fluid to other influences such as putting it in an electric or magnetic field. If the system is wholly defined by two properties,

Figure 6.1. Joule's apparatus for the 'paddle wheel' experiments used in determining the mechanical equivalent of heat. (From J.P. Joule, *Phil. Trans. Roy. Soc.*, 1850, **140**, opposite p. 64.)

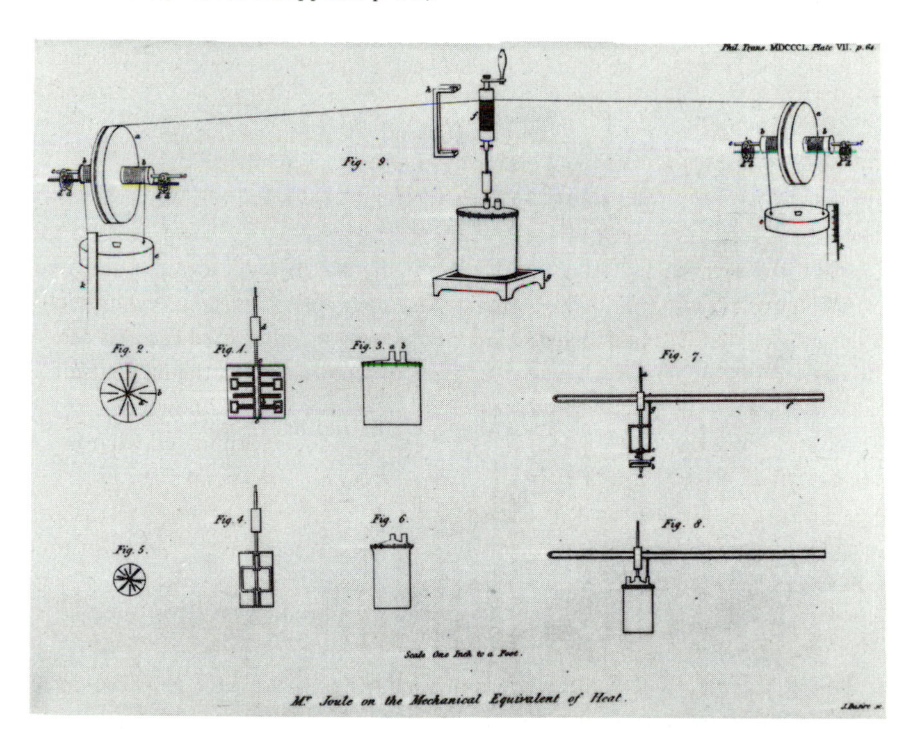

it is known as a two-coordinate system. Most of the systems we have to deal with are two-coordinate systems but it is relatively straightforward to develop the formalism for multi-coordinate systems.

Notice carefully what this assertion means. Suppose we take some fluid through a sequence of processes such that it ends up having pressure p_1 and volume V_1. Then suppose we take some more fluid and perform a totally different set of operations so that it also ends up with coordinates p_1 and V_1. We have just asserted that these two fluids will be totally indistinguishable in their physical properties.

Now let us take two isolated systems consisting of fluids with coordinates p_1, V_1 and p_2, V_2. We then bring them into thermal contact. If we leave them for a very long time, they will change their properties such that they reach a state of *thermodynamic equilibrium.* By this term, we mean that all components which make up the system are allowed to interact thermally until after a very long time no further changes in the bulk properties of the system are found. In attaining this state, generally heat will be exchanged and work done. Eventually, the two fluids come into thermodynamic equilibrium such that their thermodynamic coordinates are p_1, V_1 and p_2, V_2. Now it is clear that these four values cannot be arbitrary. It is a fact of experience that you cannot have two fluids with arbitrary values for $p_1 V_1$ and $p_2 V_2$ and expect them to be in thermodynamic equilibrium. There must be some mathematical relation between the four quantities and we can write

$$F(p_1, V_1, p_2, V_2) = 0$$

This equation tells us the value of the fourth quantity if the other three are fixed.

So far we have avoided using the word *temperature*. We will find a suitable definition by developing another common fact of experience which is so central to the subject that it is included as a law of thermodynamics – the *zeroth law.* The formal statement is as follows:

> If two systems, 1 and 2, are separately in thermal equilibrium with a third, 3, then they must also be in thermal equilibrium with one another.

We can write this down mathematically and see the consequences:

System 1 being in equilibrium with system 3 means that

$$F(p_1, V_1, p_3, V_3) = 0$$

or expressing the pressure p_3 in terms of the other variables

$$p_3 = f(p_1, V_1, V_3)$$

System 2 being in equilibrium with system 3 means, in the same way, that

$$p_3 = g(p_2, V_2, V_3)$$

therefore,

$$f(p_1, V_1, V_3) = g(p_2, V_2, V_3) \tag{6.1}$$

But the zeroth law tells us that 1 and 2 must also be in equilibrium and consequently there must exist a function such that

$$h(p_1, V_1, p_2, V_2) = 0 \tag{6.2}$$

The relation (6.2) tells us that equation (6.1) must be of such a form that the dependence on V_3 cancels out on either side, e.g. $f(p_1, V_1, V_3) = \phi_1(p_1, V_1)\zeta(V_3) + \eta(V_3)$. Therefore, cancelling out the term V_3, we find that

$$\phi_1(p_1, V_1) = \phi_2(p_2, V_2) = \phi_3(p_3, V_3) = \theta = \text{constant} \tag{6.3}$$

in thermodynamic equilibrium. This is the logical consequence of the zeroth law – there exists a function of p and V which takes a constant value for all systems in thermodynamic equilibrium with each other. Different equilibrium states will be characterised by different constants. Thus, this constant which characterises the equilibrium is what we call a *function of state*, i.e. it is a quantity which takes a definite value for a particular equilibrium state. It is called the *empirical temperature* θ. We also define *an equation of state* which relates p and V to the empirical temperature

$$\phi(p, V) = \theta \tag{6.4}$$

We can now find from experiment all the combinations of p and V which correspond to a fixed value of the empirical temperature θ. We have three quantities, p, V and θ, which define the equilibrium state and any two of them are sufficient to define it completely. Lines of constant θ are called *isotherms.*

At this stage the empirical temperature looks nothing like what we customarily call temperature and in fact we could devise horridly complicated temperature scales. To put the whole thing on a firm experimental foundation, we have to decide upon a *thermometric scale.* Once we have fixed this scale for one system, it will be fixed for all others by virtue of the fact that all systems have the same value of the empirical temperature when they are in thermodynamic equilibrium.

The subject of thermometric scales is a vast one. We note only the *constant volume gas thermometer* which is illustrated in Figure 6.2. Although bulky and unwieldy, the constant volume gas thermometer is of special importance because it was found empirically that all gases have the same variation of pressure with volume at low pressures, i.e. they give the same temperature at low enough pressures. In addition, such gases approximate very closely to what we will define to be *ideal or perfect gases* and one can then go further and relate the pressure and volume measures of these gases to the thermodynamic temperature T. The relation between these quantities is the *perfect gas law*,

$$pV = RT \tag{6.5}$$

where R is the gas constant for 1 mole of gas. We will show in Section 6.5.2 that this is not an empirically defined temperature but one which can be related to basic thermodynamic principles. Thus, in the limit of low pressures, we can use gas thermometers to measure *thermodynamic temperatures* directly. Symbolically, we can write

$$T = \lim_{p \to 0} (pV)/R \tag{6.6}$$

Notice that, in general, p and V are much more complicated functions of T, in particular, at high pressures and close to phase transitions.

6.3.2 *The mathematical expression of the first law of thermodynamics*

We have already made a simple statement of the law 'Energy is conserved if heat is taken into account.' To give quantitative substance to the law, we must define clearly what we mean by heat, energy and work. The last two are straightforward. The basic definition is that the work done is given by

$$W = \int_{\mathbf{r}_1}^{\mathbf{r}_2} \mathbf{F} \cdot d\mathbf{r} \tag{6.7}$$

When we do a certain amount of work on a body, we increase its *energy*. Let us note some of the ways in which we can do work on a system.

Work done in compressing a fluid. When we do work on a fluid, we regard the work done as a positive quantity. Therefore, since the volume decreases when we do mechanical work on the fluid, the amount of (positive) work done is

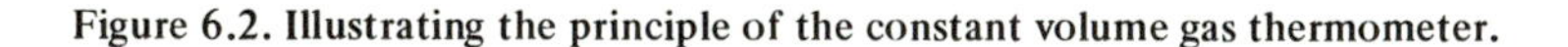
Figure 6.2. Illustrating the principle of the constant volume gas thermometer.

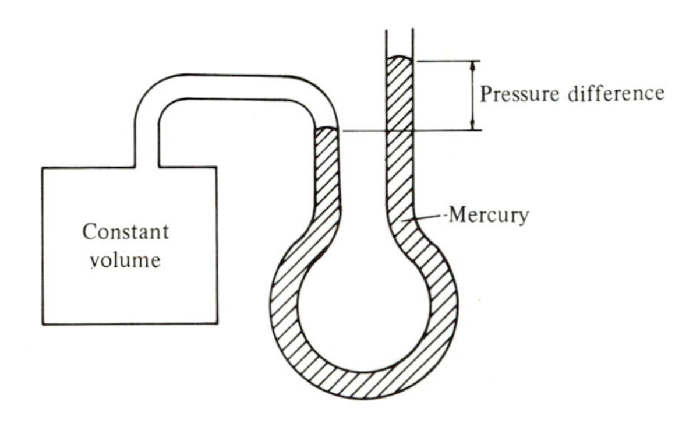

$$\mathrm{d}W = -\int p\,\mathrm{d}V$$

If the fluid does work by expanding, the work done on the surroundings is positive and the work done by the system negative, i.e. the sign of the volume increment is important.

Work done in stretching a wire by $\mathrm{d}\mathbf{l}$: $\mathrm{d}W = \mathbf{F}\cdot\mathrm{d}\mathbf{l}$.

Work done by an electric field on a charge q: $\mathrm{d}W = q\mathbf{E}\cdot\mathrm{d}\mathbf{r}$.

Work done against surface tension in increasing the surface area $\mathrm{d}A$: $\mathrm{d}W = \gamma\mathrm{d}A$, where γ is the surface tension.

Work done by a couple C: $\mathrm{d}W = C\,\mathrm{d}\theta$.

Work done on a dielectric by an electric field: $\mathrm{d}W = \mathbf{E}\cdot\mathrm{d}\mathbf{p}$ where $\mathbf{p}$ is the total dipole moment.

Work done by a magnetic field: $\mathrm{d}W = \mathbf{B}\cdot\mathrm{d}\mathbf{m}$ where $\mathbf{m}$ is the magnetic dipole moment.

We see that, in general, the work done is the product of a generalised force $\mathbf{X}$ times a generalised displacement $\mathrm{d}\mathbf{x}$, $\mathrm{d}W = \mathbf{X}\cdot\mathrm{d}\mathbf{x}$. We note that the work done is always the product of an *intensive variable* $\mathbf{X}$, by which we mean a property defined at a particular point in the fluid, and an *extensive variable* $\mathrm{d}\mathbf{x}$ which tells us about the 'extent' of the system.

Now let us consider an isolated system in which there is no thermal interaction with the surroundings. It is a fact of experience that if we do work on the system in some fashion, the system attains a new equilibrium state and that it does not matter how we do the work. For example, we may compress a gas, or stir it with a paddle wheel, or pass an electric current through it for a certain time. It was Joule's great contribution to thermodynamics to demonstrate precisely by experiment that this is indeed the case. The result is that we give energy to the system. We say that there is an increase in the internal energy U of the system by virtue of work having been done. Since it does not matter how the work is done, U must be a *function of state*. For this isolated system,

$$W = U_2 - U_1 \quad \text{or} \quad W = \Delta U \tag{6.8}$$

Now suppose the system is not isolated but that there is a thermal interaction between the surroundings and the system. Then the system will reach a new internal energy state and this will not all be due to work. We then define the heat supplied to be

$$Q = \Delta U - W \tag{6.9}$$

This is our definition of *heat*. It may look like a rather roundabout way of proceeding but it has the great value of logical consistency. It completely avoids the problem of specifying exactly what heat is, the problem which was at the root of the difficulties which existed up till the 1840s in incorporating heat into the conservation laws.

It is convenient to write this in differential form

$$\mathrm{d}Q = \mathrm{d}U - \mathrm{d}W \tag{6.10}$$

It is also useful to distinguish between those differentials which refer to functions of state and those which do not. Clearly, $\mathrm{d}U$ is a differential of a function of state as are $\mathrm{d}p$, $\mathrm{d}V$ and $\mathrm{d}T$ but $\mathrm{d}Q$ and $\mathrm{d}W$ are not because we can get from U_1 to U_2 by adding together different amounts of $\mathrm{d}Q$ and $\mathrm{d}W$. We write these differentials $đQ$, $đW$. Thus,

$$đQ = \mathrm{d}U - đW \tag{6.11}$$

The expression (6.11) is the formal mathematical expression of the *first law of thermodynamics*. We are now in a position to apply the conservation law to all sorts of diverse problems. We will consider some of these briefly in the next subsection.

6.3.3 Some applications of the first law of thermodynamics

(*i*) *Specific heat capacities.* U is a function of state and we know we can describe the properties of a gas entirely in terms of two other functions of state. So let us express U in terms of T and V: $U = U(T, V)$. Then the differential of U is

$$\mathrm{d}U = \left(\frac{\partial U}{\partial T}\right)_V \mathrm{d}T + \left(\frac{\partial U}{\partial V}\right)_T \mathrm{d}V \tag{6.12}$$

Hence,

$$đQ = \left(\frac{\partial U}{\partial T}\right)_V \mathrm{d}T + \left[\left(\frac{\partial U}{\partial V}\right)_T + p\right] \mathrm{d}V \tag{6.13}$$

We can now introduce the concepts of *heat capacity C*. At constant volume,

$$C_V = \left(\frac{đQ}{\mathrm{d}T}\right)_V = \left(\frac{\mathrm{d}U}{\mathrm{d}T}\right)_V$$

At constant pressure

$$C_p = \left(\frac{đQ}{\mathrm{d}T}\right)_p = \left(\frac{\partial U}{\partial T}\right)_V + \left[\left(\frac{\partial U}{\partial V}\right)_T + p\right]\left(\frac{\partial V}{\partial T}\right)_p \tag{6.14}$$

These expressions tell us by how much the temperature rises for a given input of heat. Notice that these heat capacities do not refer to any particular volume or mass. It is conventional to use *specific heat capacities* or *specific heats* where the word specific takes its usual meaning of 'per unit mass'. Conventionally, specific quantities are written in lower case letters, i.e.

$$c_V = C_V/m; \quad c_p = C_p/m \tag{6.15}$$

Subtracting, we find

$$C_p - C_V = \left[\left(\frac{\partial U}{\partial V}\right)_T + p\right]\left(\frac{dV}{dT}\right)_p \tag{6.16}$$

The interpretation of this equation is straightforward. The second term on the right-hand side clearly tells us how much work is done in pushing back the surroundings at constant p. The first term is clearly something to do with the internal properties of the gas because it tells us how the total internal energy changes with volume. Obviously, it must be associated with work done against intermolecular forces in the gas. Thus $C_p - C_V$ gives us information about $(dU/dV)_T$.

(*ii*) *The Joule expansion.* One way of finding out the relation between C_p and C_V is to perform what is known as a *Joule expansion*, i.e. a free expansion of the gas into a larger volume (Figure 6.3). There is no heat inflow and no work of the form $p\,dV$ is done in the free expansion, i.e. all the walls are fixed. Therefore, if $(dU/dV)_T = 0$ there should not be a change in U. This is the way in which we can define classically a *perfect gas.* It is defined by the following two properties.

(*a*) Its equation of state is the perfect gas law, $pV = RT$.

(*b*) In a Joule expansion, there is no change in internal energy U.

Thus, for a perfect gas, we find a simple relation between C_p and C_V. We will consider only 1 mole of gas. From (*b*),

$$\left(\frac{dU}{dV}\right)_T = 0$$

From (*a*),

$$\left(\frac{\partial V}{\partial T}\right)_p = \left(\frac{\partial}{\partial T}\left(\frac{RT}{p}\right)\right)_p = \frac{R}{p}$$

$$C_p - C_V = [0 + p]\,\frac{R}{p}$$

$$C_p - C_V = R \tag{6.17}$$

The importance of this result is that it demonstrates that the internal energy of a perfect gas is a function only of temperature. You will remember that the internal energy is a function of state and therefore may be expressed as a function of

Figure 6.3. Illustrating a Joule expansion.

the only two variables we need to describe the system: p and V. What the Joule expansion of the perfect gas tells us is that the internal energy is independent of volume. It must also be independent of pressure because the pressure certainly decreases in a Joule expansion. This proves that the internal energy is only a function of temperature for a perfect gas.

For real gases, of course, there actually is a change in U with volume – physically this is because work is done against the intermolecular forces, e.g. the van der Waals forces. Also, at very high pressures, there is an effective repulsive force because the molecules cannot be squeezed – hard core repulsion. Thus, in practice, there will be a small change in temperature in the expansion. The Joule coefficient is defined to be $(\partial T/\partial V)_U$, i.e. the change in T as volume increases at constant U, and we can relate this to other properties of the gas.

(*iii*) *The enthalpy and the Joule–Kelvin expansion.* Now let us look again at the heat capacities. You will note that the heat capacity at constant volume is the differential of a function of state and we may ask if there is a function of state which corresponds to C_p. Let us start again and write $U = U(p, T)$ instead of $U(V, T)$. Remember that in the general case we always need two coordinates to specify the state of the gas.

$$\mathrm{d}U = \left(\frac{\partial U}{\partial p}\right)_T \mathrm{d}p + \left(\frac{\partial U}{\partial T}\right)_p \mathrm{d}T$$

Now we proceed as before.

$$\begin{aligned} đQ &= \mathrm{d}U + p\,\mathrm{d}V \\ &= \left(\frac{\partial U}{\partial p}\right)_T \mathrm{d}p + p\,\mathrm{d}V + \left(\frac{\partial U}{\partial T}\right)_p \mathrm{d}T \\ \left(\frac{đQ}{\mathrm{d}T}\right)_p &= p\left(\frac{\partial V}{\partial T}\right)_p + \left(\frac{\partial U}{\partial T}\right)_p \\ &= \left[\frac{\partial}{\partial T}(pV + U)\right]_p \end{aligned} \tag{6.19}$$

But the combination $pV + U$ is entirely composed of functions of state and hence this new function must also be a function of state. This quantity is known as the *enthalpy* H. Thus,

$$H = U + pV$$

$$C_p = \left(\frac{đQ}{\mathrm{d}T}\right)_p = \left(\frac{\partial H}{\partial T}\right)_p = \left[\frac{\partial}{\partial T}(U + pV)\right]_p \tag{6.20}$$

The enthalpy often appears in flow processes and, in particular, in another type of expansion process known as a Joule–Kelvin expansion. In this case, gas is transferred from one cylinder to another, the pressures in both cylinders being maintained constant at values p_1 and p_2. Suppose we push a given number of moles of gas through a nozzle or a porous plug. What starts off as gas with internal energy U_1, in volume V_1, under pressure p_1, ends up on the other side as gas with pressure p_2, volume V_2 and temperature T_2. The system is completely isolated thermally and hence we can apply the law of conservation of energy to the gas. The energy consists of internal energy U_1 plus work done on the gas on one side $p_1 V_1$ and this must equal the internal energy U_2 plus the work done by the gas on the other side $p_2 V_2$.

$$p_1 V_1 + U_1 = p_2 V_2 + U_2$$

or

$$H_1 = H_2 \tag{6.21}$$

Now, again, if we are dealing with a perfect gas, we must have $H = pV + U = RT + U(T)$. But $U(T) + RT$ is a unique function of temperature and hence T must be the same in both volumes. Thus, for a perfect gas, there is no change in temperature in a Joule–Kelvin expansion. In real gases, however, there is a temperature change, again, just because of the internal forces between molecules. The temperature change may be either positive or negative depending upon the pressure and temperature. The Joule–Kelvin coefficient is defined to be $(\partial T/\partial p)_H$.

We have now come very close to deriving a yet more general flow equation in which we take into account other contributions to the flow, e.g. the kinetic energy and the potential energy of the gas if it is in a gravitational field. Let us consider the flow through a black box and add in these other energies as well (Figure 6.4).

We consider only steady flow of a given mass of gas or fluid as it enters and leaves the black box. The energy conservation equation then reads

$$H_1 + \tfrac{1}{2}mv_1^2 + m\phi_1 = H_2 + \tfrac{1}{2}mv_2^2 + m\phi_2$$

$$p_1 V_1 + U_1 + \tfrac{1}{2}mv_1^2 + m\phi_1 = p_2 V_2 + U_2 + \tfrac{1}{2}mv_2^2 + m\phi_2 \tag{6.22}$$

i.e.

$$\frac{p}{m/V} + \frac{U}{m} + \tfrac{1}{2}v^2 + \phi = \text{constant}$$

Figure 6.4. Illustrating the conservation of energy in fluid flow in the presence of a gravitational field.

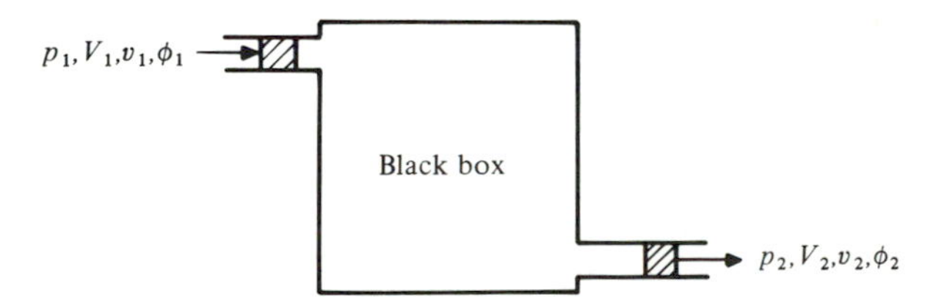

$$\frac{p}{\rho} + u + \tfrac{1}{2}v^2 + \phi = \text{constant} \tag{6.23}$$

This is leading us to the equations of fluid flow. In particular, for an incompressible fluid, $U =$ constant and we obtain Bernoulli's equation,

$$\frac{p}{\rho} + \tfrac{1}{2}v^2 + \phi = \text{constant} \tag{6.24}$$

which we derived fluid mechanically in Appendix A5.3. Notice also that we assume that all the additional terms present in Bernoulli's equation are absent when we treat the simple Joule–Kelvin expansion. The full Bernoulli equation indicates that we must assume that the Joule–Kelvin expansion takes place very slowly so that the kinetic energy terms can be neglected.

(*iv*) *Adiabatic expansion.* In an adiabatic expansion, the volume of the gas changes without any thermal contact between the system and its surroundings. The classical method of illustration is in the expansion or compression of the gas within a perfectly insulated piston (Figure 6.5). An important point about expansions of this type is that they take place very slowly so that the system passes through an infinite number of equilibrium states between the initial and final thermodynamic coordinates. We will return to this key concept in our discussion of reversible processes in Section 6.5.1. We can therefore write

$$đQ = dU + p\,dV = 0 \tag{6.25}$$

We consider n moles of gas and therefore, if $C_V = (\partial U/\partial T)_V$ is taken to refer to 1 mole of gas, $dU = nC_V dT$. During the expansion, the gas passes through an infinite series of equilibrium states for which the perfect gas law is assumed to apply, $pV = nRT$, and hence, from equation (6.25),

$$nC_V\, dT + \frac{nRT}{V}\, dV = 0 \tag{6.26}$$

$$\frac{C_V}{R}\frac{dT}{T} = -\frac{dV}{V}$$

Integrating, we find that

Figure 6.5

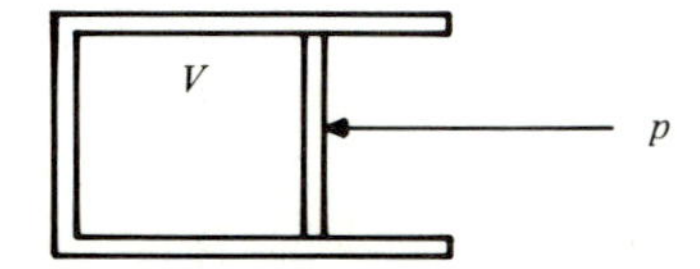

$$\frac{V_2}{V_1} = \left(\frac{T_2}{T_1}\right)^{-C_V/R} \quad \text{or} \quad VT^{C_V/R} = \text{constant} \tag{6.27}$$

Since $pV = nRT$ at all stages in the expansion, this result can also be written

$$pV^{\gamma} = \text{constant}$$

where

$$\gamma = 1 + \frac{R}{C_V}$$

We have already shown that, for one mole of gas, $C_V + R = C_p$ and consequently

$$1 + \frac{R}{C_V} = \frac{C_p}{C_V} = \gamma \tag{6.28}$$

γ is the ratio of specific heats or the adiabatic index. For a monatomic gas, $C_V = \frac{3}{2}R$ and consequently $\gamma = \frac{5}{3}$.

(*v*) *Isothermal expansion.* In this case, there must be heat exchange with the surroundings so that the gas within the piston remains at the same temperature, $T =$ constant. In the expansion, work is done in pushing back the piston and this must be made up by a corresponding inflow of heat, i.e. the work done is

$$\int_{V_1}^{V_2} p \, dV = \int_{V_1}^{V_2} \frac{RT}{V} \, dV = RT \ln\left(\frac{V_2}{V_1}\right) \tag{6.29}$$

This is the amount of heat which must be supplied from the surroundings to maintain an isothermal expansion. This result will find important applications in the understanding of heat engines.

(*vi*) *Expansions of different types.* We should notice carefully the four different types of expansion which we have described in this section.

Isothermal expansion, $\Delta T = 0$. Heat must be supplied or removed from the system to maintain $\Delta T = 0$ (see item (v)).

Adiabatic expansion, $\Delta Q = 0$. No heat exchange takes place with the surroundings (see item (iv)).

Joule expansion, $\Delta U = 0$. *For a perfect gas*, the free expansion to a larger volume with fixed walls involves no change in internal energy (item (ii)).

Joule–Kelvin expansion, $\Delta H = 0$. When gas passes from one volume to another and the pressures in the two vessels are maintained at p_1 and p_2 during the transfer, enthalpy is conserved if the gas is perfect (item (iii)).

The basic point of principle in all these apparently different phenomena is simply the conservation of energy, taking account of heat, i.e. they are no more than simple applications of the first law of thermodynamics.

6.4 The origin of the second law of thermodynamics

The origin of the second law of thermodynamics is historically linked with the name of Sadi Carnot. He was the eldest son of Lazare Carnot who was a member of the Directory after the French Revolution and later, during the Hundred Days in 1815, Napoleon's Minister of the Interior. After 1807, Lazare Carnot devoted much of his energies to the education of his sons. Sadi Carnot was educated at the elite École Polytechnique where his teachers included Poisson, Gay-Lussac and Ampère. After a period as a military engineer, from 1819 he was able to devote himself wholly to research.

His great work *Réflexions sur la Puissance Motrice du Feu et sur les Machines Propres à Developper cette Puissance*[5] was published in 1824; is is normally translated as *Reflections on the Motive Power of Fire.* The treatise concerned the question of the maximum efficiency of heat engines. In his approach, he was strongly influenced by his father's work on steam engines. The treatise is, however, of much greater generality and is an intellectual achievement of the greatest originality.

Most previous work on the maximum efficiency of steam engines involved empirical studies such as comparing the fuel input with the work output or else theoretical studies on the basis of specific models for the behaviour of the gases in heat engines. Carnot's aims were unquestionably practical in intent but his basic perceptions were entirely novel. In my view, the imaginative leap involved is one of genius.

In seeking to derive a completely general theory of heat engines, he was guided by his father's basic premise in his study of steam engines of the impossibility of perpetual motion. In the *Réflexions*, he adopted the caloric theory of heat and assumed that caloric was conserved in the cyclic operation of heat engines. He postulated that it was the transfer of caloric from a hotter to a colder body which is the origin of the work done by a heat engine. The flow of caloric was envisaged as being analogous to the flow of fluid which, as in a waterwheel, can produce work when it falls down a potential gradient.

Carnot's basic insights into the operation of heat engines are twofold. First, he recognised that a heat engine works most efficiently if the transfer of heat occurs as part of a cyclic process. The second is the fact that the crucial factor in determining the amount of work which can be extracted from a heat engine is the temperature difference between the source of heat and the sink into which the caloric flows. It turns out that these basic ideas are independent of the particular model of the heat flow process.

By another stroke of imaginative insight, he devised the cycle of operations which we now know as the *Carnot cycle* as an idealisation of the behaviour of any heat engine. We will discuss the cycle in more detail in Section 6.5.2. A key feature of the Carnot cycle is that it is *reversible* so that by reversing the sequence of operations work can be done on the system and caloric transferred from the cooler to the hotter body. By joining together an arbitrary heat engine and a reversed Carnot heat engine, he was able to demonstrate that no heat engine can ever produce more work than a Carnot heat engine. If it were otherwise, by joining the two engines together, we can either transfer heat from the colder to the hotter body without doing any work, or we could produce a net amount of work without any net heat transfer, both phenomena being in violation of common experience. The influence of Lazare Carnot's premise about the impossibility of perpetual motion becomes apparent. We will demonstrate the results formally in Section 6.5.2.

It is tragic that Sadi Carnot died of cholera at the age of 36 in August 1832, before the profound significance of his work was appreciated by anyone. However, in 1834, Emile Clapeyron reformulated Carnot's arguments analytically and related the ideal Carnot engine to the standard pressure–volume indicator diagram. There the matter rested until William Thomson began work on certain aspects of Clapeyron's paper and went back to the original version in the *Réflexions*. The big problem for Thomson and others at that time was to reconcile Carnot's work in which caloric is conserved with Joule's work which demonstrated the interconvertibility of heat and work. The matter was resolved by Rudolf Clausius who showed that Carnot's theorem concerning the maximum efficiency of heat engines was correct but that the assumption of no heat loss was wrong. In fact, there is a conversion of heat into work in the Carnot cycle. This reformulation by Clausius constitutes the bare bones of the second law of thermodynamics. However, as we will show, the law goes far beyond the efficiency of heat engines. It serves not only to define a proper thermodynamic temperature scale but also to resolve the problem of the way in which systems evolve thermodynamically. Let us demonstrate this now more formally, bringing out the basic assumptions made in formulating the second law mathematically.

6.5 The second law of thermodynamics

What we have done so far is to conserve energy and define precisely what we mean by heat. However, we observe that there must be other restrictions upon thermodynamic processes. For example, we do not have any rules about the direction in which heat flows or about the way in which thermodynamic systems evolve. We will build up the second law in such a way that we establish these rules but let us first discuss the crucial distinction between reversible and irreversible processes.

6.5.1 Reversible and irreversible processes

A reversible process is one which is carried out infinitely slowly so that, in passing from state A to B, the system passes through an infinite number of equilibrium states. Since the process takes place infinitely slowly, there is no friction or turbulence and no sound waves are generated. At no stage are there any unbalanced forces. At each stage, we make only an infinitesimal change. The implication is that, by reversing the process precisely, we can get back to the point from which we started and nothing will have changed in either the system or its surroundings. Clearly, if there were frictional losses, we could not get back to where we started without extracting some energy from the surroundings.

Let us emphasise this point by considering in detail how we could carry out a reversible isothermal expansion. Suppose we have a large heat reservoir at temperature T and a cylinder with gas also at temperature T (Figure 6.6). Now no heat flows if the two are at the same temperature. But if we make an infinitesimally small movement of the piston outwards, the gas in the cylinder cools infinitesimally and so an infinitesimal amount of heat flows into the gas by virtue of the temperature difference. This small amount of energy brings the gas back to T. The system is reversible because, if we compress the gas at T slightly, it heats up and the heat flows from the gas back into the reservoir. Thus, provided we consider only infinitesimal changes, the heat flow process occurs reversibly.

Clearly, this is not possible if the reservoir and the piston are at different temperatures. In this case, we cannot reverse the direction of heat flow by making an infinitesimal change in the temperature of the cooler object. This makes the point clear that in reversible processes, we have to be able to get from one state to another by going through an infinite set of equilibrium states which we join up by infinitesimal increments of work and energy flow.

Just to make this abundantly clear, let us repeat the argument for an adiabatic expansion. Again, we perform each step infinitesimally slowly. There is no flow of heat in or out of the system and there is no friction. Thus, since

Figure 6.6

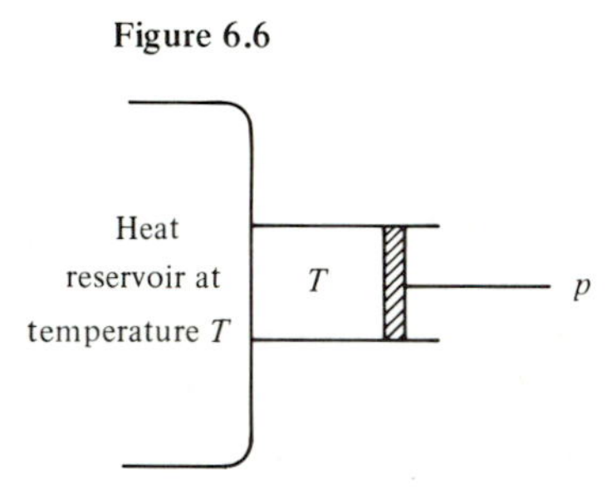

each infinitesimal step is reversible, we can perform the whole expansion by adding lots of them together.

Now let us contrast this behaviour with the other two expansions we have described. In the Joule expansion, the gas expands into a large volume and, clearly, this cannot happen without all sorts of non-equilibrium processes taking place. Unlike the adiabatic and isothermal expansions, there is no way in which we can design a series of equilibrium states through which the final state is reached.

The Joule–Kelvin expansion is a case in which there is a discontinuity in the properties of the gas on passing into the second cylinder. We do not reach the final state by passing through a series of equilibrium states infinitely slowly.

This long introduction stresses the point that reversible processes are highly idealised but they provide the norm against which all other real processes can be measured.

6.5.2 The Carnot cycle and the definition of thermodynamic temperature

We will restate the second law in the form due to Clausius: 'No process is possible whose sole result is the transfer of heat from a colder to a hotter body'. Notice that this is another 'fact of experience' which we assert without proof. The implication of this law is that you cannot transfer heat from a cold to a hot body without some sort of change in the surroundings of the system. Notice also that it assumes that we can define what we mean by the terms hotter and colder. We have not done this so far. What we have done is to set up an empirical temperature scale which I asked you to believe would eventually be shown to be identical with the thermodynamic temperature scale. We will prove this in a moment.

It is a matter of common experience that it is easy to convert work into heat but it is much more difficult to devise a means of achieving the opposite, i.e. to convert heat into work. This is where heat engines come in because we define them to be devices which convert heat into work. We have already described Carnot's great insights but they are worth repeating. First of all, in any efficient heat engine, there is a working substance which is used *cyclically*. For example, in a steam turbine, the working substance is steam which undergoes a cyclic process (Figure 6.7). Heat is put into the working substance at a high temperature in the boiler, the steam goes into the turbine where it does work in turning the rotor of the electricity generator and then the cooled vapour is condensed into water which is recycled by the pump.

Second, in all real heat engines, heat is put into the working substance at a high temperature, the working substance then does work and gets rid of the rest of the heat at a lower temperature. Third, the best we can ever do is to build a heat engine in which all the stages of operation of the engine are reversible, i.e. all the changes are carried out infinitely slowly and there are no dissipative losses such as friction or turbulence.

Carnot's famous cycle is illustrated in Figure 6.8(*a*). I strongly recommend Feynman's careful description of the Carnot cycle in Chapter 44 of Volume 1 of his *Lectures on Physics.*[6] Our exposition is modelled on his.

We have two very large heat reservoirs 1 and 2 which are maintained at temperatures T_1 and T_2. The working substance is gas which is contained within the cylinder which has a piston. We now go through the following reversible sequence of operations which simulate how real machines do work.

(1) We place the cylinder in thermal contact with the reservoir at temperature T_1 and then perform a very slow reversible isothermal expansion. As we described above, in a reversible isothermal expansion, heat must flow from the reservoir into the gas in the cylinder. We will stop when a quantity of heat Q_1 has been absorbed by the gas in the cylinder. Let us indicate this change as A to B on a p–V indicator diagram (Figure 6.8(*b*)). Note that work is done by the piston on the surroundings.

(2) Now we perform an infinitely slow adiabatic expansion of the piston so that the temperature falls from T_1 to T_2, the temperature of the second reservoir. Again this process is reversible. Work is again done on the surroundings.

Figure 6.7. On reading the children's book *What Do People Do All Day*? by Richard Scarry to my son, I was delighted to find that the electric turbine illustrated clearly Carnot's basic insight that in an efficient heat engine the working substance (steam and water) is used cyclically (R. Scarry, 1968, *What Do People Do All Day*?, p. 50–1, Collins).

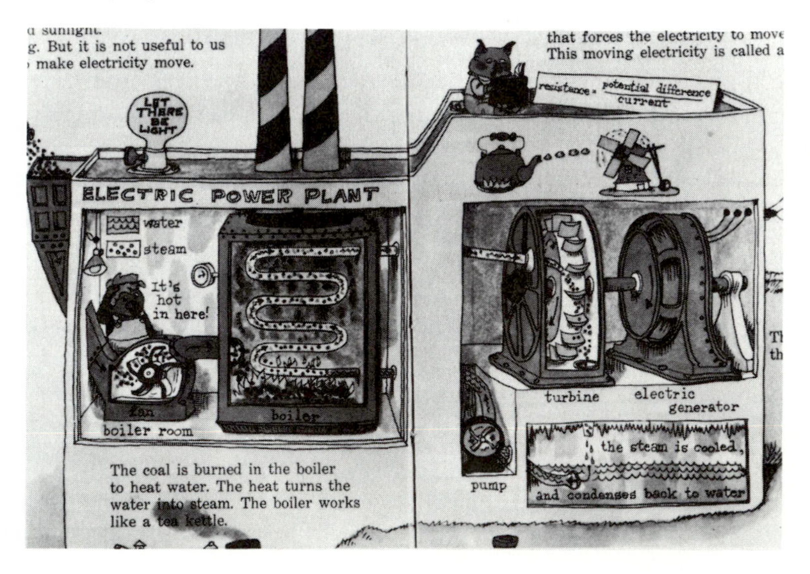

(3) Now we begin to compress the gas in the cylinder, again infinitely slowly and reversibly at temperature T_2. In this process, we continuously increase the temperature of the gas infinitesimally above T_2 and so heat flows into the second reservoir. We continue this process until the isotherm intersects the adiabatic curve which will take the working substance back to the beginning. Heat Q_2 is given up to the reservoir and work is done on the system.

Figure 6.8. (*a*) Illustrating schematically the four stages in the Carnot cycle of an ideal machine. (*b*) The four stages of operation of an ideal Carnot engine indicated on a *p*–*V* or indicator diagram.

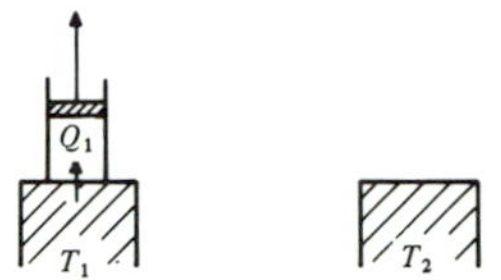

Stage 1: Reversible isothermal expansion at T_1

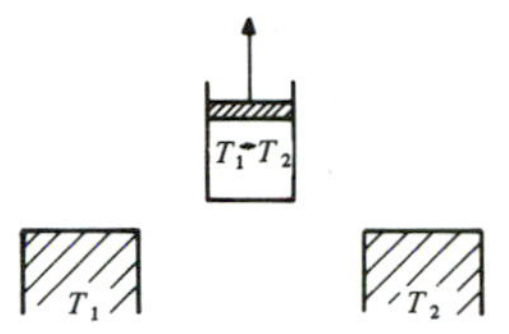

Stage 2: Reversible adiabatic expansion $T_1 \to T_2$

Stage 3: Reversible isothermal compression at T_2

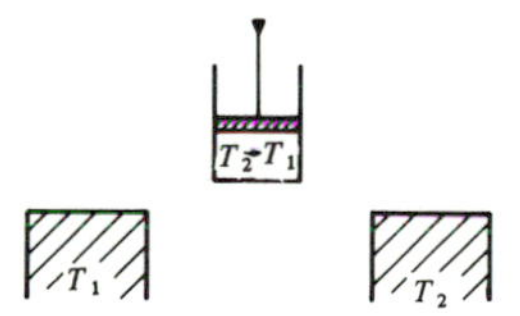

Stage 4: Reversible adiabatic compression, $T_2 \to T_1$

(*b*)

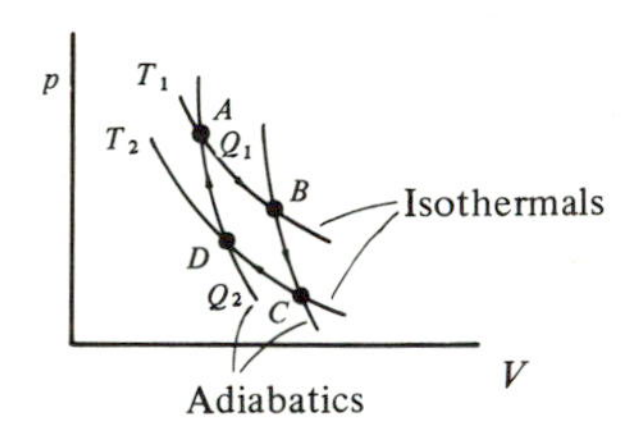

(4) We perform another infinitely slow compression of the gas under adiabatic conditions bringing the gas back to its original state.

This is the operation of a 'perfect engine' in the sense which we will define precisely. The net effect of this cycle is that we have extracted heat Q_1 at T_1 and put back Q_2 at T_2. In the process, the engine has done work. Now the amount of work is easily calculated. It is just $W = \int p\,dV$ taken round the cycle. From the diagram it is clear that the cyclic integral is just equal to the area of the closed curve described by the cycle of the engine in the $p-V$ indicator diagram (Figure 6.9). We also know from the first law that this work must be just $Q_1 - Q_2$, i.e.

$$W = Q_1 - Q_2 \tag{6.30}$$

We note incidentally that the adiabatic curves must be steeper than the isotherms because for the former $pV^\gamma = \text{constant}$ and $\gamma > 1$ for all gases. We can draw this machine schematically as shown in Figure 6.10.

Now the really beautiful aspect of this engine is that it is completely reversible, i.e. we can run the whole system backwards so that the sequence of events and their functions are reversed (Figure 6.11). These would be:

(1) Adiabatic expansion from A to D, reducing the temperature of the working substance from T_1 to T_2.

(2) Isothermal expansion at T_2 in which heat Q_2 is withdrawn from the reservoir.

(3) Adiabatic compression from C to B so that the working substance returns to temperature T_1.

Figure 6.9

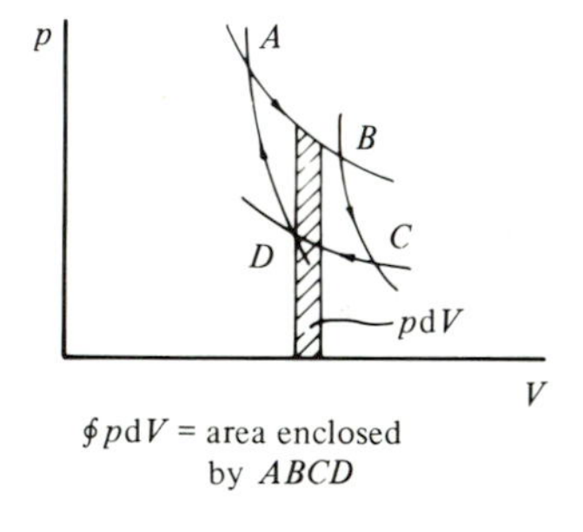

Figure 6.10. A representation of a reversible Carnot heat engine.

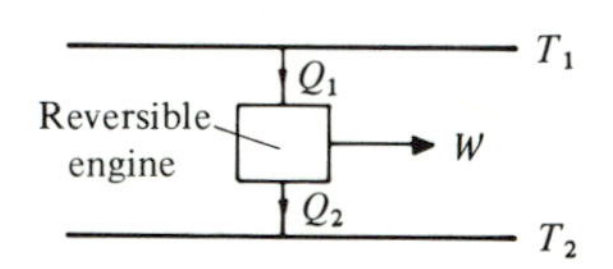

(4) Isothermal compression at T_1 so that the working substance gives up heat Q_1 to the reservoir at T_1.

Thus the cycle working in the reverse direction is acting as an ideal refrigerator or heat pump in which heat is extracted from the reservoir at the lower temperature and passed to the reservoir at the higher temperature. We show this refrigerator or heat pump schematically in Figure 6.12. Notice that, in this cycle, we have to do work in order to extract the heat from T_2 and deliver it to T_1.

We now see how we can define the efficiencies of heat engines. For the standard heat engine run forwards, the efficiency

$$\eta = \frac{\text{work done in the cycle}}{\text{heat input}} = \frac{W}{Q_1} = \frac{Q_1 - Q_2}{Q_1}$$

For a refrigerator,

$$\eta = \frac{\text{heat extracted from reservoir 2}}{\text{work done}} = \frac{Q_2}{W} = \frac{Q_2}{Q_1 - Q_2}$$

For a heat pump, the cycle is run backwards but used to supply Q_1 at T_1 by doing work W

$$\eta = \frac{\text{heat supplied to reservoir 1}}{\text{work done}} = \frac{Q_1}{W} = \frac{Q_1}{Q_1 - Q_2}$$

We can now do three things – prove Carnot's theorem, show the equivalence of the Clausius and Kelvin statements of the second law and derive the concept of thermodynamic temperature.

Figure 6.11. A reversed Carnot cycle representing a refrigerator or a heat pump.

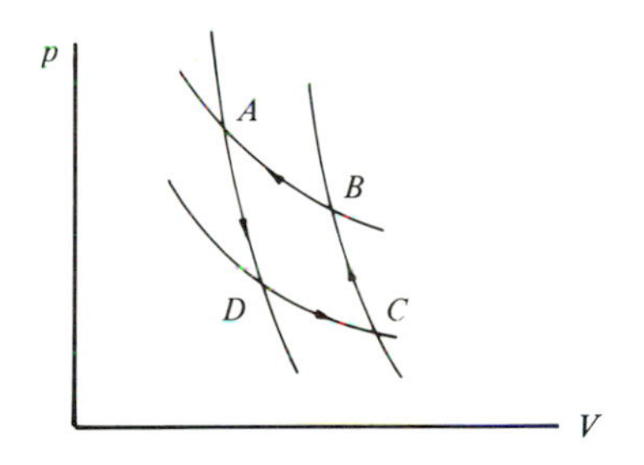

Figure 6.12. A representation of a reversed Carnot cycle acting as a refrigerator or heat pump.

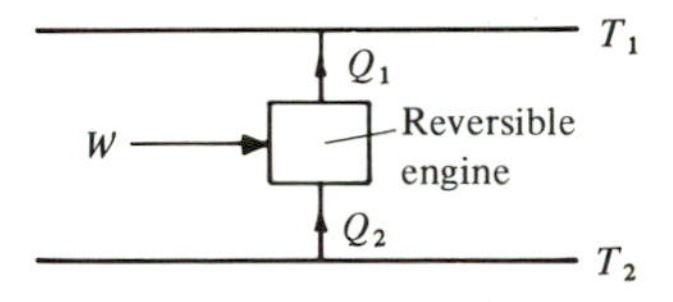

Carnot's theorem states that 'Of all heat engines working between two given temperatures, none can be more efficient than a reversible heat engine.'

Let us suppose the opposite is true – that we have an irreversible heat engine which has greater efficiency than a reversible engine working between the same temperatures. Then we can use the work produced by the irreversible engine to drive the reversible engine backwards (Figure 6.13). Let us now regard the system consisting of the two combined engines to be a single engine. We use all the work produced by the irreversible engine to drive the reversible engine. Therefore, since we assume that

$$\eta_{irr} > \eta_{rev}$$

we must have $W/Q_1' > W/Q_1$ and consequently $Q_1 > Q_1'$, i.e. $Q_1 - Q_1' > 0$. Thus, looked at overall, the only net effect of this combined engine is to produce a transfer of energy from the lower to the higher temperature which is forbidden by the second law. Therefore, the irreversible engine cannot have an efficiency greater than that of the reversible engine. This is the theorem which Carnot published in his *Réflexions* of 1824, 26 years before the formal statement of the laws of thermodynamics by Clausius in 1850.

Kelvin's statement of the second law is 'No process is possible whose sole result is the complete conversion of heat into work.' Again, we suppose the opposite to be true and that the engine completely converts heat Q_1 into work W. Again, we use this work to drive a reversible Carnot engine backwards as a heat pump (Figure 6.14). Then, regarded as a single system, no net work is done but the total amount of heat delivered to the reservoir at T_1 is

$$Q_1 - Q_1^K = Q_1 - W = Q_1 - (Q_1 - Q_2) = Q_2$$

where Q_1^K means the heat supplied to the hypothetical Kelvin engine which is completely converted into work W. There is, therefore, a net transfer of heat Q_2

Figure 6.13. Illustrating the proof of Carnot's theorem.

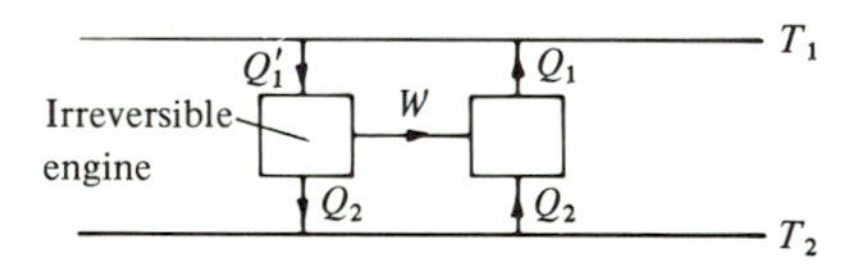

Figure 6.14. Illustrating the equivalence of Kelvin's and Clausius's statements of the second law using a hypothetical Kelvin engine which converts all the heat supplied at T_1 into work.

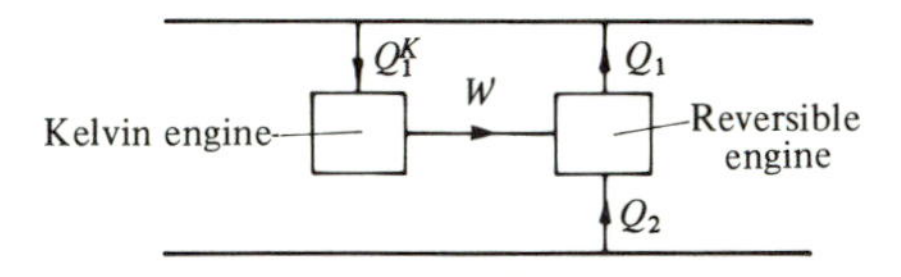

to the reservoir T_1 without anything else changing in the Universe and this is forbidden by the Clausius statement. Therefore, the Clausius and Kelvin statements are equivalent.

At last we can define *thermodynamic temperature.* The clue to the thermodynamic definition comes from Carnot's theorem. A reversible heat engine working between two temperatures has the maximum possible efficiency. Expressing this in terms of the heat input and output,

$$\eta = \frac{Q_1 - Q_2}{Q_1}; \quad \frac{Q_1}{Q_2} = \frac{1}{1-\eta} \tag{6.31}$$

According to the theorem, this must be a unique function of the two temperatures which so far we have called rather sloppily T_1 and T_2. Now let us proceed with the argument mathematically. We let the ratio Q_1/Q_2 be given by some function $f(\theta_1, \theta_2)$ where we call the temperatures θ_1 and θ_2 to emphasise that they are really empirical temperatures. Now let us join together two heat engines in series as shown in Figure 6.15. The engines are so connected that the heat delivered at tempeature θ_2 goes into the second heat engine which transfers the energy to the reservoir at θ_3. Thus, we have

$$\frac{Q_1}{Q_2} = f(\theta_1, \theta_2); \quad \frac{Q_2}{Q_3} = f(\theta_2, \theta_3) \tag{6.32}$$

for each stage. On the other hand, we can look at the combined system as a single engine operating between θ_1 and θ_3, in which case we find that

$$\frac{Q_1}{Q_3} = f(\theta_1, \theta_3) \tag{6.33}$$

Thus, since

$$\frac{Q_1}{Q_3} = \frac{Q_1}{Q_2}\frac{Q_2}{Q_3}$$

Figure 6.15. Illustrating the origin of the definition of thermodynamic temperature.

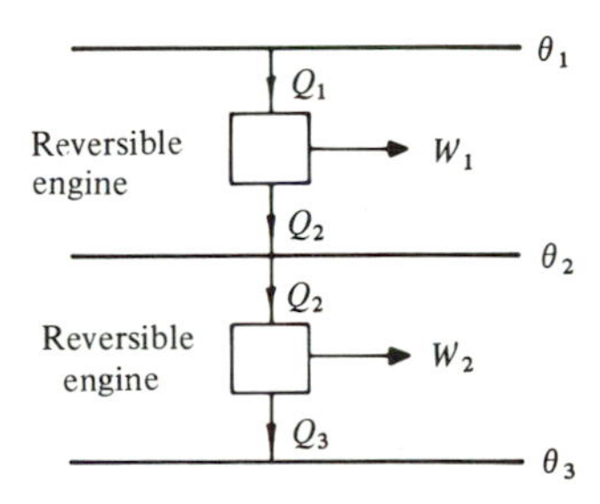

$$f(\theta_1, \theta_3) = f(\theta_1, \theta_2) f(\theta_2, \theta_3) \tag{6.34}$$

Consequently, the function f must have the form

$$f(\theta_1, \theta_3) = \frac{g(\theta_1)}{g(\theta_3)} = \frac{g(\theta_1)}{g(\theta_2)} \frac{g(\theta_2)}{g(\theta_3)} \tag{6.35}$$

We then adopt a definition of *thermodynamic temperature* consistent with this requirement, i.e.

$$\frac{g(\theta_1)}{g(\theta_3)} = \frac{T_1}{T_3} \quad \text{i.e.} \quad \frac{Q_1}{Q_3} = \frac{T_1}{T_3} \tag{6.36}$$

All we have to do now is to show that this definition is identical with that corresponding to the perfect gas temperature scale. We will write the perfect gas law temperature scale as T^{P} and hence $pV = RT^{\mathrm{P}}$. Thus, for the Carnot cycle with a perfect gas we have

$$\left.\begin{aligned} Q_1 &= \int_A^B p\,\mathrm{d}V = R\,T_1^{\mathrm{P}} \ln\left(\frac{V_B}{V_A}\right) \\ Q_2 &= \int_D^C p\,\mathrm{d}V = R\,T_2^{\mathrm{P}} \ln\left(\frac{V_C}{V_D}\right) \end{aligned}\right\} \tag{6.37}$$

Along the adiabatic legs of the cycle we have

$$pV^{\gamma} = \text{constant} \quad \text{and} \quad T^{\mathrm{P}} V^{\gamma-1} = \text{constant}$$

Therefore,

$$(V_B/V_C)^{\gamma-1} = (T_2^{\mathrm{P}}/T_1^{\mathrm{P}}) \tag{6.38}$$

$$(V_D/V_A)^{\gamma-1} = (T_1^{\mathrm{P}}/T_2^{\mathrm{P}}) \tag{6.39}$$

Multiplying (6.38) and (6.39) together, we find that

$$\left(\frac{V_B V_D}{V_A V_C}\right) = \text{constant}$$

i.e.

$$\frac{V_B}{V_A} = \frac{V_C}{V_D} \tag{6.40}$$

and, consequently, from the relations (6.37) and (6.36) we see that

$$\frac{Q_1}{Q_2} = \frac{T_1^{\mathrm{P}}}{T_2^{\mathrm{P}}} = \frac{T_1}{T_2}$$

At last, we have a rigorous thermodynamic definition of temperature and it all derives from the operation of perfect heat engines, the line of thought initiated by Carnot.

Now we can rewrite the maximum efficiencies of heat engines, refrigerators and heat pumps in terms of the temperatures between which they work.

$$\left.\begin{aligned} &\textit{Heat engine:} \quad \eta = \frac{Q_1 - Q_2}{Q_1} = \frac{T_1 - T_2}{T_1} \\ &\textit{Refrigerator:} \quad \eta = \frac{Q_2}{Q_1 - Q_2} = \frac{T_2}{T_1 - T_2} \\ &\textit{Heat pump:} \quad \eta = \frac{Q_1}{Q_1 - Q_2} = \frac{T_1}{T_1 - T_2} \end{aligned}\right\} \tag{6.41}$$

6.6 Entropy

You may have noticed something rather remarkable about the relation we developed in our definition of thermodynamic temperature. When we go round a Carnot cycle, taking the heats Q_1 and Q_2 to be positive quantities,

$$\frac{Q_1}{T_1} - \frac{Q_2}{T_2} = 0 \tag{6.42}$$

We can rewrite this in the following form

$$\int_A^C \frac{đQ}{T} - \int_C^A \frac{đQ}{T} = 0 \tag{6.43}$$

Since both legs of the cycle are reversible, this means that if we go from A to C down either leg of the cycle, we have

$$\int_{\substack{A \\ \text{via } B}}^{C} \frac{đQ}{T} = \int_{\substack{A \\ \text{via } D}}^{C} \frac{đQ}{T} \tag{6.44}$$

This strongly suggests that we have discovered another function of state. It does not matter how we get from A to C; this integral will always remain the same.

We see that for any two-coordinate system we can always move between any two points by an infinite number of infinitesimally small Carnot cycles, all reversible (Figure 6.16). Whichever path we take, we will always get the same result. Mathematically, for two points A and B,

$$\sum_A^B \frac{đQ}{T} = \text{constant} \tag{6.45}$$

Figure 6.16

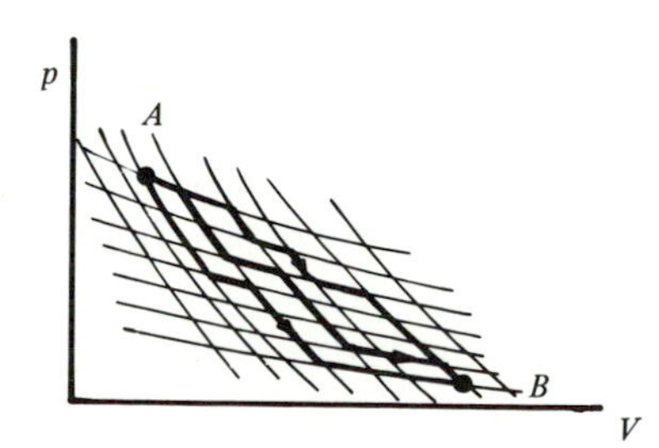

Writing this in integral form, we see that

$$\int_A^B \frac{\text{đ}Q}{T} = \text{constant} = S_B - S_A \tag{6.46}$$

This new function of state defined by $\int_A^B \text{đ}Q/T$ is called the *entropy* of the system. Notice that it is defined for reversible processes connecting the states A and B. By T, we mean the temperature at which the heat is supplied to the system which in this case is the same as the temperature of the system itself because the processes involving heat exchange are performed reversibly.

Now we remember that, for any engine, the efficiency must be less than that of the perfect Carnot engine working between the same temperatures and therefore

$$\eta_{\text{irr}} \leqslant \eta_{\text{rev}}$$

$$\frac{Q_1 - Q_2}{Q_1} \leqslant \frac{Q_{1\,\text{rev}} - Q_{2\,\text{rev}}}{Q_{1\,\text{rev}}}$$

Therefore,

$$\frac{Q_2}{Q_1} \geqslant \frac{Q_{2\,\text{rev}}}{Q_{1\,\text{rev}}} = \frac{T_2}{T_1}$$

$$\frac{Q_2}{T_2} \geqslant \frac{Q_1}{T_1} \tag{6.47}$$

i.e. round the cycle

$$\frac{Q_1}{T_1} - \frac{Q_2}{T_1} \leqslant 0$$

This suggests that in general we must have

$$\oint \frac{\text{đ}Q}{T} \leqslant 0 \tag{6.48}$$

where the Qs are the heat inputs and extractions in the real cycle. Notice that only in the case when the heat engine is reversible does the equals sign apply. Notice also our sign convention for heat. When we put heat into the system, we take a positive sign; when it is taken out, it has a negative sign.

This relation $\oint \text{đ}Q/T \leqslant 0$ is very fundamental to the whole of thermodynamics and it is known as *Clausius's theorem.* I regard this proof of Clausius's theorem as adequate for our present purposes. The only limitation is that it applies to two-coordinate systems and of course, in general, one may deal with multi-coordinate systems. This question is treated nicely in Pippard's book where it is shown that the same result is true, in general, for multi-coordinate systems.

6.7 The law of increase of entropy

Notice that we can only define a change in entropy when the system moves from state A to state B along a reversible path. In practice, this means that when we take the system from state A to state B in real life we can never effect precisely a reversible change and so the heat transfer is not directly related to the entropy difference between the initial and final states. Let us compare what happens in a reversible and irreversible change between two states. Suppose the irreversible change is that from A to B (Figure 6.17). Then we can complete the cycle by taking any reversible path back from B to A. Then, according to Clausius's theorem, we must have

$$\oint \frac{đQ}{T} \leqslant 0$$

$$\int_{\substack{A\\\text{irrev}}}^{B} \frac{đQ}{T} + \int_{\substack{B\\\text{rev}}}^{A} \frac{đQ}{T} \leqslant 0$$

i.e.

$$\int_{\substack{A\\\text{irrev}}}^{B} \frac{đQ}{T} \leqslant \int_{\substack{A\\\text{rev}}}^{B} \frac{đQ}{T}$$

But, because the second change is reversible, we must have

$$\int_{A_{\text{rev}}}^{B} \frac{đQ}{T} = S_B - S_A$$

according to the definition of entropy. Consequently,

$$\int_{A_{\text{irrev}}}^{B} \frac{đQ}{T} \leqslant S_B - S_A$$

or writing this for a differential irreversible change, we have

$$\frac{đQ}{T} \leqslant \Delta S \tag{6.49}$$

Figure 6.17

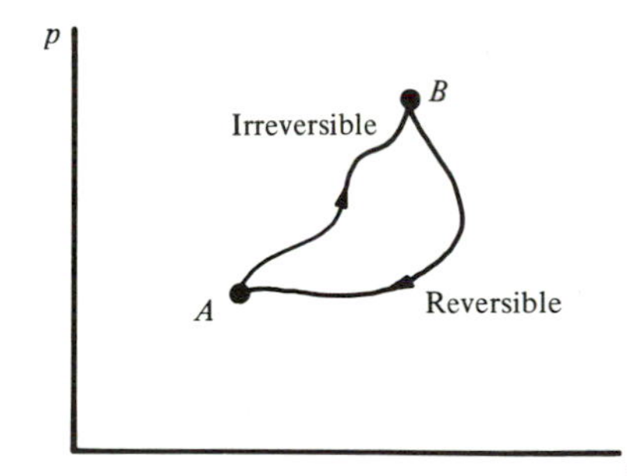

Thus, we obtain the general result that for any differential change $\Delta S \geqslant đQ/T$ where the equality holds only in the case of a reversible change.

This is an extremely important result because at last we are beginning to see where we can find quantitatively what it is that determines the direction in which physical processes must evolve. Notice that, if we are dealing with reversible processes, the temperature at which heat is supplied is that of the system itself. This is not necessarily the case in irreversible processes.

For an *isolated system*, there is no thermal contact with the outside universe and hence in the above inequality we have $đQ = 0$. Therefore

$$\Delta S \geqslant 0 \tag{6.50}$$

Thus, the entropy of an isolated system cannot decrease. A corollary of this is that, in approaching equilibrium, the entropy of the isolated system must tend towards a maximum and the final equilibrium configuration will be that for which the entropy is the greatest.

For example, consider two bodies at temperatures T_1 and T_2 and suppose that a small amount of heat ΔQ is exchanged between them. Then heat flows from the hotter to the colder body and we find an entropy decrease of the hotter body $\Delta Q/T_1$ and an entropy increase of the colder body $\Delta Q/T_2$ whilst the total entropy change is $(\Delta Q/T_2)-(\Delta Q/T_1) > 0$ which is positive, meaning that we were correct in our statement about the direction of flow of the heat.

So far we have only dealt with isolated systems. Now let us consider the opposite case in which the system has thermal contact with the surroundings.

As a first example, let us take our system round a complete cycle of operations so that it ends up in exactly the same state as at the beginning. The cycle may well involve irreversible processes as is the case in most real systems. Since the system ends up in exactly the same state in which it started, all the functions of state are exactly the same and hence, for the system, $\Delta S = 0$. According to Clausius' theorem, this must be greater than or equal to $\oint đQ_{sys}/T$, i.e.

$$0 = \Delta S \geqslant \oint \frac{đQ_{sys}}{T}$$

In order to return to the initial state, heat must have been communicated to the system from its surroundings. The very best we can do is to transfer the heat reversibly to and from the surroundings. Then, in each stage of the cycle, $đQ_{sys} = -\, đQ_{surr}$ and hence

$$0 = \Delta S \geqslant \oint \frac{đQ_{sys}}{T} = -\oint \frac{đQ_{surr}}{T}$$

Only if the heat transfer is reversible can we equate the last quantity $\oint đQ_{surr}/T$ to ΔS_{surr}. Thus, we find that

$$0 \geqslant -\Delta S_{\text{surr}}$$

or

$$\Delta S_{\text{surr}} \geqslant 0$$

As a second example, consider an irreversible change in going from states 1 to 2. Again we assume that the heat exchange with the surroundings takes place reversibly. Then, as above,

$$\Delta S_{\text{sys}} \geqslant \int_1^2 \frac{đQ_{\text{irr}}}{T} = -\int \frac{đQ_{\text{surr}}}{T} = -\Delta S_{\text{surr}}$$

$$\Delta S_{\text{sys}} + \Delta S_{\text{surr}} \geqslant 0$$

You will have noticed the implication of these examples. When we have irreversible processes present in the cycle or changes of state, although there might not be a change in the entropy of the system itself, the entropy of the whole Universe has increased. It is always important to consider the surroundings as well as the system itself.

These examples indicate the origin of the popular expression of the two laws of thermodynamics due to Clausius, who in 1865 invented the word entropy (from the Greek word for transformation):

> The energy of the Universe is constant; the entropy of the Universe tends to a maximum.

The entropy change need not necessarily involve heat exchange. It is a measure of the irreversibility of the process. Let us look again at the Joule expansion of a perfect gas as an example. We recall that in this case the gas expands to fill a larger volume whilst doing no work, i.e. the internal energy U is a constant. The entropy change of the perfect gas at temperature T is $\int_A^B đQ/T$. Now

$$đQ = \mathrm{d}U + p\,\mathrm{d}V$$

and hence for 1 mole of gas

$$đQ = C_V\,\mathrm{d}T + \frac{RT}{V}\,\mathrm{d}V$$

Therefore,

$$\int_A^B \frac{đQ}{T} = C_V \int_A^B \frac{\mathrm{d}T}{T} + R\int_A^B \frac{\mathrm{d}V}{V}$$

i.e.

$$S - S_0 = C_V \ln\frac{T}{T_0} + R\ln\frac{V}{V_0} \tag{6.51}$$

Since $T =$ constant in this expansion,

$$S - S_0 = R\ln V/V_0 \tag{6.52}$$

Thus, although there is no heat flow in the conventional sense, the system has undergone a patently irreversible change and as a result there must be an increase in entropy in reaching the new equilibrium state from the old.

Finally, let us write down the entropy change of a perfect gas in an interesting form. From equation (6.51), we can write:

$$S - S_0 = C_V \ln\left(\frac{pV}{p_0 V_0}\right) + R \ln \frac{V}{V_0}$$

$$= C_V \ln \frac{p}{p_0} + (C_V + R) \ln \frac{V}{V_0}$$

$$= C_V \ln\left(\frac{pV^\gamma}{p_0 V_0^\gamma}\right) \tag{6.53}$$

Thus, if the expansion is adiabatic, $pV^\gamma = \text{constant}$, there is no entropy change. For this reason, adiabatic expansions are sometimes called *isentropic* expansions. Also, we note that this provides us with an interpretation of the entropy function – isotherms are curves of constant temperature, adiabats are curves of constant entropy.

6.8 The differential form of the combined first and second laws of thermodynamics

At last, we have come to the central equation which unites the first and second laws. For reversible processes,

$$\mathrm{d}S = \frac{đQ}{T}$$

and hence, combining this with the relation $đQ = \mathrm{d}U - \mathrm{d}W$, we find that

$$T\mathrm{d}S = \mathrm{d}U + p\mathrm{d}V \tag{6.54}$$

More generally, if we write the work done on the system as

$$\sum_i X_i \,\mathrm{d}x_i$$

we find that

$$T\mathrm{d}S = \mathrm{d}U - \sum_i X_i \,\mathrm{d}x_i \tag{6.55}$$

The remarkable thing about this formula is that it is a way of writing the combined first and second laws entirely in terms of functions of state and hence the relation must be true for all changes.

We shall not take this story much further, but we should note that the relation (6.54) is the origin of some very powerful results which we will need later. In particular, for any gas, we find that

$$\left(\frac{\partial S}{\partial U}\right)_V = \frac{1}{T} \tag{6.56}$$

i.e. the partial derivative of the entropy with respect to U at constant volume defines the thermodynamic temperature T. This is of particular significance in statistical mechanics because the concept of entropy S is easy in statistical mechanics and the internal energy U is one of the first things we need to determine statistically. Thus, we have already the key to define the concept of temperature in statistical mechanics.

APPENDIX TO CHAPTER 6

MAXWELL'S RELATIONS AND JACOBIANS

A6.1 Perfect differentials in thermodynamics

It is convenient to introduce the idea of *perfect differentials* in thermodynamics. The reason is that we have emphasised that the systems with which we have been most concerned are two-coordinate systems, i.e. a thermodynamic equilibrium state can be defined by only two functions of state. When we move between one equilibrium state and another, the change in any function of state only depends on the initial and final coordinates. Thus, in general, if z is a function of state and x and y two suitable coordinates, the change of state can be written

$$\mathrm{d}z = \left(\frac{\partial z}{\partial x}\right)_y \mathrm{d}x + \left(\frac{\partial z}{\partial y}\right)_x \mathrm{d}y \tag{A6.1}$$

Now the fact that the differential change $\mathrm{d}z$ does not depend upon the way in which we make the incremental steps in $\mathrm{d}x$ and $\mathrm{d}y$ enables us to set restrictions upon the functional dependence of z upon x and y. The relation is most simply demonstrated by the two ways in which we can make the differential change $\mathrm{d}z$ illustrated in Figure A6.1. We can go from A to C either via D or B. Since z is a function of state, the change $\mathrm{d}z$ must be the same along the paths of ABC and ADC. Along the path ADC, we first move $\mathrm{d}x$ in the x direction and then $\mathrm{d}y$ from the point $x + \mathrm{d}x$, i.e.

Figure A6.1

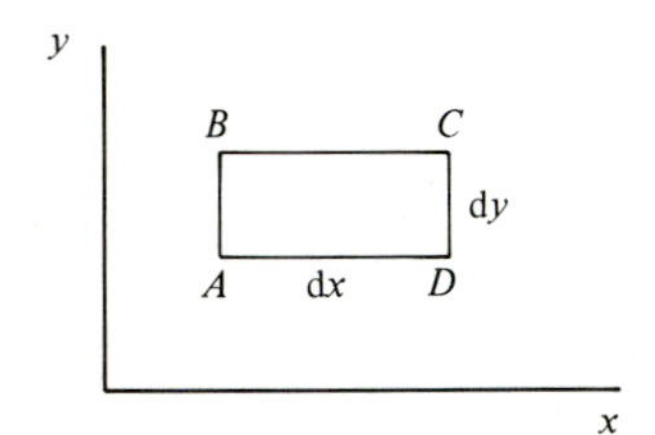

$$z(C) = z(A) + \left(\frac{\partial z}{\partial x}\right)_y \mathrm{d}x + \left\{\frac{\partial}{\partial y}\left[z + \left(\frac{\partial z}{\partial x}\right)_y \mathrm{d}x\right]\right\}_x \mathrm{d}y$$

$$= z(A) + \left(\frac{\partial z}{\partial x}\right)_y \mathrm{d}x + \left(\frac{\partial z}{\partial y}\right)_x \mathrm{d}y + \frac{\partial^2 z}{\partial y \partial x}\,\mathrm{d}x\,\mathrm{d}y \qquad \text{(A6.2)}$$

where by $\partial^2 z/\partial y \partial x$ we mean

$$\left[\frac{\partial}{\partial y}\left(\frac{\partial z}{\partial x}\right)_y\right]_x$$

Along the path ABC, the value of $z(C)$ is

$$z(C) = z(A) + \left(\frac{\partial z}{\partial y}\right)_x \mathrm{d}y + \left\{\frac{\partial}{\partial x}\left[z + \left(\frac{\partial z}{\partial y}\right)_x \mathrm{d}y\right]\right\}_y \mathrm{d}x$$

$$= z(A) + \left(\frac{\partial z}{\partial y}\right)_x \mathrm{d}y + \left(\frac{\partial z}{\partial x}\right)_y \mathrm{d}x + \frac{\partial^2 z}{\partial x \partial y}\,\mathrm{d}x\,\mathrm{d}y \qquad \text{(A6.3)}$$

Notice that in equations (A6.2) and (A6.3), the order in which we take the double partial derivatives is important. Since equations (A6.2) and (A6.3) must be identical, the function z must have the property that

$$\frac{\partial^2 z}{\partial y \partial x} = \frac{\partial^2 z}{\partial x \partial y} \quad \text{i.e.} \quad \frac{\partial}{\partial y}\left(\frac{\partial z}{\partial x}\right) = \frac{\partial}{\partial x}\left(\frac{\partial z}{\partial y}\right) \qquad \text{(A6.4)}$$

This is the mathematical condition that z be a perfect differential of x and y.

A6.2 Maxwell's relations

You will have noticed that, although we have been dealing with two-coordinate systems, we have, in fact, introduced a much greater number of functions of state: p, V, T, S, U, H, etc. Indeed, we can go on to define an infinite number of functions of state by combinations of these functions. Therefore, when we make differential changes in the state of the system, we expect that there will be corresponding changes in the other functions of state. The four most important of these differential relations are called *Maxwell's relations.* Let us derive two of them.

Consider first the equation (6.54) which relates differential changes in S, U and V.

$$\left.\begin{aligned} T\mathrm{d}S &= \mathrm{d}U + p\,\mathrm{d}V \\ \mathrm{d}U &= T\mathrm{d}S - p\,\mathrm{d}V \end{aligned}\right\} \qquad \text{(A6.5)}$$

$\mathrm{d}U$ being the differential of a function of state must be a perfect differential and hence can be written

$$\mathrm{d}U = \left(\frac{-\partial U}{\partial S}\right)_V \mathrm{d}S + \left(\frac{\partial U}{\partial V}\right)_S \mathrm{d}V \qquad \text{(A6.6)}$$

Comparison with equation (A6.5) shows that

$$T = \left(\frac{\partial U}{\partial S}\right)_V; \quad p = -\left(\frac{\partial U}{\partial V}\right)_S \tag{A6.7}$$

However, since $\mathrm{d}U$ is a perfect differential, we also require from equation (A6.4) that

$$\frac{\partial}{\partial V}\left(\frac{\partial U}{\partial S}\right) = \frac{\partial}{\partial S}\left(\frac{\partial U}{\partial V}\right) \tag{A6.8}$$

Substituting relations (A6.7) into (A6.8) we find that

$$\left(\frac{\partial T}{\partial V}\right)_S = -\left(\frac{\partial p}{\partial S}\right)_V \tag{A6.9}$$

This is the first of four similar relations between T, S, p and V. We can go through exactly the same procedure for the enthalpy H:

$$\begin{aligned} H &= U + pV \\ \mathrm{d}H &= \mathrm{d}U + p\mathrm{d}V + V\mathrm{d}p \\ &= T\mathrm{d}S + V\mathrm{d}p \end{aligned} \tag{A6.10}$$

By exactly the same mathematical procedure, we find that

$$\left(\frac{\partial T}{\partial p}\right)_S = \left(\frac{\partial V}{\partial S}\right)_p \tag{A6.11}$$

We can go through the analyses for other functions of state, the *Helmholtz free energy* $F = U - TS$ and the *Gibbs free energy* $G = U - TS + pV$, to find the other two relations

$$\left(\frac{\partial S}{\partial V}\right)_T = \left(\frac{\partial p}{\partial T}\right)_V \tag{A6.12}$$

$$\left(\frac{\partial V}{\partial T}\right)_p = -\left(\frac{\partial S}{\partial p}\right)_T \tag{A6.13}$$

The functions F and G are particularly useful in studying processes which occur at constant temperature and constant pressure respectively.

The set of relations (A6.9), (A6.11), (A6.12) and (A6.13) are known as *Maxwell's relations* and are very useful in many thermodynamic problems. We will use them in subsequent lectures. It is useful to have a mnemonic by which to remember them and I find that Jacobians provide an easy way of doing this.

A6.3 Jacobians in thermodynamics

If the variables x, y, and η, ξ, are related by

$$\begin{aligned} x &= x(\eta, \xi) \\ y &= y(\eta, \xi) \end{aligned} \tag{A6.14}$$

the Jacobian is defined to be the determinant

$$\frac{\partial(x, y)}{\partial(\eta, \xi)} = \begin{vmatrix} \dfrac{\partial x}{\partial \eta} & \dfrac{\partial y}{\partial \eta} \\[2ex] \dfrac{\partial x}{\partial \xi} & \dfrac{\partial y}{\partial \xi} \end{vmatrix} \tag{A6.15}$$

Then we can show from the properties of determinants that

$$\left.\begin{aligned} &\frac{\partial(x, y)}{\partial(x, y)} = -\frac{\partial(y, x)}{\partial(x, y)} = 1 \\ &\frac{\partial(v, v)}{\partial(x, y)} = 0 = \frac{\partial(k, v)}{\partial(x, y)} \quad \text{if } k \text{ is a constant} \\ &\frac{\partial(u, v)}{\partial(x, y)} = -\frac{\partial(v, u)}{\partial(x, y)} = \frac{\partial(-v, u)}{\partial(x, y)} = \frac{\partial(v, -u)}{\partial(x, y)} \end{aligned}\right\} \tag{A6.16}$$

etc. and similar rules for changes 'downstairs'. Notice, in particular, that

$$\frac{\partial(u, y)}{\partial(x, y)} = \left(\frac{\partial u}{\partial x}\right)_y \tag{A6.17}$$

$$\frac{\partial(u, v)}{\partial(x, y)} = \frac{\partial(u, v)}{\partial(r, s)} \times \frac{\partial(r, s)}{\partial(x, y)} = \frac{1}{\partial(x, y)/\partial(u, v)} \tag{A6.18}$$

Note also that

$$\left(\frac{\partial y}{\partial x}\right)_z \left(\frac{\partial z}{\partial y}\right)_x \left(\frac{\partial x}{\partial z}\right)_y = -1 \tag{A6.19}$$

The value of the Jacobian notation is that we can write all four of Maxwell's equations in the very neat form

$$\frac{\partial(T, S)}{\partial(x, y)} = \frac{\partial(p, V)}{\partial(x, y)} \tag{A6.20}$$

where x and y are different variables. Let us give one example of how this works. Because of the relation (A6.17), the Maxwell relation

$$\left(\frac{\partial T}{\partial V}\right)_S = -\left(\frac{\partial p}{\partial S}\right)_V$$

is exactly the same as

$$\frac{\partial(T, S)}{\partial(V, S)} = -\frac{\partial(p, V)}{\partial(S, V)} = \frac{\partial(p, V)}{\partial(V, S)}$$

according to (A6.16). We can see how we can generate all the relations by introducing the four possible combinations of T, S, p and V into (A6.20). By virtue of (A6.18), equation (A6.20) is exactly the same as

$$\frac{\partial(T, S)}{\partial(p, V)} = 1 \tag{A6.21}$$

The Jacobian (A6.21) is the key to remembering all four relations. We need only remember that the Jacobian is + 1 if T, S, p and V are written in the above order. One way of remembering this is that the ***intensive variables***, T and p, and the ***extensive variables***, S and V, should appear in the same order on the top and bottom if the sign is positive.

Besides the mnemonic function, the Jacobian formalism is a rather elegant way of finding relations between thermodynamic quantities in terms of partial derivatives.

7

THE KINETIC THEORY OF GASES AND THE ORIGIN OF STATISTICAL MECHANICS

7.1 Introduction

In Chapter 6, we emphasised that classical thermodynamics is independent of the particular physical model chosen to explain the properties of matter or radiation. In a nutshell, the subject consists of a set of definitions of thermodynamic quantities and the two laws of thermodynamics. On occasion, we did mention the forces between molecules and gave interpretation of the various quantities we discussed in terms of physical models for the behaviour of the molecules which make up the gas. However, that was not really part of the story at all. The relations between the variables in classical thermodynamics have absolute validity and there is no need to interpret them if we are only interested in the bulk properties of matter and radiation.

Nonetheless, it comes as no surprise to learn that, in fact, the science of classical thermodynamics developed in parallel with advances in the atomic and molecular interpretation of the properties of matter. Indeed, we find the same pioneers in the field of classical thermodynamics, Thomson, Clausius, Maxwell and Boltzmann, reappearing in the history of the understanding of the microscopic structure of solids, liquids and gases. We will tell two stories in this chapter, that of the kinetic theory of gases and the way in which it leads to Boltzmann's statistical mechanics. In fact, they form part of one single story because, despite the fact that there were good reasons why the kinetic theory was by no means readily accepted, it clearly suggested that the law of increase of entropy is basically a statistical result and this concept profoundly influenced Boltzmann's thinking.

7.2 The kinetic theory of gases

The controversy between the caloric theory and the kinetic or dynamic theory of heat had apparently been resolved by Joule's experiments of the 1840s and by the establishment of the two laws of thermodynamics by Clausius in the early 1850s. Before 1850, however, various kinetic theories of gases had been proposed, in particular, by John Herapath and John James Waterston. We will return to the case of Waterston at the end of this section. Joule had noted how his discovery of the equivalence of heat and work could be interpreted in a kinetic theory and Clausius, in his first formulation of the two laws of thermodynamics, had described how the laws could be interpreted in terms of the kinetic theory, although he emphasised that the laws are quite independent of the theory. The first systematic account of the theory was by Clausius in 1857 in a paper entitled 'The Nature of the Motion which we call Heat'.[1] This contained a simple derivation of the perfect gas law assuming that the atoms of a gas consist of elastic spheres which exert a pressure on the walls of the containing vessel by virtue of bouncing off them. At the risk of insulting my readers, who, I am told, learn this material nowadays 'at the same time as the alphabet', let me outline in brief the simple derivation of Clausius's theory. In his original paper, Clausius worked solely in terms of the average velocity of particles but we can do a bit better than that. We will assume that there is a distribution of velocities such that the probability of the velocity lying in the range u to $u + \mathrm{d}u$ is $f(u)\mathrm{d}u$ and $\int_0^\infty f(u)\mathrm{d}u = 1$. Then the average velocity is

$$\bar{u} = \int_0^\infty uf(u)\mathrm{d}u \tag{7.1}$$

and the mean square velocity

$$\overline{u^2} = \int_0^\infty u^2 f(u)\mathrm{d}u \tag{7.2}$$

Let us consider first elastic collisions of atoms of the gas with the walls of the vessel (Figure 7.1). Kinetic energy (or *vis viva*) is conserved in the collision but the momentum vector of the atom perpendicular to the wall is reversed. Thus, the change in momentum is $\Delta p = 2mu \cos\theta$. All we need do now is to work out the rate at which particles arrive at unit area of the wall and the total rate of change of momentum on that area will be the pressure of the gas.

Figure 7.1

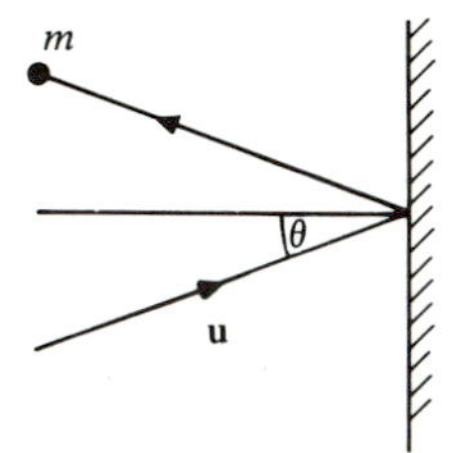

Locating ourselves at a particular point on the wall, we ask 'What is the probability of an atom arriving in the angular range θ to $\theta + \mathrm{d}\theta$, where θ is the angle measured from the direction normal to the wall.' By considering the solid angle subtended by an annulus of sphere in the range θ to $\theta + \mathrm{d}\theta$ at distance r (Figure 7.2), the probability distribution is

$$P(\theta)\mathrm{d}\theta = \frac{2\pi r \sin\theta \, r\mathrm{d}\theta}{4\pi r^2}$$

$$= \tfrac{1}{2}\sin\theta \, \mathrm{d}\theta \tag{7.3}$$

Now we also know that the velocity distribution of the particles arriving in the angular range θ to $\theta + \mathrm{d}\theta$ is $f(u)\mathrm{d}u$. Those with velocity u are approaching the surface with normal velocity $u\cos\theta$. Therefore, if we consider unit area of the surface, the number arriving at the surface per second is

$$n(u,\theta)\,\mathrm{d}u\,\mathrm{d}\theta = \tfrac{1}{2}f(u)\sin\theta\,\mathrm{d}u\,\mathrm{d}\theta \times n_0\,u\cos\theta \tag{7.4}$$

The factor $\frac{1}{2}f(u)\sin\theta\,\mathrm{d}u\mathrm{d}\theta$ is the probability that a single particle has velocity in the range u to $u + \mathrm{d}u$ and angle in the range θ to $\theta + \mathrm{d}\theta$ and the factor $n_0 u\cos\theta$ is the number of particles in a rectangular box of unit cross section and of height such that in one second all the particles with velocity u and angle θ will reach the surface. From this calculation, we can do two things.

First, the total number of particles arriving per unit area per second at the surface is simply the integral of expression (7.4), i.e.

$$\text{flux of particles per unit area} = \int_0^\infty \int_0^{\frac{\pi}{2}} \tfrac{1}{2} n_0 f(u) u \cos\theta \sin\theta \,\mathrm{d}\theta\,\mathrm{d}u$$

$$= \tfrac{1}{4} n_0 \int_0^\infty f(u) u \,\mathrm{d}u$$

$$= \tfrac{1}{4} n_0 \bar{u} \tag{7.5}$$

Figure 7.2

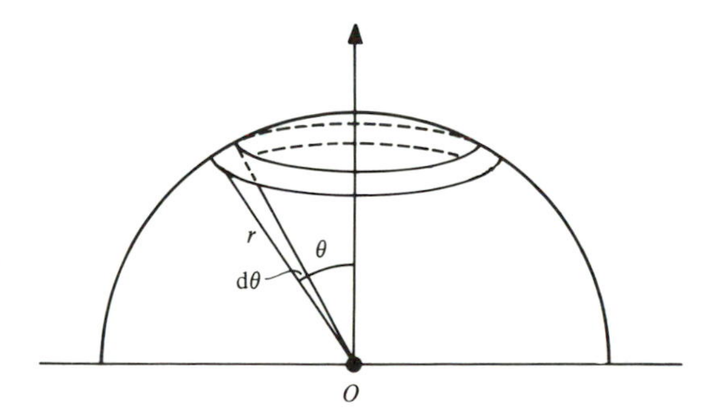

Second, the change of momentum perpendicular to the wall of the vessel for the particles arriving at angle θ with velocity u is $2mu\cos\theta$ and hence the pressure is

$$p = \int_0^\infty \int_0^{\frac{\pi}{2}} (2mu \cos\theta)\, \tfrac{1}{2} n_0 f(u) u \cos\theta \sin\theta \,\mathrm{d}\theta\, \mathrm{d}u$$

$$= n_0 m \int_0^\infty u^2 f(u)\,\mathrm{d}u \int_0^{\frac{\pi}{2}} \sin\theta \cos^2\theta \,\mathrm{d}\theta$$

$$= \tfrac{1}{3} n_0 m \overline{u^2} \tag{7.6}$$

Then Clausius noted that this is just the expression for a perfect gas. If we consider 1 mole of gas, N is the total number of atoms and hence $n_0 V = N$, i.e.

$$pV = \tfrac{1}{3} Nm\overline{u^2} \tag{7.7}$$

For a perfect gas, we must identify the right-hand side of equation (7.7) with RT, i.e. the temperature of the gas is proportional to $\overline{u^2}$, the mean square velocity of the gas atoms. Thus,

$$RT = \tfrac{1}{3} Nm\overline{u^2} \tag{7.8}$$

But we can go further and note that the internal energy of the gas is just the kinetic energy of the gas molecules, i.e. for 1 mole

$$U = \tfrac{1}{2} Nm\overline{u^2} \tag{7.9}$$

We find that

$$U = \tfrac{3}{2} RT \tag{7.10}$$

and hence the heat capacity at constant volume

$$C_V = \left(\frac{\partial U}{\partial T}\right)_V = \tfrac{3}{2} R \tag{7.11}$$

The ratio of specific heats is

$$\gamma = \frac{C_V + R}{C_V} = \tfrac{5}{3} \tag{7.12}$$

Correspondingly, the average energy per atom is

$$\tfrac{1}{2} m\overline{u^2} = \frac{U}{N} = \tfrac{3}{2} \frac{R}{N} T = \tfrac{3}{2} kT \tag{7.13}$$

where $k = R/N = 1.38 \times 10^{-23}\ \mathrm{J\,K^{-1}}$ is Boltzmann's constant.

This remarkably elegant argument is so taken for granted nowadays that one forgets what an achievement is represented in 1857. Whilst accounting for the perfect gas law admirably, it did not give good agreement with the known values of γ for molecular gases which was ≈ 1.4. There must therefore exist other ways of storing kinetic energy within molecular gases which can increase the internal energy per molecule, an important point clearly recognised by Clausius in the last sentences of his paper.

Two aspects of this argument are worthy of note. First, Clausius knew that there had to be a dispersion in velocity $f(u)$ of the atoms of the gas but he had no way of knowing what it was and so worked in terms of mean values only. Second, the idea that the pressure of the gas should be proportional to $Nm\overline{u^2}$ had been worked out 15 years earlier by Waterston. As early as 1843, Waterston wrote in a book published in Edinburgh, 'A medium constituted of elastic spherical atoms that are continually impinging against each other with the same velocity, will exert against a vacuum an elastic force that is proportional to the square of this velocity and to its density'.[2] In December 1845, Waterston developed a more systematic version of his theory and gave the first statement of the equipartition theorem for a mixture of gases of different atomic species. He also worked out the ratio of specific heats although his calculation contains a numerical error.

Waterston's work was submitted to the Royal Society for publication in 1845 but was harshly refereed and remained unpublished until 1892, eight years after his death. The paper was discovered by Lord Rayleigh in the Royal Society archives and he then had it published with an introduction which he himself wrote. In the introduction Rayleigh wrote

> Impressed by the above passage [from a later work by Waterston] and with the general ingenuity and soundness of Waterston's views, I took the first opportunity of consulting the Archives, and saw at once that the memoir justified the large claims made for it, and that it makes an immense advance in the direction of the now generally received theory. The omission to publish it was a misfortune, which probably retarded the development of the subject by ten or fifteen years.[3]

Later he notes

> The character of the advance to be dated from this paper will be at once understood when it is realised that Waterston was the first to introduce into the theory the conception that heat and temperature are to be measured by 'vis viva' (i.e. kinetic energy) . . . . In the second section, the great feature is the statement (VII) that in mixed media the mean square velocity is inversely proportional to the specific weight of the molecules. The proof which Waterston gave is doubtless not satisfactory; but the same may be said of that advanced by Maxwell fifteen years later.[4]

One final quotation is particularly memorable.

> The history of this paper suggests that highly speculative investigations, especially by an unknown author, are best brought before the scientific

> world through some other channel than a scientific society, which naturally hesistates to admit into its printed records matter of uncertain value. Perhaps one may go further and say that a young author who believes himself capable of great things would usually do well to secure the favourable recognition of the scientific world by work whose scope is limited, and whose value is easily judged, before embarking on greater flights.[5]

What is one to say? The situation is a rather tragic one and yet it is so easy to see how it can come about. How often have we dismissed the ideas of an unknown scientist or student whose work we did not fully comprehend? I wish I could believe I had not but I will probably never know for certain.

7.3 Maxwell's velocity distribution

Before proceeding to Maxwell's novel approach to the kinetic theory of gases, we should note one other feature of Clausius's work which was of particular significance for Maxwell. From the kinetic theory, Clausius was able to work out the typical velocities of the atoms and molecules of air from his formula $RT = \frac{1}{3}Nm\overline{u^2}$. For oxygen and nitrogen, he deduced velocities of 461 and 492 m s^{-1} respectively. The Dutch meteorologist Buys Ballot criticised this aspect of the theory, since it is well known that pungent odours take minutes to permeate a room. Clausius's response was to point out that the molecules of air collide and therefore the particles diffuse from one part of a volume to another rather than propagate in straight lines. In his response published in 1858, Clausius introduced the concept of *mean free path* for the atoms and molecules of gases for the first time. Thus, in the kinetic theory of gases, it must be supposed that there are continually collisions between the molecules.

Both papers by Clausius were known to Maxwell when he turned to the problem of the kinetic theory of gases in 1859 and 1860. His work was published in 1860 in another characteristically novel and profound paper entitled 'Illustrations of the dynamical theory of gases'.[6] The quite amazing achievement was that, in one paper, he derived the correct formula for the velocity distribution $f(u)$ and introduced statistical concepts into the kinetic theory of gases and thermodynamics. C.W.F. Everitt writes that this derivation of what we now know as Maxwell's velocity distribution marks the beginning of a new epoch in physics.[7] From it come directly the concepts of the statistical nature of the laws of thermodynamics, which is the key to Boltzmann's statistics, and the modern theory of statistical mechanics.

Maxwell's derivation of the distribution occupies no more than half a dozen short paragraphs. He states the problem as Proposition IV of his paper 'To find the average number of particles whose velocities lie between given limits, after a great number of collisions among a great number of equal particles'. The total number of particles is N and we will write the x, y and z components of the velocities of the particles as u_x, u_y and u_z. He then supposes that the velocity distribution in the three orthogonal directions will be the same after a great number of collisions, i.e.

$$Nf(u_x)\,\mathrm{d}u_x = Nf(u_y)\,\mathrm{d}u_y = Nf(u_z)\,\mathrm{d}u_z \tag{7.14}$$

where f is always the same function. Now the three perpendicular components of the velocity are entirely independent and hence the number of particles with velocities in the range u_x to $u_x + \mathrm{d}u_x$, u_y to $u_y + \mathrm{d}u_y$, u_z to $u_z + \mathrm{d}u_z$ is

$$Nf(u_x)f(u_y)f(u_z)\,\mathrm{d}u_x\,\mathrm{d}u_y\,\mathrm{d}u_z \tag{7.15}$$

But, the total velocity of a particle with components u_x, u_y, u_z is $u^2 = u_x^2 + u_y^2 + u_z^2$ and, because we assume large numbers of collisions have taken place, the velocity distribution must be isotropic and hence depend only on u^2, i.e.

$$f(u_x)f(u_y)f(u_z) = \phi(u^2) = \phi(u_x^2 + u_y^2 + u_z^2) \tag{7.16}$$

This is what is known as a functional equation and we have to ask what forms of the function $f(u_x)$ are consistent with (7.14) and (7.16). The straightforward solution is

$$f(x) = C\,\mathrm{e}^{Au_x^2}, \quad f(y) = C\,\mathrm{e}^{Au_y^2}, \quad f(z) = C\,\mathrm{e}^{Au_z^2}$$

so that

$$\begin{aligned}\phi(u^2) &= f(u_x)f(u_y)f(u_z) = C^3\,\mathrm{e}^{A(u_x^2+u_y^2+u_z^2)}\\ &= C^3\,\mathrm{e}^{Au^2}\end{aligned} \tag{7.17}$$

Now the distribution must converge as $u \to \infty$ and hence A must be negative. Maxwell writes this as

$$\phi(u^2) = C^3\,\mathrm{e}^{-u^2/\alpha^2} \tag{7.18}$$

We know that the total number of particles is N and hence we must have for each velocity component

$$N = N\int_{-\infty}^{\infty} C\,\mathrm{e}^{-u_x^2/\alpha^2}\,\mathrm{d}u_x \tag{7.19}$$

This is a standard integral, $\int_{-\infty}^{\infty}\mathrm{e}^{-x^2}\,\mathrm{d}x = \pi^{\frac{1}{2}}$ and hence

$$C = \frac{1}{\alpha\pi^{\frac{1}{2}}} \tag{7.20}$$

This leads directly to four conclusions which we quote in Maxwell's words (but using our notation).

> 1st. The number of particles whose velocity, resolved in a certain direction, lies between u_x and $u_x + \mathrm{d}u_x$ is

$$N\frac{1}{\alpha\pi^{\frac{1}{2}}}e^{-u_x^2/\alpha^2}\,du_x$$

2nd. The number whose actual velocity lies between u and $u + du$ is

$$N\frac{4}{\alpha^3\pi^{\frac{1}{2}}}u^2\,e^{-u^2/\alpha^2}\,du$$

This is because

$$N(u)\,du = N\frac{1}{\alpha^3\pi^{\frac{3}{2}}}\,e^{-u^2/\alpha^2}\,du_x\,du_y\,du_z$$

But we obtain the velocity u^2 from all the volume elements $du_x\,du_y\,du_z$ which lie within the spherical shell of radius u and width du (Figure 7.3), i.e. the total volume element for velocities u to $u + du$ is $4\pi u^2 du$. Therefore,

$$N(u)\,du = N\frac{4\pi u^2}{\alpha^3\pi^{\frac{3}{2}}}\,e^{-u^2/\alpha^2}\,du$$

$$= N\frac{4}{\alpha^3\pi^{\frac{1}{2}}}\,u^2 e^{-u^2/\alpha^2}\,du \qquad (7.21)$$

3rd. To find the mean value of u, add the velocities of all the particles together and divide by the number of particles; the result is

$$\text{mean velocity} = \frac{2\alpha}{\pi^{\frac{1}{2}}} \qquad (7.22)$$

This is the standard procedure:

$$\bar{u} = \frac{\int_0^\infty uN(u)du}{\int_0^\infty N(u)du} = \int_0^\infty \frac{4}{\alpha^3\pi^{\frac{1}{2}}}\,u^3 e^{-u^2/\alpha^2}\,du$$

Figure 7.3. Conversion of the volume integral over $du_x\,du_y\,du_z$ over velocity space to one over the total velocity u and the element of velocity space $4\pi u^2 du$.

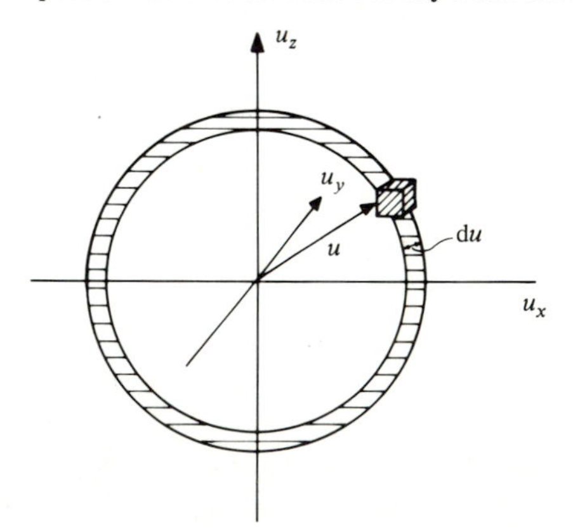

This is another standard integral $\int_0^\infty x^3 \, \mathrm{e}^{-x^2} \, \mathrm{d}x = \frac{1}{2}$. Therefore,

$$\bar{u} = \frac{4}{\alpha^3 \pi^{\frac{1}{2}}} \, \frac{\alpha^4}{2} = \frac{2\alpha}{\pi^{\frac{1}{2}}}.$$

4th. To find the mean value of u^2, add all the values together and divide by N,

mean value of $u^2 = \frac{3}{2}\alpha^2$

i.e. we form

$$\overline{u^2} = \int_0^\infty \frac{4}{\alpha^3 \pi^{\frac{1}{2}}} u^4 \, \mathrm{e}^{-u^2/\alpha^2} \, \mathrm{d}u \tag{7.23}$$

This can again be reduced to the above standard forms by integration by parts. Maxwell immediately goes on to note 'that the velocities are distributed among the particles according to the same law as the errors are distributed among the observations in the theory of the "method of least squares" ', i.e. in this very first paper, the direct relation to statistical procedures is established.

To put Maxwell's distribution into its standard form, we need only compare the result (7.23) with the value deduced for $\overline{u^2}$ from Clausius's theory (equation (7.13)) and we find that

$$\overline{u^2} = \tfrac{3}{2}\alpha^2 = \frac{3kT}{m} \tag{7.24}$$

We can therefore write Maxwell's distribution (7.21) in its final form:

$$N(u)\,\mathrm{d}u = 4\pi N \left(\frac{m}{2\pi kT}\right)^{\frac{3}{2}} u^2 \, \mathrm{e}^{-mu^2/2kT} \, \mathrm{d}u \tag{7.25}$$

This is quite an astonishing achievement. It is by far the simplest derivation of Maxwell's distribution and is in contrast to the normal derivation which proceeds through the Boltzmann distribution.

Maxwell then proceeds to apply this law in a variety of circumstances and, in particular, he shows that if two different types of particle are present in the same volume, the mean *vis viva*, or kinetic energy, will be the same for each type of particle. In the final section of the paper, Maxwell addresses the problem of accounting for the ratio of specific heats of gases, which for most gases had measured values $\gamma \approx 1.4$. Clausius had already noted that some additional means of storing *vis viva* was needed and Maxwell proposed that it be stored in the kinetic energy of rotation of the atoms or molecules which he modelled as rough particles. He found that, in equilibrium, as much energy could be stored in the rotational motion as in translation motion, i.e. $\frac{1}{2}I\omega^2 = \frac{1}{2}mu^2$ and, consequently, he could derive a value for the ratio of specific heats. Instead of $U = \frac{3}{2}RT$ (equation 7.10)), he wrote $U = 3RT$ since as much energy is stored in rotation as translation and consequently, following through the analysis of equations (7.10) to (7.12), we find that

$$\gamma = \tfrac{4}{3} = 1.333$$

This value is just as bad as the value 1.667 which follows if only translational motion is considered. This profoundly depressed Maxwell. The last sentence of this great paper reads 'Finally, by establishing a necessary relation between the motions of translation and rotation of all particles not spherical, we proved that a system of such particles could not possibly satisfy the known relation between the two specific heats of all gases'. His inability to explain the value $\gamma \approx 1.4$ was a grave disappointment to Maxwell who, in his report to the British Association for the Advancement of Science of 1860, states that this discrepancy 'overturned the whole hypothesis'.[8]

The second key concept which results directly from this paper is the *principle of the equipartition of energy*. This was to prove to be a highly contentious issue until it was finally resolved by Einstein's application of quantum concepts to the average energy of an oscillator (see Section 11.4). The principle states that in equilibrium energy is shared equally among the separate modes in which energy can be stored by atoms and molecules. The problem of accounting for values of $\gamma \approx 1.4$ was aggravated by the discovery of spectral lines in the spectra of gases. These were interpreted as molecular resonances associated with the internal structure of molecules and there were so many of them that if each of them were to acquire its share of the internal energy of the gas, the ratio of specific heats of gases would tend to 1. These two fundamental problems cast grave doubt upon the validity of the principle of equipartition of energy and in consequence upon the whole idea of a kinetic theory of gases, despite its success in accounting for the perfect gas law.

These issues were at the centre of much of Maxwell's subsequent work. In 1867, he produced another derivation of the velocity distribution related directly to molecular collisions. To maintain equilibrium in collisions, the distribution functions of the velocities in a collision in which

$$u_1 + u_2 \rightarrow u_1' + u_2'$$

must satisfy

$$f(u_1)\,f(u_2) = f(u_1')\,f(u_2') \tag{7.26}$$

By virtue of the conservation of energy, we must also have

$$\tfrac{1}{2}mu_1^2 + \tfrac{1}{2}mu_2^2 = \tfrac{1}{2}mu_1'^2 + \tfrac{1}{2}mu_2'^2 \tag{7.27}$$

Inspection of Maxwell's velocity distribution shows that indeed it satisfies this criterion, i.e.

$$\exp\left[-\frac{m}{2kT}(u_1^2+u_2^2)\right] = \exp\left[-\frac{m}{2kT}(u_1'^2+u_2'^2)\right] \tag{7.28}$$

Despite the problems with the kinetic theory, Maxwell never seriously doubted its validity or that of the velocity distribution he had derived, either from his first general argument or from the second which involved molecular collisions. Towards the end of his life, the full significance of what he had done began to become apparent with Boltzmann's radical new approach to thermodynamics.

7.4 The statistical nature of the second law of thermodynamics

In 1867, Maxwell first presented his famous argument by which he demonstrated how it is possible to transfer heat from a colder to a hotter body on the basis of the kinetic theory of gases. He had by now established the velocity distribution which describes the range of velocities present in the gas. He considered a vessel divided into two halves, A and B, the gas in A being hotter than that in B. Maxwell proposed that a small hole be drilled in the partition between A and B and a 'finite being' watch the molecules as they approach the hole. The finite being has a shutter and he adopts the strategy of allowing only fast molecules to pass from B to A through the hole and slow molecules from A into B. By this means, the hot molecules in the tail of the Maxwell distribution of the gas in B heat up the gas in A and the cold molecules in the low energy tail of the distribution of A cool the cool gas in B. The finite being thus enables the system to violate the second law of thermodynamics. Thomson referred to the finite being as 'Maxwell's demon', a name objected to by Maxwell who asked Tait to 'call him no more a demon but a valve'.[9]

Maxwell's last remark brings out the key point about the statistical nature of the second law of thermodynamics. Quite independent of finite beings or demons, there is a small but finite probability that, from time to time, exactly what Maxwell describes does indeed occur. Whenever one fast molecule moves from B to A, heat is transferred from the colder to the hotter body without the influence of any external agency. Now, of course, it is overwhelmingly more likely that hot molecules move from A to B and in this process heat flows from the hotter to the colder body with the consequence that the entropy of the whole system increases. However, there is no question but that, according to the kinetic theory of gases, there is a very small but finite probability that the reverse will happen spontaneously and entropy will decrease in this natural process. Maxwell was quite clear about the significance of this argument. He remarked to Tait that his argument was designed 'to show that the second law of thermodynamics has only a statistical certainty'.[10] I find this a brilliant and compelling argument but it does depend upon a specific model for the gas, i.e. the kinetic theory of gases.

The essentially statistical nature of the second law was emphasised by another argument of Maxwell's. In the late 1860s, both Clausius and Boltzmann attempted to derive the second law of thermodynamics from mechanics, an approach which was known as the dynamical interpretation of the second law. In this approach, the dynamics of individual particles were followed in the hope that they would ultimately lead to an understanding of the origin of the second law. Maxwell refuted this approach as a matter of principle because of the simple but powerful argument that Newton's laws of motion, and indeed Maxwell's equations for the electromagnetic field, are completely time reversible and consequently the irreversibility implicit in the second law cannot be explained by a dynamical theory. The second law could only be understood as a statement based upon a statistical analysis of an immense number of particles.

Boltzmann originally belonged to the dynamical school of thought but he was fully aware of Maxwell's work. Among the most significant of his contributions during these years was a reworking of Maxwell's analysis of the equilibrium distribution of velocities in a gas, including the presence of a potential term $\phi(\mathbf{r})$ which describes the potential energy of the particle in the field. The conservation of energy requires that

$$\tfrac{1}{2}mu_1^2 + \phi(\mathbf{r}_1) = \tfrac{1}{2}mu_2^2 + \phi(\mathbf{r}_2)$$

and the corresponding probability distribution has the form

$$f(u) \propto \exp\left[-\frac{\tfrac{1}{2}mu^2 + \phi(\mathbf{r})}{kT}\right]$$

We recognise the primitive form of the Boltzmann factor $e^{-E/kT}$ in this analysis.

Eventually, he accepted Maxwell's doctrine concerning the statistical nature of the second law and set about working out the formal relationship between entropy and probability,

$$S = k \ln p \tag{7.29}$$

where S is the entropy, p the probability of that state and k is Boltzmann's constant. In fact, in Boltzmann's analysis, the exact value of the constant k was not known. Boltzmann's analysis is of considerable mathematical complexity and, indeed, this was one of the problems which stood in the way of the scientists of his day fully appreciating the significance of what he had achieved. We do not wish to begin the formal development of statistical mechanics but we can indicate the reasonableness of Boltzmann's theorem which is the basis of modern statistical physics.

7.5 Entropy and probability

The law of entropy increase tells us that systems evolve in such a way as to become more uniform. To put it another way, the particles which make up the system become more randomised – there is less organised structure. We have already met examples of this in Chapter 6. When bodies at different temperatures are brought together, heat is exchanged so that they come to the same temperature, i.e. temperature inequalities are evened out. In a Joule expansion, the gas expands to fill a greater volume and thus produces uniformity throughout a greater volume. In both cases, there is an increase in entropy.

Maxwell's arguments strongly suggest that the entropy increase is a statistical phenomenon although he did not attempt to quantify the nature of the relation between the second law and statistics. The advance made by Boltzmann was to describe the degree of disorder in systems in terms of the probability that it could arise by chance and relate this probability to the entropy of the system.

Now suppose we have two systems and we work out the probabilities p_1 and p_2 that they arise by chance. Then the probability that they occur together is the product of the two probabilities $p = p_1 p_2$. On the other hand, if we associate entropies S_1 and S_2 with each system, we know that entropies are additive and consequently the total entropy

$$S = S_1 + S_2 \tag{7.30}$$

Thus, if entropy and probability are related, it must be a logarithmic relation of the form $S = C \ln p$ where C is some constant.

Let us apply this statistical definition of entropy to the Joule expansion of a perfect gas and see if it works. We consider the Joule expansion of the gas from volume V to $2V$. We will suppose initially that there is 1 mole of gas in V. We now ask 'What is the probability that, if the molecules were free to occupy $2V$, they would all occupy only a volume V?' For each molecule, the probability is $\frac{1}{2}$. Therefore, if we have only 2 molecules, the probability is $(\frac{1}{2})^2$, if 3 molecules $(\frac{1}{2})^3$, if 4 molecules $(\frac{1}{2})^4$, . . . and if there are N molecules, the probability is $(\frac{1}{2})^N$. Since $N = 6 \times 10^{23}$, this probability is very small indeed. However, let us press on and apply the definition $S = C \ln p$ to $p(V_1/V_2) = (\frac{1}{2})^N$. We therefore expect the entropy change to be

$$S_2 - S_1 = \Delta S = C \ln (p_2/p_1) = C \ln 2^N = CN \ln 2 \tag{7.31}$$

However, we have already worked out this expression for the Joule expansion according to classical thermodynamics (equation 6.52),

$$\Delta S = R \ln V_2/V_1 = R \ln 2$$

We immediately see that we must have

$$R = CN$$

i.e.

$$C = R/N = k$$

We find Boltzmann's fundamental relation between entropy and probability

$$S = k \ln p$$

or

$$S_2 - S_1 = k \ln (p_2/p_1) \tag{7.32}$$

The signs are chosen so that the more likely event corresponds to an increase in entropy.

Let us perform another illustrative example which shows how the statistical approach works. Let us consider two equal volumes of gas of different pressures p_1 and p_2. On one side of the partition between the volumes, there are r moles of gas and in the other $1 - r$ moles. We assume that the volume of each is $V/2$. We can now write down the entropy of 1 mole of gas when it changes volume and temperature from V_0, T_0 to V, T. According to equation (6.51), for one mole of gas

$$S_1(T, V) = C_V \ln \left(\frac{T}{T_0}\right) + R \ln \left(\frac{V}{V_0}\right) + S_1 (T_0, V_0)$$

The subscript 1 refers to 1 mole of gas. Now suppose we have m moles of gas. Then, since entropies are additive,

$$S_m (T, V) = mC_V \ln \left(\frac{T}{T_0}\right) + mR \ln \left(\frac{V}{V_0}\right) + mS_1 (T_0, V_0)$$

if we keep all the variables the same as we add on volumes. The only catch is that V and V_0 still refer to only 1 mole of gas. The volume of the m moles is $mV = V_m$. Let us therefore rewrite the expression for the entropy:

$$S_m (T, V_m) = mC_V \ln \left(\frac{T}{T_0}\right) + mR \ln \left(\frac{V_m}{mV_0}\right) + mS_1 (T_0, V_0) \tag{7.33}$$

We can now apply this result to the two volumes $V_m = V/2$ which contain r and $1 - r$ moles, i.e.

$$\left.\begin{aligned} S_1 &= rC_V \ln \left(\frac{T}{T_0}\right) + rR \ln \left(\frac{V}{2rV_0}\right) + rS_1 (T_0, V_0) \\ S_2 &= (1 - r)\, C_V \ln \left(\frac{T}{T_0}\right) + (1 - r) R \ln \left(\frac{V}{2(1 - r) V_0}\right) \\ &\quad + (1 - r)\, S_1 (T_0, V_0) \end{aligned}\right\} \tag{7.34}$$

Adding these entropies, the entropy of the system is

$$S = S_1 + S_2 = C_V \ln \left(\frac{T}{T_0}\right) + R \ln \left(\frac{V}{2V_0}\right) + S_1 (T_0, V_0)$$

$$+ (1 - r) R \ln \left(\frac{1}{1 - r}\right) + rR \ln \frac{1}{r}$$

$$= C_V \ln\left(\frac{T}{T_0}\right) + R\ln\left(\frac{V}{2V_0}\right) + S_1(T_0, V_0) + \Delta S(r) \tag{7.35}$$

where we have absorbed all the terms in r in $\Delta S(r)$, i.e.

$$\Delta S(r) = -R[(1-r)\ln(1-r) + r\ln r] \tag{7.36}$$

It is the term $\Delta S(r)$ which tells us how the entropy depends upon how much gas there is on either side of the partition.

Now let us look at the problem from the statistical point of view. We can use simple statistical procedures to work out the number of ways in which we can distribute N identical objects between two boxes. We develop in Section 10.2 the necessary tools. Reference to that section shows that the number of ways in which we can arrange m objects in one box and $N-m$ in the other is

$$g(N, m) = \frac{N!}{(N-m)!\, m!}$$

Let us refer these numbers to the case of equal numbers of objects in each box, i.e.

$$g(N, x) = \frac{N!}{[(N/2)-x]!\,[(N/2)+x]!}$$

where $x = m-(N/2)$. Now this is not exactly a probability but actually the exact number of microscopic ways in which we can end up with $[(N/2)-x]$ in one box and $[(N/2)+x]$ in the other, but it is clearly the probability if we divide by the total number of possible ways of distributing the objects between the boxes, and this is just a constant. The part of the probability which includes x is only the above expression for $g(N, x)$. We can therefore adopt a definition of entropy

$$S = k\ln g(N, x) = k\{\ln N! - \ln[(N/2)-x]! - \ln[(N/2)+x]!\} \tag{7.37}$$

We now use Stirling's approximation in the form

$$\ln M! \approx M\ln M - M \tag{7.38}$$

After a bit of simple manipulation of equation (7.37), we find that

$$S/k = N\ln 2 - \left[\left(\frac{N}{2}-x\right)\ln\left(1-\frac{2x}{N}\right)\right] - \left[\left(\frac{N}{2}+x\right)\ln\left(1+\frac{2x}{N}\right)\right] \tag{7.39}$$

$$= -N\left[\left(\frac{1}{2}-\frac{x}{N}\right)\ln\left(\frac{1}{2}-\frac{x}{N}\right) + \left(\frac{1}{2}+\frac{x}{N}\right)\ln\left(\frac{1}{2}+\frac{x}{N}\right)\right] \tag{7.40}$$

Now by analogy with our classical derivation let us suppose that N corresponds to the number of atoms in 1 mole. Then

$$r = \frac{1}{2} - \frac{x}{N}; \quad 1 - r = \frac{1}{2} + \frac{x}{N}$$

and

$$S = -Nk\,[r\ln r + (1-r)\ln(1-r)] \tag{7.41}$$

This is exactly the result we obtained from classical thermodynamics with the correct value for k, i.e. $k = R/N$. We notice that the definition $S = k\ln p$ can account rather beautifully for the classical result by statistical arguments.

Let us look at the expansion of our function $g(N, x)$ for small values of x. From expression (7.39),

$$\frac{S}{k} = \ln g(N,x) = N\ln 2 - \frac{N}{2}\left[\left(1 - \frac{x}{2N}\right)\ln\left(1 - \frac{2x}{N}\right) + \left(1 + \frac{x}{2N}\right)\ln\left(1 + \frac{2x}{N}\right)\right]$$

Expanding to order x^2, since

$$\ln(1+x) = x - x^2/2 + \ldots$$

$$\ln g(N,x) = N\ln 2 - \frac{N}{2}\left[\left(1 - \frac{x}{2N}\right)\left(-\frac{2x}{N} - \frac{1}{2}\frac{4x^2}{N^2}\right) + \left(1 + \frac{x}{2N}\right)\left(\frac{2x}{N} - \frac{1}{2}\frac{4x^2}{N^2}\right)\right]$$

$$= N\ln 2 - \frac{2x^2}{N} + \ldots$$

$$g(N,x) = 2^N \exp\left(-\frac{2x^2}{N}\right) \tag{7.42}$$

Now the total number of ways in which the objects can be divided between the boxes is 2^N and hence in this approximation the probability distribution is

$$p(N,x) = \exp\left(-\frac{2x^2}{N}\right) \tag{7.43}$$

i.e. it is a Gaussian distribution with mean value 0 with probability 1! We note that the standard deviation of the distribution about the value 0 is $N^{\frac{1}{2}}/2$, i.e. to order of magnitude, $N^{\frac{1}{2}}$. This tells us that the likely deviation about the value $x = 0$ corresponds to very tiny fluctuations indeed. Since $N \sim 10^{23}$, $N^{\frac{1}{2}} \sim 10^{11.5}$ and in terms of the fractional fluctuations $N^{\frac{1}{2}}/N \sim 10^{-11.5}$. Our analysis of Section 12.2.1 shows that this is no more than the statistical fluctuation about the mean value.

This example illustrates the way in which statistical mechanics works. We are dealing with huge ensembles of particles $N \sim 10^{23}$ and hence, although it is true in principle that we can find statistical deviations from the mean behaviour, in practice they are very small. Indeed, although it is possible for the statistical entropy we have defined to decrease spontaneously, the likelihood of this happening is absolutely negligibly small because N is so large.

Let us look at one final example which shows how entropies depend upon the volume of the system not only in physical three dimensional space but also in terms of velocity (or phase) space. We have derived classically the entropy of the perfect gas

$$\Delta S(T, V) = C_V \ln\left(\frac{T}{T_0}\right) + R \ln\left(\frac{V}{V_0}\right) \tag{7.44}$$

Now let us express the first expression on the right-hand side in terms of molecular velocities rather than temperatures. *If* we assume the kinetic theory, $T \propto \overline{u^2}$ and hence, since $C_V = \frac{3}{2}R$,

$$\begin{aligned}\Delta S(T, V) &= \tfrac{3}{2}R \ln \frac{\overline{u^2}}{\overline{u_0^2}} + R \ln\left(\frac{V}{V_0}\right) \\ &= R\left[\ln\left(\frac{\overline{u^2}}{\overline{u_0^2}}\right)^{\frac{3}{2}} + \ln \frac{V}{V_0}\right]\end{aligned} \tag{7.45}$$

We can interpret this formula as saying that when we change T and V, the available physical volume changes in the ratio V/V_0. In addition, the first term indicates that the particles occupy a larger volume of velocity space by a factor $(\overline{u^2}/\overline{u_0^2})^{\frac{3}{2}}$. We see that we can interpret the formula for the entropy increase in terms of the increase in the available volume in both real and velocity space.

7.6 Concluding remarks

With this introduction, the way is now clear for the construction of the full statistical mechanical interpretation of classical thermodynamics. This development is traced in such standard texts as Kittel's *Thermal Physics*[11] and Mandl's *Statistical Physics.*[12]

Boltzmann's great discovery was appreciated by relatively few physicists at the time. The stumbling blocks were the fact that the kinetic theory of gases was not properly understood since it failed to account for the specific heats of gases and it was not clear how to incorporate the internal vibrations of atoms and molecules, as exhibited by spectral lines, into the equipartition theorem. In fact, towards the end of the 19th century, a reaction against atomic and molecular theories of the properties of matter gained currency in continental Europe. It was considered that one should only deal with the bulk properties of systems, i.e. with classical thermodynamics, and abolish the atomic and molecular concepts as unnecessary. These were profoundly discouraging developments for Boltzmann. It is probable that they contributed to his suicide in 1906. It is a tragedy that he should have been driven to this just at that time when the correctness of his fundamental insights were appreciated by Einstein who resolved the problem of the specific heats in his classical papers of 1905 and 1906.

We will return to Boltzmann's procedures when we survey Einstein's great contributions to the discovery of quanta.

Case Study 5

THE ORIGINS OF THE CONCEPT OF QUANTA

Max Planck (1858–1947).
(From *Introduction to Concepts and Theories in Physical Science*, G. Holton & S.G. Brush, p. 431, Addison-Wesley, 1973.)

Albert Einstein (1879–1955)
(From *Einstein: A Centenary Volume*, ed. A.P. French, p. 69, Heinemann, 1979, reproduced by permission of the Einstein Estate.)

Quanta and relativity are the two phenomena of physics which are quite outside our everyday experience – they are also perhaps the greatest discoveries of modern physics. In this case study, I want to look in some detail into the origins of the concept of quanta. For me, this is one of the most marvellous stories in intellectual history. It is very exciting and catches the flavour of an epoch when, within 25 years, physicists' view of nature changed totally and completely new perspectives were opened up. The story illustrates many important points about how physics and theoretical physics work in practice. We find the greatest physicists making mistakes, individuals having to struggle against the accepted views of virtually all physicists, and, most of all, a level of inspiration and scientific creativity which I find dazzling. If only everyone, and not only those who have had two years of training as physicists, could appreciate the intellectual beauty of this story.

In addition to telling a fascinating and compelling story, I want to prove everything essential to it using the physics and mathematics as understood at the time. This will provide some excellent revision material of basic parts of physics. We will find a striking contrast between those things which can be proved *classically* and those which are necessarily *quantum* in nature. The story will cover the years 1890 to about 1920 when matters came to a head and, at the end, all physicists were faced with a new view of the whole of physics in which all the fundamental entities have to be quantised.

The story will centre upon two very great physicists – Planck and Einstein. Planck is properly given credit for the discovery of quanta and we will trace how he came to this discovery. Einstein's contribution was perhaps even greater in that, long before anyone else, he inferred that all natural phenomena are quantum in nature and he was the first to put the subject on a firm theoretical basis.

I have based my telling of this story upon a set of lectures by M.J. Klein entitled *The Beginnings of the Quantum Theory*[1] published in the proceedings of the fifty seventh Varenna summer school. When I first read them, I found these lectures a revelation and felt cheated at not having known of this story before. I acknowledge fully my indebtedness to Dr Klein in inspiring what I consider in many ways to be the core of the present volume.

8

BLACK BODY RADIATION UP TO 1895

8.1 Physics and theoretical physics in 1890

Over the last four case studies, we have been building up a picture of the state of physics and theoretical physics towards the end of the 19th century. The achievements had been immense. In mechanics and dynamics, all the approaches described in Chapter 5 were understood. In thermodynamics, the first and second laws were firmly established, largely through the work of Clausius, and the full ramifications of the concept of entropy for classical thermodynamics were being elaborated. In Chapters 3 and 4, we described how Maxwell derived the basic equations of electromagnetism which were completely validated by Hertz's experiments of 1889. It was now known that light and electromagnetic waves are the same thing. This discovery provided a firm theoretical foundation for the wave theory of light which could account for virtually all the known phenomena of optics.

The impression is sometimes given that most physicists of the 1890s believed that the combination of thermodynamics, electromagnetism and classical mechanics could account for all known physical phenomena and that all that remained to be done was to work out the consequences of these recently won achievements. In fact, it was a period of ferment when there were still many fundamental unresolved problems which exercised the greatest minds of the period.

We have described the ambiguous status of the kinetic theory of gases as expounded by Clausius, Maxwell and Boltzmann. The fact that it could not satisfactorily account for all the properties of gases was a major barrier to its acceptance. The whole status of atomic and molecular theories of the structure of matter came under attack both for the technical reason outlined above and because of a movement away from mechanistic atomic models for physical phenomena in favour of empirical or phenomenological theories. The origin of the 'resonances' within molecules which were presumed to be the source of spectral lines had no clear interpretation and they were an embarrassment to the supporters of the kinetic theory. Boltzmann had discovered the statistical basis of thermodynamics but the theory had won little success, particularly in the face of a movement which denied that kinetic theories had any value, even as hypotheses.

Among these basic problems was the origin of the spectrum of black body radiation which proved to be the key, not only to the discovery of quanta, but also to the resolution of many of the basic problems identified above. The discovery of quanta was the precursor of the modern quantum theory of matter and radiation.

8.2 The Stefan–Boltzmann law

Stefan deduced the law which bears his name empirically from some experiments by Tyndall on the radiation from platinum strip heated to different known temperatures. He found that the total energy radiated by a black body over all wavelengths (or frequencies) is proportional to the fourth power of temperature

$$-\left(\frac{\mathrm{d}E}{\mathrm{d}t}\right) = \text{total radiant energy per second} \propto T^4 \tag{8.1}$$

In 1884, Boltzmann deduced this law from considerations of classical thermodynamics. It is important to note that his analysis was entirely classical and I think we should demonstrate how this can be done without taking any short cuts.

The simplest approach is to begin with a volume consisting only of electromagnetic radiation and as usual we suppose that the volume contains a piston so that the 'gas' of radiation can be compressed or expanded. Now suppose we add some heat $\mathrm{d}Q$ to the system and, as a result, the total internal energy increases by $\mathrm{d}U$ and the piston is allowed to be pushed out slightly so that the volume increases by $\mathrm{d}V$. Then, by conservation of energy,

$$\mathrm{d}Q = \mathrm{d}U + p\mathrm{d}V \tag{8.2}$$

Now let us rearrange this relation so that we can introduce the increase in entropy $\mathrm{d}S = \mathrm{d}Q/T$:

$$T\mathrm{d}S = \mathrm{d}U + p\mathrm{d}V \tag{8.3}$$

Now we convert this into a partial differential equation by dividing through by $\mathrm{d}V$ at constant T

$$T\left(\frac{\partial S}{\partial V}\right)_T = \left(\frac{\partial U}{\partial V}\right)_T + p \tag{8.4}$$

We can now use one of the Maxwell's relations which are derived in the Appendix to Chapter 6 to recast this relation:

$$\left(\frac{\partial p}{\partial T}\right)_V = \left(\frac{\partial S}{\partial V}\right)_T \tag{A6.12}$$

Therefore,

$$T\left(\frac{\partial p}{\partial T}\right)_V = \left(\frac{\partial U}{\partial V}\right)_T + p \tag{8.5}$$

This is the relation we were seeking because we can find the relation between U and T, provided we know the equation of state for the gas, i.e. the relation between p, V and U. Now this last relation is one which comes directly from Maxwell's theory of electromagnetism. Let us derive the radiation pressure of a 'gas' of electromagnetic radiation from Maxwell's equations. If you already know the answer that $p = \frac{1}{3}\epsilon$, where ϵ is the energy density of radiation, and you can prove it classically, you may wish to advance to Section 8.2.3.

8.2.1 The reflection of electromagnetic waves by a conducting plane

There are a number of ways of deriving the expression for radiation pressure but I will give a rather simple one based upon the idea that, when electromagnetic waves are reflected, they exert a pressure. Since it is well known that the waves are perfectly reflected from a perfectly conducting surface, let us consider the case of waves normally incident on a sheet of large but finite conductivity (Figure 8.1). This is an excellent revision example using some of the tools we have already developed in our study of electromagnetism (Chapters 3 and 4).

Figure 8.1. Illustrating the boundary conditions at the interface between a vacuum and a highly conducting medium for a normally incident electromagnetic wave.

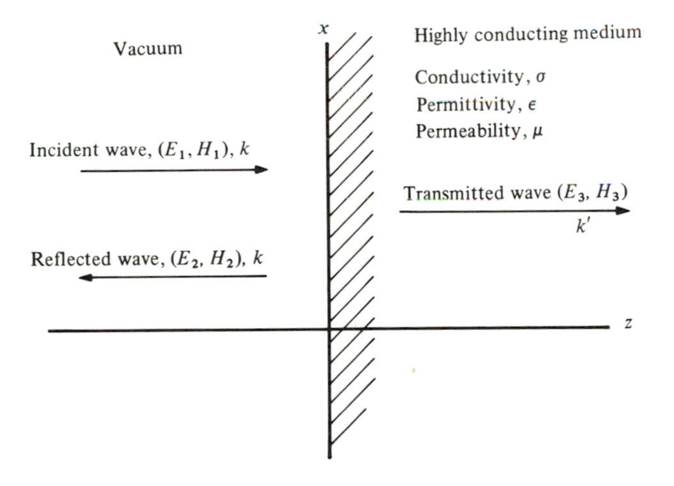

Medium 1 is a vacuum and medium 2 has high conductivity σ. We show in Figure 8.1 incident and reflected waves in the vacuum and a wave transmitted through the conductor. Now from our studies of Chapters 3 and 4, we can write down the *dispersion relations* for waves propagating in the vacuum and for waves in the conducting medium.

$$\left.\begin{array}{lll} \textit{Medium 1} & k^2 = \omega^2/c^2 & \\ \textit{Medium 2} & \left.\begin{array}{l} \nabla\times\mathbf{H} = \dfrac{\partial \mathbf{D}}{\partial t} + \mathbf{J} \\ \nabla\times\mathbf{E} = -\dfrac{\partial \mathbf{B}}{\partial t} \end{array}\right\} & \begin{array}{l} \mathbf{J} = \sigma\mathbf{E} \\ \mathbf{D} = \epsilon\epsilon_0\mathbf{E} \\ \mathbf{B} = \mu\mu_0\mathbf{H} \end{array} \end{array}\right\} \quad (8.6)$$

Therefore, using the relations developed in Appendix A3.6

$$\nabla\times \rightarrow \mathrm{i}\mathbf{k}\times$$

$$\frac{\partial}{\partial t} \rightarrow -\mathrm{i}\omega$$

we find that

$$\left.\begin{array}{l} (\mathbf{k}\times\mathbf{H}) = -(\omega\epsilon\epsilon_0 + \mathrm{i}\sigma)\mathbf{E} \\ (\mathbf{k}\times\mathbf{E}) = \omega\mathbf{B} \end{array}\right\} \quad (8.7)$$

Thus, we find the same answer as in Section 3.2, but with $\omega\epsilon\epsilon_0$ replaced by $(\omega\epsilon\epsilon_0 + \mathrm{i}\sigma)$, i.e.

$$k^2 = \epsilon\mu\frac{\omega^2}{c^2}\left(1 + \frac{\mathrm{i}\sigma}{\epsilon\epsilon_0\omega}\right) \quad (8.8)$$

Let us consider only the case in which the conductivity is very high, $\sigma/\epsilon\epsilon_0\omega \gg 1$. Then

$$k^2 = \mathrm{i}\frac{\mu\omega\sigma}{\epsilon_0 c^2}$$

Since $\mathrm{i}^{\frac{1}{2}} = (1/\sqrt{2})(1+\mathrm{i})$, the solution for k is

$$\left.\begin{array}{ll} & k = \pm(\mu\omega\sigma/2\epsilon_0 c^2)^{\frac{1}{2}}(1+\mathrm{i}) \\ \text{or} & \\ & k = \pm(\mu\omega\sigma/\epsilon_0 c^2)^{\frac{1}{2}}\,\mathrm{e}^{\mathrm{i}\pi/4} \end{array}\right\} \quad (8.9)$$

Now the phase relations between **E** and **B** in the wave can be found from the second relation of (8.7), $\mathbf{k}\times\mathbf{E} = \omega\mathbf{B}$.

In free space, **k** is real and hence **E** and **B** oscillate in phase with constant amplitude. However, *in the conductor*, there is a difference of $\pi/4$ between the phases of **E** and **B** (equations (8.9)) and both fields decrease exponentially into the conductor, i.e.

$$\begin{aligned} E &= \text{constant } \mathrm{e}^{\mathrm{i}(kz-\omega t)} \\ &= \text{constant } \mathrm{e}^{-z/l}\,\mathrm{e}^{\mathrm{i}[(z/l)-\omega t]} \end{aligned} \quad (8.10)$$

where $l = (2\epsilon_0 c^2/\mu\omega\sigma)^{\frac{1}{2}}$. The amplitude of the wave decreases by a factor 1/e in this length l which is called the *skin depth* of the conductor. This is the typical depth to which electromagnetic fields can penetrate into the conductor.

There is a nice general feature of the solution represented by the relation (8.9) which is worth noting. If we trace back the steps involved in arriving at the solution, we find that the assumption that the conductivity is high corresponds to neglecting the displacement current $\partial \mathbf{D}/\partial t$ in comparison with $\mathbf{J}$. Then the equation which we have solved is just a diffusion equation,

$$\nabla^2 \mathbf{H} = \sigma\mu\mu_0 \frac{\partial \mathbf{H}}{\partial t} \tag{8.11}$$

In general, wave solutions of diffusion equations have dispersion relations of the form $k = A(1 + \mathrm{i})$, i.e. the real and imaginary parts of the wave vector are equal, corresponding to waves which decay in amplitude by $\mathrm{e}^{-2\pi}$ in each complete cycle of the wave into the medium. This is true for equations such as

(*a*) the *heat diffusion equation*

$$\kappa\nabla^2 T - \frac{\partial T}{\partial t} = 0, \quad \kappa = \frac{K}{\rho C} \tag{8.12}$$

where K is the thermal conductivity of the medium, ρ its density, C the specific heat and T its temperature;

(*b*) the *diffusion equation*

$$D\nabla^2 N - \frac{\partial N}{\partial t} = 0 \tag{8.13}$$

where D is the diffusion coefficient and N the number density of particles.

(*c*) *viscous waves*

$$\frac{\mu}{\rho}\nabla^2 \mathbf{u} - \frac{\partial \mathbf{u}}{\partial t} = 0 \tag{8.14}$$

where μ is the viscosity, ρ the density of the fluid and $\mathbf{u}$ the fluid velocity. This equation is derived from the Navier–Stokes equation for fluid flow in a viscous medium (see the Appendix to Chapter 5, Section A5.4, equation A5.12).

Returning to our story, we seek to match the $\mathbf{E}$ and $\mathbf{H}$ vectors of the waves at the interface between the two media. Taking z to be the direction normal to the interface, we introduce the following

Incident wave

$$\left.\begin{aligned} E_x &= E_1\, \mathrm{e}^{\mathrm{i}(kz-\omega t)} \\ H_y &= E_1 Z_0 \mathrm{e}^{\mathrm{i}(kz-\omega t)} \end{aligned}\right\} \tag{8.15}$$

where $Z_0 = \mu_0/\epsilon_0$ is the impedance of free space.

Reflected wave

$$\left.\begin{aligned} E_x &= E_2\, e^{-i(kz+\omega t)} \\ H_y &= -E_2 Z_0\, e^{-i(kz+\omega t)} \end{aligned}\right\} \tag{8.16}$$

Transmitted wave

$$\left.\begin{aligned} E_x &= E_3\, e^{i(k'z-\omega t)} \\ H_y &= E_3\, \frac{(\mu\omega\sigma/2\epsilon_0 c^2)^{\frac{1}{2}}}{\omega\mu\mu_0}(1+i)\, e^{i(k'z-\omega t)} \end{aligned}\right\} \tag{8.17}$$

where k' is given by the value of k in the relation (8.9). The values of H_y are found by substituting for k in the relation between $\mathbf{E}$ and $\mathbf{B}$, $\mathbf{k} \times \mathbf{E} = \omega\mathbf{B}$.

For simplicity, let us write $q = [(\mu\omega\sigma/2\epsilon_0 c^2)^{\frac{1}{2}}/\omega\mu\mu_0](1+i)$. Therefore

$$H_y = qE_3\, e^{i(k'z-\omega t)} \tag{8.18}$$

The boundary conditions are that E_x and H_y are continuous at the interface (see Section 4.5), i.e. at $z = 0$, we find that

$$\left.\begin{aligned} E_1 + E_2 &= E_3 \\ E_1 Z_0 - E_2 Z_0 &= qE_3 \end{aligned}\right\} \tag{8.19}$$

Therefore,

$$\frac{E_1}{[1+(q/Z_0)]} = \frac{E_2}{[1-(q/Z_0)]} = \frac{E_3}{2} \tag{8.20}$$

In general, q is a complex number and hence there are phase differences between E_1, E_2 and E_3. However, we are interested in the case in which the conductivity is very large, i.e. $|q|/Z_0 \gg 1$ and hence

$$\frac{E_1}{q/Z_0} = -\frac{E_2}{q/Z_0} = \frac{E_3}{2} \tag{8.21}$$

i.e.

$$E_1 = -E_2 = (q/2Z_0)E_3$$

Therefore, on the vacuum side of the interface the total electric field strength $E_1 + E_2$ is zero and the magnetic field $H_1 + H_2 = 2H_1$.

It may seem as though we have strayed rather far from the thermodynamics of radiation but we are now set to work out the pressure exerted by the incident wave upon the surface.

8.2.2 *The formula for radiation pressure*

Let us suppose that we confine the radiation in a box with rectangular sides and let the waves bounce back and forth between the walls at $z = \pm z_1$. If we assume that the walls of the box are highly conducting as in the preceding subsection, we now have values of the electric and magnetic field strengths in the vicinity of the walls for normal incidence.

Part of the origin of the phenomenon of radiation pressure may be understood as follows. The electric field in the conductor E_x causes a current density to flow in the $+x$ direction,

$$J_x = \sigma E_x \tag{8.22}$$

But the force per unit volume acting on this electric current in the presence of a magnetic field is

$$\begin{aligned} \mathbf{F} &= N_q q(\mathbf{v} \times \mathbf{B}) \\ &= \mathbf{J} \times \mathbf{B} \end{aligned} \tag{8.23}$$

where N_q is the number of conduction electrons per unit volume and q their charge. Since $\mathbf{B}$ is in the $+y$ direction, this force acts in the $\mathbf{i}_x \times \mathbf{i}_y$ direction, i.e. in the $\mathbf{k}$ direction of the incident wave. Therefore, the pressure acting on a layer of thickness $\mathrm{d}z$ in the conductor is

$$\mathrm{d}p = J_x B_y \,\mathrm{d}z \tag{8.24}$$

However, we also know that, in the conductor, curl $\mathbf{H} = \mathbf{J}$ because the conductivity is very high and hence we can relate J_x and B_y by

$$\left(\frac{\partial H_z}{\partial y} - \frac{\partial H_y}{\partial z}\right) = J_x$$

Since $H_z = 0, -\partial H_y/\partial z = J_x$. Substituting into (8.24), we find that

$$\mathrm{d}p = -B_y \frac{\partial H_y}{\partial z}\,\mathrm{d}z \tag{8.25}$$

Therefore,

$$\begin{aligned} p &= -\int_0^\infty B_y \frac{\partial H_y}{\partial z}\,\mathrm{d}z \\ &= \int_0^{H_0} B_y \,\mathrm{d}H_y \end{aligned} \tag{8.26}$$

where H_0 is the value of the magnetic field strength at the interface and, according to the analysis of Section 8.2.1, $H \to 0$ and $z \to \infty$. For a linear medium, $B_0 = \mu\mu_0 H_0$ and hence

$$p = \tfrac{1}{2}\mu\mu_0 H_0^2$$

Notice that this force is associated with currents flowing in the conducting medium.

Now we have to ask what other forces there are acting on the metal. The only forces are those associated with the stresses in the electromagnetic fields themselves. In any medium, these are given by the appropriate components of the

Maxwell stress tensor, but, in our simple case, we can derive the stresses by an argument based on Faraday's concept of lines of force. Suppose we have a uniform longitudinal magnetic field confined within a certain region, which we will take to be a long rectangular perfectly conducting tube (Figure 8.2). Then we know that, if the medium is linear, $B = \mu\mu_0 H$, the energy per unit length of tube is

$$E = \tfrac{1}{2}BHxl$$

where x is the breadth and l the height of the tube. Now let us squash the rectangle by $\mathrm{d}x$ whilst maintaining the same number of lines of force through it. Then the magnetic flux density increases to $Bx/(x - \mathrm{d}x)$ because of conservation of lines of force and correspondingly, because the field is linear, H becomes $Hx/(x - \mathrm{d}x)$. Thus, the energy in the volume becomes

$$E + \mathrm{d}E = \tfrac{1}{2}BHlx^2(x - \mathrm{d}x)^{-1}$$

$$\mathrm{d}E = \tfrac{1}{2}BHlx\left(1 + \frac{\mathrm{d}x}{x}\right) - \tfrac{1}{2}BHlx$$

$$= \tfrac{1}{2}BHl\,\mathrm{d}x$$

But this increase in energy must be the result of work done on the field, i.e.

$$F\mathrm{d}x = \mathrm{d}E = \tfrac{1}{2}BHl\,\mathrm{d}x$$

Figure 8.2. A long perfectly conducting rectangular tube enclosing a longitudinal magnetic field. When the tube is compressed, the magnetic flux enclosed by the tube is conserved (for more details of flux freezing, see M.S. Longair *High Energy Astrophysics*, 143–9, Cambridge University Press, 1981).

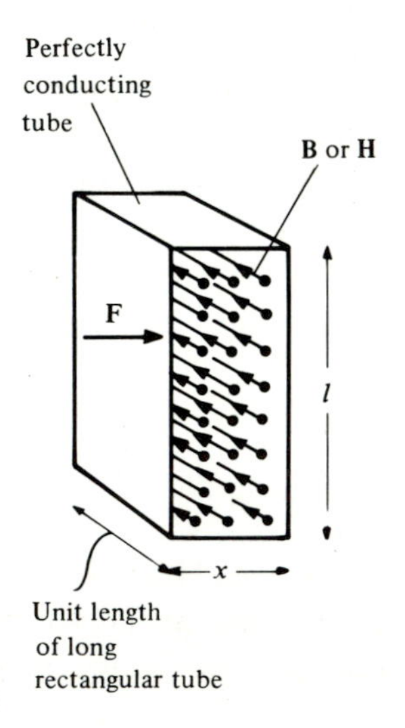

i.e. the force per unit area $F/l = \frac{1}{2}BH$. We may say that the pressure which we can associate with the field in the volume is

$$p = \tfrac{1}{2}\mu\mu_0 H^2 \tag{8.27}$$

perpendicular to the field direction. We can apply exactly the same argument to electrostatic fields in which case we would find that the total pressure associated with the electric and magnetic field is

$$p = \tfrac{1}{2}\epsilon\epsilon_0 E^2 + \tfrac{1}{2}\mu\mu_0 H^2 \tag{8.28}$$

Because the value of μ is different on either side of the interface, there is a pressure difference across it associated with the existence of the magnetic fields. In the case of the vacuum, we have shown that $E_x = 0$ and hence $p = \frac{1}{2}\mu_0 H_0^2$. Inside the conductor, the stress is $\frac{1}{2}\mu\mu_0 H_0^2$. Therefore, the total pressure on the conductor is

$$p = \underbrace{\tfrac{1}{2}\mu_0 H_0^2}_{\text{stress in vacuum}} - \underbrace{\tfrac{1}{2}\mu\mu_0 H_0^2}_{\text{stress in conductor}} + \underbrace{\tfrac{1}{2}\mu\mu_0 H_0^2}_{\text{force on conduction current}}$$

i.e.

$$p = \tfrac{1}{2}\mu_0 H_0^2 \tag{8.29}$$

Now we have shown that, when the conductivity is very high, the field H_0 is $2H_1$, where H_1 is the field strength of the wave propagating in the positive z direction in the vacuum and the energy density in this wave is $\frac{1}{2}(\epsilon_0 E_1^2 + \mu_0 H_1^2) = \mu_0 H_1^2 = \epsilon_1$. Therefore,

$$p = \tfrac{1}{2}\mu_0 H_0^2 = 2\mu_0 H_1^2 = 2\epsilon_1 = \epsilon_0 \tag{8.30}$$

where ϵ_0 is the total energy density of radiation in the vacuum in front of the conductor, i.e. the sum of the energy densities in the incident and reflected waves.

This is the basic relation for a 'one dimensional' gas, i.e. electromagnetic radiation confined between two reflecting walls. In an isotropic three dimensional volume, we have equal energy densities associated with radiation propagating in the three orthogonal directions, i.e.

$$\epsilon_x = \epsilon_y = \epsilon_z = \epsilon_0$$

and hence

$$p = \tfrac{1}{3}\epsilon \tag{8.31}$$

where ϵ is the total energy density of radiation.

This somewhat lengthy demonstration shows how it is possible to derive the pressure of a gas of electromagnetic radiation entirely by classical arguments. I have purposely given this simple treatment because I prefer to use these more physical arguments to the more mathematical treatment starting from

Maxwell's equations and involving the use of the Maxwell stress tensor for the electromagnetic field.

8.2.3 *The derivation of the Stefan–Boltzmann law*

The relation (8.31) solves the problem of the relation between p, V and U for a gas of electromagnetic radiation because we write $U = \epsilon V$ and $p = \frac{1}{3}\epsilon$ and then, from equation (8.5), we find that

$$T\left(\frac{\partial(\frac{1}{3}\epsilon)}{\partial T}\right)_V = \left(\frac{\partial(\epsilon V)}{\partial V}\right)_T + \tfrac{1}{3}\epsilon$$

$$\tfrac{1}{3}T\left(\frac{\partial \epsilon}{\partial T}\right) = \epsilon + \tfrac{1}{3}\epsilon = \tfrac{4}{3}\epsilon$$

We are now left with a relation between ϵ and T which can be solved in a straightforward manner.

$$\frac{d\epsilon}{\epsilon} = 4\frac{dT}{T}$$

$$\ln \epsilon = 4 \ln T$$

$$\epsilon \propto T^4 \tag{8.32}$$

This is what Boltzmann showed and his name is justly attached to the Stefan–Boltzmann law and the Stefan–Boltzmann constant σ. In modern form, the law is

$$I = \sigma T^4 \tag{8.33}$$

where I is the radiant energy emitted per unit area per second from the surface of a black body at temperature T. In modern notation,

$$\sigma = (\pi^2 k/60 h^3 c^2) = 5.67 \times 10^{-5} \text{ erg cm}^{-2} \text{ s}^{-1} \text{ deg}^{-4}$$

The rate of emission of radiant energy I is related to ϵ by $I = c\epsilon/4$. We can see this from the result for the rate of arrival of particles (or waves) at unit area of a surface which we derived in equation (7.5). In the present case, we can consider N to be the number density of waves of a given frequency and all of them have the same velocity c. Therefore, the rate of arrival (or departure) of waves at (or from) the surface is $\frac{1}{4}Nc$ per unit area and the total energy $\frac{1}{4}Nc\bar{E} = c\epsilon/4$ where $\bar{E}$ is the average energy per wave.

The experimental evidence for the Stefan–Boltzmann law was not particularly convincing in 1884 and it was not until 1897 that Lummer and Pringsheim undertook very careful experiments which showed that the law was indeed correct with high precision.

8.3 Wien's displacement law and the spectrum of black body radiation

The spectrum of black body radiation was even more poorly known but there had already been some important work done on the theoretical form which the radiation law should have. This was *Wien's displacement law* which he derived by a combination of electromagnetism and thermodynamics. Let us show exactly what he did. This work of singular importance in the development of the theory of black body radiation was published in 1894.

The first thing to get right is what happens to a 'gas' of radiation when it is adiabatically expanded. This comes straight out of the thermodynamic relations.

$$\mathrm{d}Q = \mathrm{d}U + p\mathrm{d}V$$

In an adiabatic expansion $\mathrm{d}Q = 0$ and we have shown that for radiation $U = \epsilon V$ and $p = \frac{1}{3}\epsilon$. Therefore,

$$\mathrm{d}(\epsilon V) + \tfrac{1}{3}\epsilon \mathrm{d}V = 0$$

$$V\mathrm{d}\epsilon + \epsilon \mathrm{d}V + \tfrac{1}{3}\epsilon \mathrm{d}V = 0$$

$$\frac{\mathrm{d}\epsilon}{\epsilon} = -\frac{4}{3}\frac{\mathrm{d}V}{V}$$

Integrating,

$$\epsilon = \text{constant} \times V^{-\frac{4}{3}} \tag{8.34}$$

But $\epsilon = aT^4$ where $a = 4\sigma/c$ and hence

$$TV^{\frac{1}{3}} = \text{constant} \tag{8.35}$$

Since V is proportional to the cube of the dimension of a spherical volume, we find that

$$T \propto r^{-1} \tag{8.36}$$

The next step is to work out the relation between the wavelength of the radiation and the volume of the enclosure. Let us do a simple sum which illustrates the answer. First of all, what is the change of wavelength of the radiation if a wave is reflected from a slowly moving mirror? This is shown in Figure 8.3. The initial position of the mirror is at X and we suppose that at that time one of the maxima of the incident waves is at A. It is then reflected along the path AC. By the time the next wavecrest comes along, the mirror has moved to X' and hence the maximum has to travel an extra distance ABN as compared with the first maximum. i.e. the distance between maxima, the wavelength, is increased by an amount $\mathrm{d}\lambda$ which is $AB + BN$. By symmetry,

$$AB + BN = A'N = AA' \cos\theta.$$

But

$$AA' = 2 \times \text{distance moved by mirror} = 2XX' = 2uT$$

where u = velocity of mirror and T = period of wave. Hence,

$$\mathrm{d}\lambda = 2uT\cos\theta$$

$$= 2\lambda\frac{u}{c}\cos\theta \quad \text{since } \mathrm{d}\lambda \ll \lambda \tag{8.37}$$

Notice that the result is correct to first order of small quantities but this is all we need since we are only interested in the result for differential changes.

Now put our wave in a spherical cavity which is expanding slowly and suppose that it makes an angle of incidence θ with the wall (Figure 8.4). Then we ask how many reflections take place and hence what will be the total change in wavelength when the cavity expands from r to $r + \mathrm{d}r$ at velocity u which is small?

If the velocity of expansion is small, θ remains the same for all reflections whilst the sphere expands by $\mathrm{d}r$. In addition, the time to expand a distance $\mathrm{d}r$ is $\mathrm{d}t = \mathrm{d}r/u$. From the geometry shown in Figure 8.4, the time between reflections for a wave propagating at velocity c is $2r \cos \theta/c$. The number of reflections in time $\mathrm{d}t$ is therefore $c\mathrm{d}t/2r \cos \theta$ and the change of wavelength $\mathrm{d}\lambda$ is

Figure 8.3. Illustrating the change in wavelength of an electromagnetic wave reflected from a moving perfectly conducting plane.

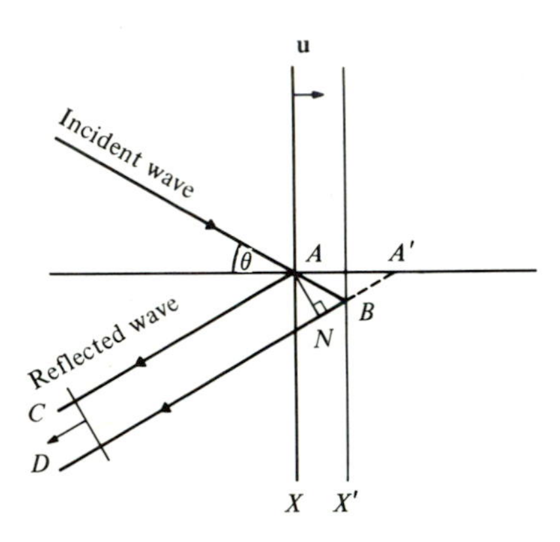

Figure 8.4. An electromagnetic wave reflected inside an expanding spherical conductor.

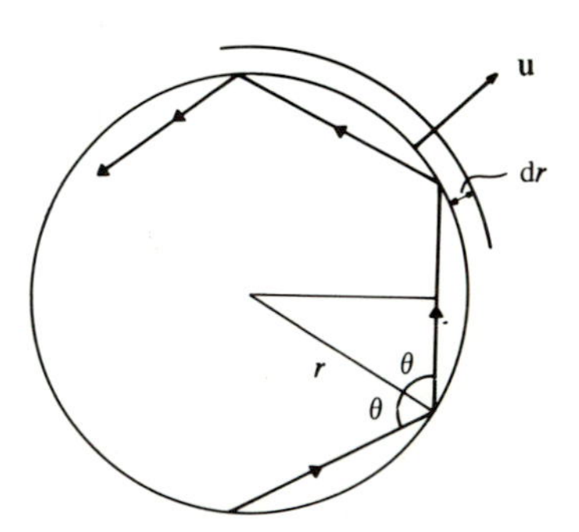

$$d\lambda = \left(\frac{2u\lambda}{c}\cos\theta\right)\left(\frac{c\,dt}{2r\cos\theta}\right)$$

i.e.

$$\frac{d\lambda}{\lambda} = \frac{u\,dt}{r} = \frac{dr}{r}$$

Integrating, we find that

$$\lambda \propto r \tag{8.38}$$

i.e. the wavelength of the radiation increases linearly proportionally to the size of the spherical volume.

We can now combine this result with relation (8.36), $T \propto r^{-1}$, and we find that

$$T \propto \lambda^{-1} \tag{8.39}$$

This is one aspect of Wien's displacement law. If radiation is adiabatically expanded, the wavelength of radiation changes inversely with temperature if we follow a particular set of waves. In other words, the wavelength of the radiation is 'displaced' as the temperature changes. In particular, if we follow the maximum of the radiation spectrum it should follow the T^{-1} law. This was found to be in agreement with experiment.

Now Wien went further than this and combined the two laws which we have just derived, the Stefan–Boltzmann law and the law $T \propto \lambda^{-1}$, to set constraints on what the spectral form of the radiation had to be. My own version of this argument is as follows. There are others, but they are much lengthier.

The first step is to note that if we enclose any system of bodies in a perfectly reflecting enclosure, then eventually they will all come to the same temperature because of the emission and absorption of radiation by them. Eventually, the radiation will come into equilibrium with the objects in the enclosure so that as much energy is radiated as absorbed by the bodies per unit time. If we wait long enough, the radiation will attain the spectrum corresponding to this equilibrium state and this will be our black body radiation spectrum. The radiation will be isotropic if we wait long enough and therefore the only parameters which can characterise the radiation will be the temperature of the enclosure and the wavelength of the radiation, T and λ respectively.

The next step is to note that if the black body radiation is initially at temperature T_1 in an enclosure at that temperature and we then expand the enclosure adiabatically, then, by the definition of an adiabatic expansion, the expansion is infinitely slow so that the radiation takes up an equilibrium spectrum at all stages in the expansion to temperature T_2. The crucial point is that in an adiabatic expansion, the radiation spectrum has black body form at the beginning and end of the expansion. The unknown law for the radiation spectrum must therefore scale appropriately with temperature.

Consider the radiation in the wavelength interval λ_1 to $\lambda_1 + d\lambda_1$ and let its energy density be $\epsilon = u(\lambda_1)\,d\lambda_1$. Then, in the expansion, according to Boltzmann's analysis we know that the energy associated with the radiation of any particular set of waves decreases as T^4 and hence

$$\frac{u(\lambda_1)d\lambda_1}{u(\lambda_2)d\lambda_2} = \left(\frac{T_1}{T_2}\right)^4 \tag{8.40}$$

But $\lambda_1 T_1 = \lambda_2 T_2$ and hence $d\lambda_1 = (T_2/T_1)\,d\lambda_2$. Therefore, we find that

$$\frac{u(\lambda_1)}{T_1^5} = \frac{u(\lambda_2)}{T_2^5}$$

i.e.

$$\frac{u(\lambda)}{T^5} = \text{constant} \tag{8.41}$$

and since $\lambda T = \text{constant}$, this can be rewritten as

$$u(\lambda)\lambda^5 = \text{constant} \tag{8.42}$$

Notice that $u(\lambda)$ is the energy density per unit bandwidth (or wavelength interval) in the radiation spectrum. Now the only combination of T and λ which is a constant is the product λT and hence we can write that, in general, the constant can only be constructed out of functions involving λT. We recall that we proved that the equilibrium spectrum should only depend upon λ and T and therefore the radiation law must have the form

$$u(\lambda)\lambda^5 = f(\lambda T)$$

or

$$u(\lambda)d\lambda = \lambda^{-5} f(\lambda T)\,d\lambda \tag{8.43}$$

This is *Wien's displacement law* in its entirety and you can see that it sets constraints on what the form of the radiation law for black body radiation should be. You may be more familiar with these laws in terms of frequencies rather than wavelength. So let us convert the law into frequency form.

$$u(\lambda)d\lambda = u(\nu)d\nu$$

$$\lambda = c/\nu, \quad d\lambda = -\frac{c}{\nu^2}d\nu$$

Hence

$$u(\nu)d\nu = \left(\frac{c}{\nu}\right)^{-5} f\left(\frac{\nu}{T}\right)\left(-\frac{c}{\nu^2}d\nu\right)$$

i.e.

$$u(\nu)d\nu = \nu^3 f\left(\frac{\nu}{T}\right)d\nu \tag{8.44}$$

This is really rather clever. Note how far Wien was able to get using only rather general thermodynamic arguments. We will see in a moment how crucial this general argument proved to be in establishing the correct form of the radiation formula.

This work was all new in 1894 when Planck first became interested in the problem of the spectrum of black body radiation. Let us now turn to Planck's huge contributions to the understanding of this problem.

9

1895–1900: PLANCK AND THE SPECTRUM OF BLACK BODY RADIATION

9.1 Planck's early career

We can learn about Planck's early career from his own short scientific autobiography.[1] He studied under Helmholtz and Kirchhoff in Berlin but in his own words 'I must confess that the lectures of these men netted me no perceptible gain. It was obvious that Helmholtz never prepared his lectures properly. . . Kirchhoff was the very opposite . . . but it would sound like a memorised text, dry and monotonous.' It is reassuring to students struggling to understand their physics lectures to note that even the greatest of physicists can be considered inadequate as university lecturers. In my own experience, although there is a general correlation between the best physicists and the best lecturers, there is a very wide dispersion about this relation. We all have to put up with the lecturers who happen to be taking our courses but, in the end, we have to understand the material ourselves rather than be spoon-fed and so it is perhaps not so bad. Indeed, one might argue that a bad lecturer requires the student to think harder for himself which is a good thing.

Planck's research interest was really stimulated by his reading of the works of Clausius and he set about investigating how the second law of thermodynamics could be applied to all sorts of physical problems, as well as elaborating as clearly as possible the basic tenets of the subject. He completed his dissertation in 1879. In his words, 'the effect of my dissertation on the physicists of those days was nil. None of my professors at the University had any understanding for its contents, as I learned for a fact in my conversations with them . . . . Helmholtz probably did not even read my paper at all. Kirchhoff expressly disapproved of its contents, with the comment that . . . the concept of entropy . . . must not be applied to irreversible processes. I did not succeed in reaching Clausius. He did not answer my letters and I did not find him at home when I tried to see him in person in Bonn.'[2]

You should note two things about these remarks. First of all, Kirchhoff was incorrect in stating that the concept of entropy cannot be applied to irreversible processes. Entropy is a function of state and therefore can be determined for any given state of the system, independent of whether or not irreversible processes are involved in getting into the state. Even the very best physicists are fallible. Second, one should not be particularly surprised that Planck had little success in reaching the great men. We are all familiar with the fact that people are very busy and it is often difficult for even a sympathetic senior scientist to have the time to devote to studies outside his immediate speciality. In addition, we are dealing with real human beings with their own personal traits, some of which are appealing and others which are not. One has to learn to take the rough with the smooth. I had a similar experience before I began my research career. A distinguished physicist gave an inspiring lecture and it was suggested I might talk with him about future research topics after it was over. It turned out, however, that he was too busy drinking sherry to talk with someone of my humble status. The moral is that, despite the set backs, do not give up or get discouraged. Keep at the problem until you are proved right or wrong.

Planck continued his work on entropy and eventually succeeded to Kirchhoff's position in the University of Berlin following the latter's death in 1889. This was an important step because he was now working in one of the most active centres of physics in the world at that time.

It was in 1894 that he turned his attention to the problem of the spectrum of thermal or black body radiation which was to dominate his subsequent work and which led to the discovery of the quantum. It is likely that his interest in this problem was stimulated by Wien's important paper which had been published in the same year. Wien's analysis, described in the last chapter, had a strong thermodynamic emphasis which must have appealed to Planck. This was the first of Planck's papers in which he diverged from his previous work in that it appeared to be about electromagnetism rather than about entropy. In 1895, he published the results of this work on the resonant scattering of plane electromagnetic waves by an oscillating dipole. However, in the last words of the paper, Planck made it clear that he regarded this as the first step towards tackling the problem of what we now call black body radiation. His aim was to set up a system of oscillators in an enclosed cavity which would radiate and interact with the radiation produced so that after a very long time the system would come into equilibrium. He could then apply the laws of thermodynamics to black body radiation with a view to understanding the origin of its spectrum. Even in this paper, he noticed one important aspect of this problem. Namely that, when an oscillator loses energy by radiation, it does not go into heat but into electromagnetic waves. In a sense, the process is conservative because, if enclosed in a box with perfectly reflecting walls, the radiation can then act back on the oscillator. In addition, the process is independent of the nature of the oscillator. In Planck's words, 'The study of conservative damping seems to me to be of great importance, since it opens up the prospect of a possible general explanation of irreversible processes by means of conservative forces – a problem that confronts research in theoretical physics more urgently every day'.[3]

Now it is important to remember that Planck was a great expert in thermodynamics and that he was a follower of Clausius as an exponent of classical thermodynamics. Planck looked upon the second law of thermodynamics as having absolute validity – processes in which the total entropy decreased, he believed, should be strictly excluded. This is a very different point of view from the interpretation which Boltzmann had set forth in his memoir of 1877. In this, the second law of thermodynamics is only statistical in nature. As we demonstrated in Chapter 7, according to Boltzmann, there is a very high probability indeed that entropy will increase in all natural processes but there remains an *extremely* small but finite probability that the system will evolve to a state of lower entropy. Planck and his students had published papers criticising some of the steps in Boltzmann's statistical approach.

Thus, Planck believed that, by studying the interaction of oscillators with electromagnetic radiation, he would be able to show that entropy would increase absolutely for a system of matter and radiation. He set to work on a series of five papers in which this idea was expounded. The idea did not work, as was pointed out by Boltzmann. One cannot obtain a monotonic approach to equilibrium without some statistical assumptions about the way in which the system approaches the equilibrium state. One can see that this must be so from Maxwell's simple but compelling arguments concerning the reversibility of all the laws of mechanics, dynamics and electromagnetism.

Finally, Planck conceded that a statistical assumption was needed and introduced the concept of 'natural radiation' which corresponds to Boltzmann's assumption of molecular chaos. This must have been somewhat galling for Planck. It is not pleasant to have the assumptions upon which you have based your researches for 20 years to be shown to be wrong. However, much worse was to come.

Once the assumption was made that there exists this state of natural radiation Planck was able to complete his programme to a certain extent. The first thing he did was to relate the energy density of the radiation to the average energy of the oscillators in an enclosure. This is a very important result and can be derived from purely classical arguments. We will go through the derivation; it is a beautiful piece of theoretical physics.

9.2 The relation between the average energy of an oscillator and its radiation in thermal equilibrium

First of all, why do we treat oscillators rather than things like atoms, molecules, lumps of rock, etc? The reason is that in thermal equilibrium everything is in equilibrium with everything else – rocks are in equilibrium with atoms and oscillators and therefore there is no advantage in treating complicated things. We may as well treat a simple oscillator for which the radiation and absorption laws can be calculated exactly. I have, in fact, just stated Kirchhoff's Law, in a somewhat unconventional guise.

9.2.1 The rate of radiation of an accelerated charged particle

Charged particles emit electromagnetic radiation when they are accelerated. Let us derive first of all a rather beautiful little formula for the rate of emission of radiation of an accelerated particle.

The normal derivation of the formula for the energy radiated by an accelerated charge proceeds from Maxwell's equations through the use of retarded potentials for the field components at a large distance. We can give a much simpler derivation of the exact results of these calculations from a remarkable argument first put forward by J.J. Thomson. This argument indicates physically, very clearly, why it is that an accelerated charge radiates electromagnetic radiation and also indicates the origin of the polar diagram and polarisation properties of the radiation.

We consider a charge q stationary at the origin O of some frame of reference S at time $t = 0$. The charge then suffers a small acceleration to velocity Δv in a short interval of time Δt. Thomson visualised what happens to the fields in terms of the field lines attached to the accelerated charge. After a time t, we can distinguish between the field configuration inside and outside a sphere of radius ct centred on the origin of S. Outside this sphere, the field lines do not yet know that the charge has moved away from the origin because information cannot travel faster than the speed of light and therefore they are radial centred on O. Inside this sphere, the field lines are radial in the frame of reference which is centred on the moving charge. Between these two regions, there is a thin shell of thickness $c\Delta t$ in which we have to join up corresponding field lines. This situation is indicated schematically in Figure 9.1(a). Geometrically, it is clear that in this shell there must be a component of the field lines in the circumferential direction, i.e. there is an E field in the $\mathbf{i}_\theta$ direction. This 'pulse' of electromagnetic field is propagated away from the charge at the speed of light. Thus, by virtue of being accelerated, there is an energy loss associated with this pulse which is propagated outwards from the charge.

Let us work out the strength of the electric field in the pulse. We assume that the increment in velocity Δv is very small, i.e. $\Delta v \ll c$, and therefore it is safe to assume that the field lines are radial at $t = 0$ and also at time t in the frame of reference S. There will, in fact, be small aberration effects associated with the velocity Δv but they are second order compared with the gross effects we are discussing. We may therefore consider a small cone of field lines at angle θ with

Figure 9.1. (*a*) Illustrating the configuration of the electric field lines at time *t* due to a charge accelerated by a velocity Δv in time Δt at time $t = 0$. (*b*) A diagram showing how to evaluate the azimuthal component of the electric field due to acceleration of the electron at the origin.

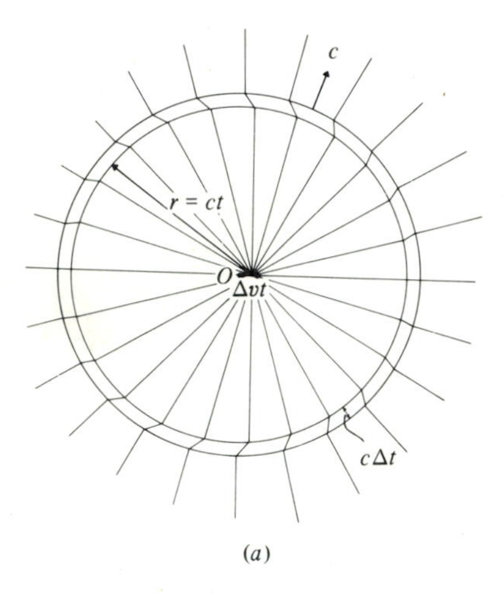

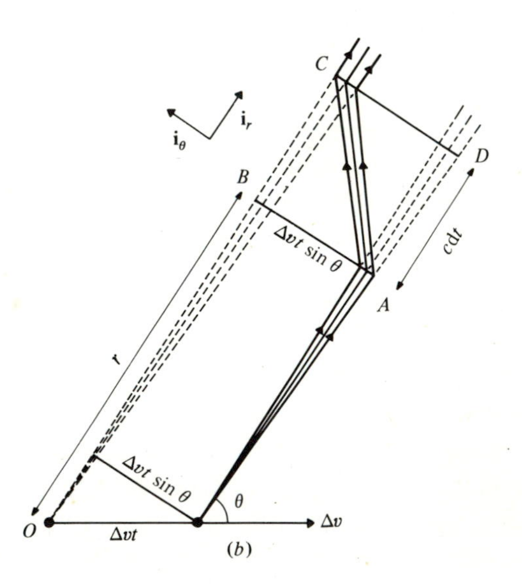

respect to the acceleration vector of the charge at $t = 0$ and at some later time t when the charge is moving with constant velocity Δv (Figure 9.1(*b*)). Now we have to join up the field lines between the two cones through the thin shell of thickness $c\Delta t$ as shown in the diagram. The strength of the E_θ component of the field is just given by the number of field lines per unit area in the $\mathbf{i}_\theta$ direction. From the geometry of Figure 9.1(*b*), this is given by the relative sizes of the sides of the rectangle *ABCD*, i.e.

$$\frac{E_\theta}{E_r} = \frac{\Delta v t \sin\theta}{c\Delta t} \tag{9.1}$$

But

$$E_r = \frac{q}{4\pi\epsilon_0 r^2}; \quad r = ct$$

and therefore

$$E_\theta = \frac{q(\Delta v/\Delta t)\sin\theta}{4\pi\epsilon_0 c^2 r}$$

$\Delta v/\Delta t$ is just the acceleration $\ddot{r}$ of the charge and hence we can write the result

$$E_\theta = \frac{q\ddot{r}\sin\theta}{4\pi\epsilon_0 c^2 r} \tag{9.2}$$

Notice how the radial component of the field decreases as r^{-2} according to Coulomb's law but the field in the pulse decreases only as r^{-1} because the field lines become more and more stretched in the E_θ direction (see relation (9.1)). Alternatively, we can write $qr = p$, where p is the dipole moment of the charge, with respect to some origin and hence

$$E_\theta = \frac{\ddot{p}\sin\theta}{4\pi\epsilon_0 c^2 r} \tag{9.3}$$

We know that this is a pulse of electromagnetic radiation and hence the energy flow per unit area per second is given by the Poynting vector $\mathbf{E} \times \mathbf{H} = Z_0 E^2$ where $Z_0 = (\epsilon_0/\mu_0)^{\frac{1}{2}}$ is the impedance of free space (see Section 8.2.1). The rate of loss of energy through solid angle $d\Omega$ at distance r from the charge is therefore

$$-\left(\frac{dE}{dt}\right)_{\text{rad},d\Omega} = \frac{Z_0|\ddot{p}|^2\sin^2\theta}{16\pi^2\epsilon_0^2c^4r^2}\,r^2\,d\Omega$$

$$= \frac{|\ddot{p}|^2\sin^2\theta}{16\pi^2\epsilon_0 c^3}\,d\Omega \tag{9.4}$$

To find the total radiation rate, we integrate over all solid angles, i.e. we integrate over θ with respect to the direction of the acceleration. We recall that, integrating over solid angle means integrating over $2\pi\sin\theta\, d\theta$.

$$-\left(\frac{\mathrm{d}E}{\mathrm{d}t}\right)_{\mathrm{rad}} = \int_0^\pi \frac{|\ddot{p}|^2 \sin^2\theta}{16\pi^2\epsilon_0 c^3} 2\pi \sin\theta \, \mathrm{d}\theta$$

Performing this integral, we find that

$$-\left(\frac{\mathrm{d}E}{\mathrm{d}t}\right)_{\mathrm{rad}} = \frac{|\ddot{p}|^2}{6\pi\epsilon_0 c^3} = \frac{q^2|\ddot{r}|^2}{6\pi\epsilon_0 c^3} \qquad (9.5)$$

Exactly the same result comes out of the full theory. Notice that it is valid for any form of acceleration $\ddot{r}$. These formulae embody the three essential properties of the radiation of an accelerated electron.

(*a*) The total rate of emission is given by the result (9.5).

(*b*) The polar diagram of the radiation, i.e. the dependence of the rate of emission of radiation on the angle θ, is of dipole form, i.e. $E_\theta \propto \sin\theta$ or $(\mathrm{d}E/\mathrm{d}t) \propto \sin^2\theta$ (expression (9.4)). There is no emission along the direction of the acceleration vector and the radiation is at a maximum perpendicular to it.

(*c*) The polarisation of the radiation, i.e. the direction of the E_θ vector, is parallel to the acceleration vector as projected onto the spherical surface at distance r.

In fact, you will find that you can understand most radiation problems of accelerated charged particles in terms of these simple rules. A number of nice examples are given in my book *High Energy Astrophysics.*[4] Although this is a rather pleasant argument, you should not regard it as a substitute for the full theory which comes out of strict application of Maxwell's equations.

9.2.2 Radiation damping of an oscillator

We now apply the result (9.5) to the case of an oscillator which performs harmonic oscillations at angular frequency ω_0 with amplitude x_0, $x = x_0 \, \mathrm{e}^{\mathrm{i}\omega_0 t}$. Therefore, $\ddot{x} = -\omega_0^2 x_0 \, \mathrm{e}^{\mathrm{i}\omega_0 t}$ and, taking the real part, $\ddot{x} = -\omega_0^2 x_0 \cos \omega t$.

The rate of loss of energy from the oscillating dipole is therefore

$$-\left(\frac{\mathrm{d}E}{\mathrm{d}t}\right) = \frac{\omega_0^4 e^2 x_0^2}{6\pi\epsilon_0 c^3} \cos^2 \omega_0 t$$

The average value of the $\cos^2 \omega_0 t$ term is $\frac{1}{2}$ and hence the average rate of loss of energy by our oscillator in the form of electromagnetic radiation is

$$-\left(\frac{\mathrm{d}E}{\mathrm{d}t}\right)_{\mathrm{average}} = \frac{\omega_0^4 x_0^2 e^2}{12\pi\epsilon_0 c^3} \qquad (9.6)$$

Now let us look at the equation for damped simple harmonic motion, since our oscillator is losing energy by virtue of emitting electromagnetic waves.

$$m\ddot{x} + a\dot{x} + kx = 0$$

where m is the mass, k the spring constant and $a\dot{x}$ is the damping force. We can find the energies associated with each of these terms by multiplying through by $\dot{x}$ and integrating with respect to time.

$$\int_0^t m\ddot{x}\dot{x}\,\mathrm{d}t + \int_0^t a\dot{x}^2\,\mathrm{d}t + \int_0^t kx\dot{x}\,\mathrm{d}t = 0$$

$$\tfrac{1}{2}m\int \mathrm{d}(\dot{x}^2) + \int a\dot{x}^2\,\mathrm{d}t + \tfrac{1}{2}\int k\,\mathrm{d}(x^2) = 0 \tag{9.7}$$

We identify these terms with the kinetic energy, the damping energy loss and the potential energy of the oscillator respectively. Now let us evaluate each term for simple harmonic motion of the form $x = x_0 \cos \omega_0 t$. The average values of the kinetic and potential energies are:

$$\text{average kinetic energy} = \tfrac{1}{4}mx_0^2\omega_0^2$$

$$\text{average potential energy} = \tfrac{1}{4}kx_0^2$$

Now, if the damping is very small, the natural frequency of oscillation of the oscillator is just $\omega_0^2 = k/m$. Thus, as we already know, the average kinetic and potential energies of the oscillator are equal and the total energy is the sum of the two which is

$$E = \tfrac{1}{2}mx_0^2\omega_0^2 \tag{9.8}$$

Now, from equation (9.7), the average rate of loss of energy by the oscillator is just

$$-\left(\frac{\mathrm{d}E}{\mathrm{d}t}\right)_{\mathrm{rad}} = \tfrac{1}{2}ax_0^2\omega_0^2 \tag{9.9}$$

Therefore, taking the ratio of equations (9.8) and (9.9), we find that

$$-\left(\frac{\mathrm{d}E}{\mathrm{d}t}\right)_{\mathrm{rad}} = \frac{a}{m}E \tag{9.10}$$

Now let us compare this relation with the expression for the loss rate by radiation of the oscillator. Substituting equation (9.8) into (9.6), we obtain

$$-\frac{\mathrm{d}E}{\mathrm{d}t} = \gamma E \tag{9.11}$$

where $\gamma = \omega_0^2 e^2/6\pi\epsilon_0 c^3 m$. This expression looks somewhat simpler if we introduce the classical electron radius $r_\mathrm{e} = e^2/4\pi\epsilon_0 m_\mathrm{e} c^2$. Thus, $\gamma = 2r_\mathrm{e}\omega_0^2/3c$. We can therefore obtain the correct expression for the decay in amplitude of the oscillator by identifying γ with a/m in equation (9.10).

We can appreciate now why Planck believed this was a profitable way of proceeding. The radiation loss does not go into heat but into electromagnetic radiation and the constant γ depends only upon fundamental constants, if, say, we take the oscillator as consisting of an oscillating electron. In contrast, if one treats ordinary frictional damping, the energy goes into heat and the loss rate formula contains constants appropriate to the material. In addition, in the case of electromagnetic waves, if we put the oscillators and waves in an enclosure with perfectly reflecting walls the waves can react back on the oscillators after they bounce off the walls and so the energy is not lost from the system. This is why Planck called the damping 'conservative damping'. The more common name for this phenomenon is *radiation damping.*

9.3 The equilibrium spectrum of radiation of an oscillator of energy E

We have now obtained the expression for the dynamics of our oscillator undergoing natural damping

$$m\ddot{x} + m\gamma\dot{x} + kx = 0$$

$$\ddot{x} + \gamma\dot{x} + \omega_0^2 x = 0$$

If an electromagnetic wave is incident on the oscillator, energy can be transferred to it and then we have to add a force term to the right-hand side,

$$\ddot{x} + \gamma\dot{x} + \omega_0^2 x = F/m \tag{9.12}$$

If the oscillator is accelerated by the E field of an incident wave, we can write $F = eE_0 \,\mathrm{e}^{\mathrm{i}\omega t}$. We find the response of the oscillator by adopting a trial solution for x of the form $x = x_0 \,\mathrm{e}^{\mathrm{i}\omega t}$. Then

$$x_0 = \frac{eE_0}{m(\omega_0^2 - \omega^2 + \mathrm{i}\gamma\omega)} \tag{9.13}$$

You will notice that there is a complex factor in the denominator and this means that the oscillator does not vibrate in phase with the incident wave. This does not matter for our calculation. We are most interested in the modulus of the amplitude as we will show.

Now we are already quite far along the track to finding the amount of energy transferred to the oscillator but let us not do that. Let us rather work out the rate of radiation of the oscillator under the influence of the incident radiation field. If we work this out and set it equal to the 'natural' radiation of the oscillator we will have satisfied our requirement of an equilibrium spectrum. This gives us a physical picture of what we mean by the oscillator being in equilibrium with the radiation field. To put it another way, the work done by the incident radiation field is just enough to supply the energy loss per second by the oscillator.

From now on the calculation is just hard work. There is really only one tricky step left.

We use the above form of the rate of radiation of an oscillator

$$-\left(\frac{\mathrm{d}E}{\mathrm{d}t}\right) = \frac{\omega^4 e^2 x_0^2}{6\pi\epsilon_0 c^3}\cos^2\omega t$$

and use the value of x_0 which we have just calculated. You will notice that we require the square of the modulus of x_0 and we get this by multiplying x_0 by its complex conjugate, i.e.

$$x_0^2 = \frac{e^2 E_0^2}{m^2[(\omega_0^2 - \omega^2)^2 + \gamma^2\omega^2]}$$

Therefore, the radiation rate is

$$-\left(\frac{dE}{dt}\right) = \frac{\omega^4 e^4 E_0^2 \cos^2\omega t}{6\pi\epsilon_0 c^3 m^2[(\omega_0^2 - \omega^2)^2 + \gamma^2\omega^2]}$$

It is now convenient to take averages over time, i.e.

$$\langle\cos^2\omega t\rangle = \tfrac{1}{2}$$

We notice that the term in E^2 is closely related to the energy in the incident wave. The incident energy per unit area per second is given by the Poynting vector $\mathbf{E} \times \mathbf{H}$. Then, since in the wave $\mathbf{B} = \mathbf{k} \times \mathbf{E}/\omega$, $|\mathbf{H}| = E/\mu_0 c$ and $|\mathbf{E} \times \mathbf{H}| = (E_0^2/\mu_0 c)\cos^2\omega t$, the average incident energy per unit area per second is

$$\frac{1}{2\mu_0 c} E_0^2 = \tfrac{1}{2}\epsilon_0 c E_0^2$$

Furthermore, we remember that when we have a superposition of waves of random phase, as we will in equilibrium radiation, we find the total incident energy by adding the energies, i.e. the E^2s. Thus we can replace the single value of E^2 in our formula by the sum over all the waves incident upon the oscillator and we will find the total reradiated power, i.e.

$$-\left(\frac{dE}{dt}\right) = \frac{\omega^4 e^4 \frac{1}{2}\sum_i E_{0i}^2}{6\pi\epsilon_0 c^3 m^2[(\omega_0^2 - \omega^2)^2 + \gamma^2\omega^2]} \tag{9.14}$$

The next step is to note that this is part of a continuum intensity distribution and we can also write this sum of energies as an incident intensity† in the frequency band ω to $\omega + d\omega$, i.e.

$$I(\omega)d\omega = \tfrac{1}{2}\epsilon_0 c \sum_i E_{0i}^2 \tag{9.15}$$

Therefore, the total radiation loss rate is

$$-\left(\frac{dE}{dt}\right) = \frac{\omega^4 e^4}{6\pi\epsilon_0^2 c^4 m^2} \frac{I(\omega)d\omega}{[(\omega_0^2 - \omega^2)^2 + \gamma^2\omega^2]}$$

† Notice that this intensity is the total power per unit area from 4π steradians. The usual definition is in terms of W m^{-2} Hz^{-1} sr^{-1}. Correspondingly, in (9.18), the relation between $I(\omega)$ and $u(\omega)$ is $I(\omega) = u(\omega)c$ rather than the usual relation $I(\omega) = u(\omega)c/4\pi$.

It will simplify the equation if we introduce the classical electron radius again, $r_e = e^2/4\pi\epsilon_0 m_e c^2$

$$-\left(\frac{dE}{dt}\right) = \frac{8\pi\omega^4}{3} r_e^2 \frac{I(\omega)d\omega}{[(\omega_0^2 - \omega^2)^2 + \gamma^2\omega^2]} \tag{9.16}$$

Now the response curve of the oscillator as described by the factor in the denominator is very sharply peaked about the value ω_0 because the radiation rate is very small in comparison with the total energy of the oscillator, i.e. $\gamma \ll 1$ (see Figure 9.2).

We can therefore make some simplifying approximations. If ω appears on its own, we can let $\omega \to \omega_0$ and $(\omega_0^2 - \omega^2) = (\omega_0 + \omega)(\omega_0 - \omega) \approx 2\omega_0(\omega_0 - \omega)$ Therefore,

$$-\left(\frac{dE}{dt}\right) = \frac{2\pi\omega_0^2 r_e^2}{3} \frac{I(\omega)d\omega}{[(\omega - \omega_0)^2 + (\gamma^2/4)]}$$

Finally, we expect that $I(\omega)$ will be a slowly varying function in comparison with the sharpness of the response curve of the oscillator and so we can set it equal to a constant over the range of values of ω of interest.

$$-\left(\frac{dE}{dt}\right) = \frac{2\pi\omega_0^2 r_e^2}{3} I(\omega_0) \int_0^\infty \frac{d\omega}{[(\omega - \omega_0)^2 + (\gamma^2/4)]}$$

The remaining integral is easy if we set the lower limit equal to minus infinity and this we can do because there is nothing left of the function as soon as we get away from the peak. Using

$$\int_{-\infty}^{\infty} \frac{dx}{x^2 + a^2} = \frac{\pi}{a}$$

the integral becomes $2\pi/\gamma$ and

Figure 9.2

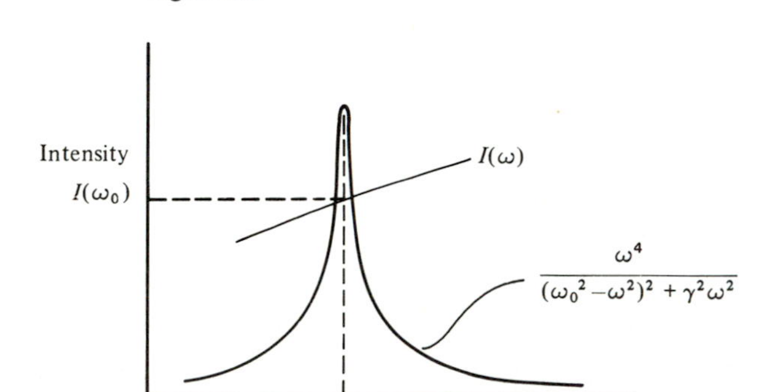

$$-\left(\frac{\mathrm{d}E}{\mathrm{d}t}\right) = \frac{2\pi\omega_0^2 r_e^2}{3}\frac{2\pi}{\gamma}I(\omega_0) = \frac{4\pi^2\omega_0^2 r_e^2}{3\gamma}I(\omega_0)$$

Now we have said that this is the rate of radiation which we will set equal to the spontaneous radiation rate of the oscillator. There is only one complication. We have assumed that the oscillator can respond equally to all possible polarisations. If we have a single oscillator, then there are directions of incidence in which there will be no effect upon it, i.e. if the electric field is perpendicular to the dipole axis of the oscillator. We can get round this problem by a clever argument which Feynman makes in his analysis of this problem. Suppose we have three oscillators mutually at right angles. Then this system will be able to respond like a completely free oscillator and follow any incident electric field. Therefore, we will obtain the correct answer if we suppose that this is the radiation which would be emitted by three oscillators, each of which oscillates at frequency ω_0. At last, we are able to arrive at the answer we have been searching for. Equating the radiation rates, we find a rather spectacular result:

$$I(\omega_0)\frac{4\pi^2\omega_0^2 r_e^2}{3\gamma} = 3\gamma E$$

$$\gamma = \tfrac{2}{3}(r_e\omega_0^2/c)$$

and hence

$$I(\omega_0) = \frac{3^2[\tfrac{2}{3}(r_e\omega_0^2/c)]^2}{4\pi^2\omega_0^2 r_e^2}E$$

$$= \frac{\omega_0^2}{\pi^2 c^2}E \tag{9.17}$$

If we write this in terms of a spectral energy density we get

$$u(\omega_0) = \frac{I(\omega_0)}{c} = \frac{\omega_0^2}{\pi^2 c^3}E \tag{9.18}$$

We may as well now drop the subscripts zero on ω since this result applies to all frequencies in equilibrium. Let us rewrite this result in terms of frequencies:

$$u(\omega)\mathrm{d}\omega = u(\nu)\mathrm{d}\nu = \frac{\omega^2}{\pi^2 c^3}E\,\mathrm{d}\omega$$

i.e.

$$u(\nu) = \frac{8\pi\nu^2}{c^3}E \tag{9.19}$$

This is the result which Planck derived in a paper published in June 1899. You can see how remarkable it is. All information about the nature of the oscillator has completely disappeared from the problem. There is no mention of its charge or mass. All that remains is the average energy of the oscillator. The meaning behind the relation is obviously very profound and fundamental in a thermodynamic sense. I find this an intriguing calculation. The whole analysis has proceeded through an analysis of the electrodynamics of oscillators and yet the final result contains no trace of the means by which we arrived at the answer. One can imagine how excited Planck must have been when he discovered this basic result.

It is also now clear that as soon as we can work out the average energy of an oscillator of frequency ν in an enclosure of temperature T, we will find the spectrum of black body radiation. Therefore, we have to find the relation between E, T and ν.

9.4 How Planck arrived at the spectrum of black body radiation

A surprising aspect of the subsequent developments was what Planck did *not* do next. We know the answer to the above problem from classical statistical mechanics. According to classical theory, in thermodynamic equilibrium, we give $\frac{1}{2}kT$ of energy to each degree of freedom and hence the mean energy of an oscillator should be kT because a harmonic oscillator has two degrees of freedom, i.e. those associated with the squared terms $\dot{x}^2$ and x^2 in the expression for the energy of the oscillator. If we set E equal to kT, we find that

$$u(\nu) = \frac{8\pi\nu^2}{c^3}kT \tag{9.20}$$

You probably know that this is actually the correct answer for the radiation law at low frequencies, the Rayleigh–Jeans law, which we will describe in its proper context in a moment. Why did Planck not do this? First of all, the equipartition theorem of Maxwell and Boltzmann is a result of statistical thermodynamics and this was the point of view which he had specifically rejected. At least, he was certainly not as familiar with statistical thermodynamics as he was with the classical theory. In addition, as we described in Chapter 7, it was not clear in 1899 how secure the equipartition theorem was. Maxwell's kinetic theory did not produce the correct answers for the ratio of the specific heats of gases.

Planck had already had his fingers burned by Boltzmann when he failed to note the necessity of statistical assumptions in order to derive the equilibrium state of black body radiation. To quote his words:

> I had no alternative than to tackle the problem once again – this time from the opposite side – namely from the side of thermodynamics, my own home territory where I felt myself to be on safer ground. In fact my previous studies of the Second Law of Thermodynamics came to stand me in good stead now, for at the very outset I hit upon the idea of correlating not the temperature of the oscillator but its entropy with

> its energy . . . . While a host of outstanding physicists worked on the problem of the spectral energy distribution both from the experimental and theoretical aspect, every one of them directed his efforts solely towards exhibiting the dependence of the intensity of radiation on the temperature. On the other hand, I suspected that the fundamental connection lies in the dependence of entropy upon energy . . . . Nobody paid any attention to the method which I adopted and I could work out my calculations completely at my leisure, with absolute thoroughness, without fear of interference or competition.[5]

What he did was to work out the following relation for the entropy of a system which is not in equilibrium – I am not going to work this one out because it will be of marginal importance in the end, although it was important to Planck in obtaining the right answer.

$$\Delta S = \Delta E \, \mathrm{d}E \frac{3}{5} \frac{\partial^2 S}{\partial E^2} \tag{9.21}$$

This equation applies to a system whose entropy deviates from the maximum entropy in that an individual resonator deviates by an amount ΔE from the equilibrium value E. The entropy change occurs when the energy of the oscillator varies by $\mathrm{d}E$. Thus, if ΔE and $\mathrm{d}E$ have opposite signs, i.e. if the system tends to return towards equilibrium, then the entropy change must be positive and hence the function $\partial^2 S/\partial E^2$ must necessarily have a negative value. We can see by inspection that a formula of this type must be correct. A negative value of $\partial^2 S/\partial E^2$ means that there is an entropy maximum and thus if ΔE and $\mathrm{d}E$ are of opposite signs, the system must approach equilibrium.

We have to look again at the state of the experimental work to appreciate what Planck did next. Wien had followed up his studies of the spectrum of thermal radiation by attempting to derive the radiation law from theory. We need not go into his ideas but he did produce a form for the radiation law which was consistent with his displacement law and which provided an excellent fit to all the data which were available in 1896. The displacement law is

$$u(\nu) = \nu^3 f(\nu/T)$$

and Wien's theory suggested that

$$u(\nu) = \frac{8\pi\alpha}{c^3} \nu^3 \, \mathrm{e}^{-\beta\nu/T} \tag{9.22}$$

This is the famous *Wien's law*, and I have written it in a form suitable for the analysis to follow. There are two unknown constants in the formula, α and β; I have included the constant $8\pi/c^3$ in front of the right-hand side for convenience. Rayleigh's comment on Wien's law was that 'Viewed from the theoretical side, the result appears to me to be little more than a conjecture'.[6] However, the importance of the formula was that it gave an excellent account of all the experimental data at the time and therefore could be used for further theoretical studies. You will notice from Figure 9.3 that one has to fit a curve which rises steeply on both sides of the maximum. Naturally, the region of the maximum is best defined by the experimental data and the errors on the energy distribution in the wings are much greater.

The next step in Planck's paper was the introduction of a definition for the entropy of the oscillator.

$$S = -\frac{E}{\beta\nu}\ln\frac{E}{\alpha\nu \mathrm{e}} \tag{9.23}$$

S is its entropy, E is its energy and α and β are constants; e is the base of natural logarithms. In fact, from his writings we know that he derived this from Wien's law. Let us see how this can be done.

We take the radiation spectrum from Wien's law (9.22) and put it into our equation relating $u(\nu)$ and E, (9.19):

$$u(\nu) = \frac{8\pi\nu^2}{c^3}E$$

Therefore,

$$E = \alpha\nu \mathrm{e}^{-\beta\nu/T} \tag{9.24}$$

Figure 9.3. Examples of the intensity spectrum of black body radiation plotted on linear scales of intensity and wavelength. Note that because of the low intensities in the wings of the curves, they are most difficult to define experimentally. (After H.S. Allen and R.S. Maxwell *A Text-book of Heat*, Part II, p. 748, MacMillan and Co. Ltd., 1952.)

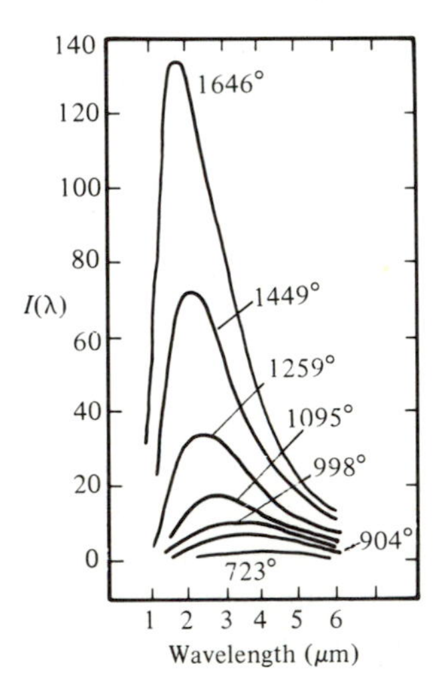

Now suppose we place the oscillator (or a set of oscillators) in a fixed volume V. The basic thermodynamic equality tells us that

$$TdS = dU + pdV$$

and hence

$$\left(\frac{\partial S}{\partial U}\right)_V = \frac{1}{T} \tag{9.25}$$

U and S are additive functions of state and hence the above relation refers to the properties of an individual oscillator as well as to an ensemble of them. Therefore, by inverting equation (9.24), we can relate $(\partial S/\partial E)_V$ to E, i.e.

$$\frac{1}{T} = \left(\frac{\partial S}{\partial E}\right)_V = -\frac{1}{\beta\nu}\ln\left(\frac{E}{\alpha\nu}\right) \tag{9.26}$$

Now let us take Planck's definition of the entropy of the oscillator (9.23) and differentiate with respect to E.

$$\begin{aligned}\frac{dS}{dE} &= -\frac{1}{\beta\nu}\ln\left(\frac{E}{\alpha\nu e}\right) - \frac{\not{E}}{\beta\nu}\frac{1}{\not{E}}\\ &= -\frac{1}{\beta\nu}\left[\ln\left(\frac{E}{\alpha\nu e}\right) + 1\right]\\ &= -\frac{1}{\beta\nu}\left[\ln\left(\frac{E}{\alpha\nu e}\right) + \ln e\right]\\ &= -\frac{1}{\beta\nu}\ln\left(\frac{E}{\alpha\nu}\right)\end{aligned}$$

We see that equation (9.26) leads directly to Planck's definition of entropy, equation (9.23).

But now we come to the step which was crucial for Planck. We take the next derivative of equation (9.26) with respect to E.

$$\frac{\partial^2 S}{\partial E^2} = -\frac{1}{\beta\nu}\frac{1}{E} \tag{9.27}$$

Notice that β, ν and E are all necessarily positive quantities and therefore $\partial^2 S/\partial E^2$ is necessarily negative. This means that Wien's law (equation (9.22)) is entirely consistent with the second law of thermodynamics. Notice also the amazing simplicity of the expression for the second derivative of the entropy with respect to energy. It is just proportional to the inverse of the energy. This profoundly impressed Planck who always sought fundamental insights in his research. In his words: 'I have repeatedly tried to change or generalise the equation for the electromagnetic entropy of an oscillator in such a way that it satisfies all theoretically sound electromagnetic and thermodynamic laws but I was unsuccessful in this endeavour'.[7] In his paper to the Prussian Academy of Sciences in May 1899 he says 'I believe that this must lead me to conclude that the definition of radiation entropy and therefore Wien's energy distribution law necessarily result from the application of the principle of the increase of entropy to the electromagnetic radiation theory and therefore the limits of validity of this law, in so far as they exist at all, coincide with those of the second law of thermodynamics'.[8] This is rather potent stuff and we now know that he was wrong but he has already made tremendous progress towards the correct theory.

An interesting aspect of the above assertion is that it is often very dangerous to use arguments of the form 'I cannot think of any other function which can do the job'. In fact any negative function of energy would have satisfied Planck's requirement so far as the second law of thermodynamics was concerned. There only remained the problem of ensuring consistency with the measurements of the black body spectrum.

These calculations were presented to the Prussian Academy of Sciences in June 1900. By October 1900, things changed again. Rubens and Kurlbaum showed beyond any doubt that Wien's law was inadequate to explain the distribution of black body radiation at low frequencies and high temperatures. You will remember that the radiation law depends only upon ν/T and thus the law is inadequate for small values of this parameter. Their experiments were made over a wide range of temperatures and were carried out with the greatest care. In particular, they showed that, at low frequencies and high temperatures, the intensity of radiation was proportional to temperature. This is clearly inconsistent with Wien's law because if $u(\nu) \propto \nu^3 \mathrm{e}^{-\beta\nu/T}$, then for $\beta\nu/T \ll 1$, $u(\nu) \propto \nu^3$ and is independent of temperature.

Rubens and Kurlbaum told Planck about their results before they presented them in October 1900 and he was offered the opportunity to make some remarks about their implications. The result was his paper entitled 'An Improvement of the Wien Distribution'[9] and it is the first time that the Planck formula appears in its primitive form.

Here is what he did. He now knew that he had to find a law which would result in the relation $u \propto T$ in the limit $\nu/T \to 0$. Let us run through the relations he had already derived and see where this will lead. We know from (9.19) that

$$u(\nu) = \frac{8\pi\nu^2}{c^3} E$$

and, because at low frequencies we must have $u(\nu) \propto T$,

$$E \propto T$$

$$\frac{\mathrm{d}S}{\mathrm{d}E} = \frac{1}{T}$$

and hence

$$\frac{dS}{dE} \propto \frac{1}{E}$$

$$\frac{d^2S}{dE^2} \propto \frac{1}{E^2} \tag{9.28}$$

Therefore, the function d^2S/dE^2 must change its functional dependence upon E between high and low values of ν/T. From Wien's law, which remains good for large values of ν/T, we require that

$$\frac{d^2S}{dE^2} \propto \frac{1}{E} \tag{9.29}$$

and now for low values

$$\frac{d^2S}{dE^2} \propto \frac{1}{E^2}$$

The standard technique in this situation is to write

$$\frac{d^2S}{dE^2} = -\frac{a}{E(b+E)} \tag{9.30}$$

which has exactly the required properties for large and small values of E. Then the analysis is straightforward. Integrating,

$$\frac{dS}{dE} = -\int \frac{a}{E(b+E)}\,dE$$

$$= -\frac{a}{b}\left[\ln E - \ln(b+E)\right] \tag{9.31}$$

But

$$\frac{dS}{dE} = \frac{1}{T}$$

and hence

$$\frac{1}{T} = -\frac{a}{b}\ln\left(\frac{E}{b+E}\right)$$

$$e^{b/aT} = \frac{b+E}{E}$$

$$E = \frac{b}{e^{b/aT}-1} \tag{9.32}$$

Now we can find the radiation spectrum.

$$u(\nu) = \frac{8\pi\nu^2}{c^3}E = \frac{8\pi\nu^2}{c^3}\frac{b}{e^{b/aT}-1} \tag{9.33}$$

By looking at the high frequency, low temperature limit we can compare the constants with those which appear in Wien's formula

$$u(\nu) = \frac{8\pi\nu^2 b}{c^3 \, e^{b/aT}} \equiv \frac{8\pi\alpha}{c^3} \frac{\nu^3}{e^{\beta\nu/T}}$$

Thus b must be proportional to frequency ν. We can therefore write the primitive form of the Planck formula.

$$u(\nu) = \frac{A\nu^3}{(e^{\beta\nu/T} - 1)} \tag{9.34}$$

Notice that this is in agreement with Wien's displacement law. Of equal importance for the rest of the story is the fact that Planck was also able to find the expression for the entropy of his oscillator by integrating dS/dE. This results in the relation

$$\frac{dS}{dE} = -\frac{a}{b}\,[\ln E - \ln (b + E)]$$

Adding $-\ln b$ to both logarithms since E/b is dimensionless

$$\frac{dS}{dE} = -\frac{a}{b}\left[\ln \frac{E}{b} - \ln \left(1 + \frac{E}{b}\right)\right]$$

Integrating,

$$\begin{aligned} S &= -\frac{a}{b}\left\{\left(E \ln \frac{E}{b} - E\right) - \left[E \ln \left(1 + \frac{E}{b}\right) - E + b \ln \left(1 + \frac{E}{b}\right)\right]\right\} \\ &= -\frac{a}{b}\left[E \ln \frac{E}{b} - (E + b) \ln \left(1 + \frac{E}{b}\right)\right] \\ &= -a\left[\frac{E}{b} \ln \frac{E}{b} - \left(1 + \frac{E}{b}\right) \ln \left(1 + \frac{E}{b}\right)\right] \end{aligned} \tag{9.35}$$

with $b \propto \nu$.

Planck's new formula was a remarkably elegant result and could now be confronted with the experimental evidence. Before doing that, let us look at the origins of another of the formulae which had been proposed at about the same time, the Rayleigh–Jeans law.

9.5 Rayleigh's derivation of the Rayleigh–Jeans law

Lord Rayleigh was the author of the famous book, *The Theory of Sound*[10] and thus a leading exponent of the theory of waves in general. His contribution to the theory of black body radiation was stimulated by the inadequacies of Wien's law in accounting for the low frequency behaviour of black body radiation as a function of temperature. His original paper[11] is a short and elegant exposition of the application of the theory of waves to black body radiation and it is reproduced in full on pages 208 and 209. The reason for the inclusion of Jeans' name in the law is that there is a numerical error in Rayleigh's analysis which was corrected by Jeans in a paper to *Nature* in 1906. We will work through Rayleigh's calculations below but let us just look at the paper itself for a moment.

The first paragraph is an exposition of the state of knowledge of the black body spectrum in 1900. In the second, he acknowledges the success of Wien's formula in accounting for the maximum in the black body spectrum but then expresses his worries about the long wavelength behaviour of the formula. The third and fourth paragraphs set out his proposal and the fifth his preferred form for the radiation spectrum which, in the sixth paragraph, he hopes will be compared with experiment by 'the distinguished experimenters who have been occupied with this subject'.

Let us now rework the essence of paragraphs three, four and five in modern notation, getting the numerical factors right. We begin with the problem of waves in a box. Suppose the box is a cube with sides of length L. Inside, we have all possible sorts of wave 'banging' around. We can therefore write down the wave equation for this system.

$$\nabla^2 \psi = \frac{\partial^2 \psi}{\partial x^2} + \frac{\partial^2 \psi}{\partial y^2} + \frac{\partial^2 \psi}{\partial z^2} = \frac{1}{c_s^2}\frac{\partial^2 \psi}{\partial t^2} \tag{9.36}$$

where c_s is the velocity of the waves. The walls are fixed and so we require the waves to have zero amplitude there, i.e. at $x, y, z = 0$ and $x, y, z = L$, $\psi = 0$. You should by now know intuitively that the solution of this problem is

$$\psi = C\,\mathrm{e}^{-\mathrm{i}\omega t} \sin\frac{l\pi x}{L} \sin\frac{m\pi y}{L} \sin\frac{n\pi z}{L} \tag{9.37}$$

This form will guarantee that the waves will fit into the box provided l, m and n are integers. Now each combination of l, m and n is called a *mode of oscillation* of the waves in the box. This means that, for any mode, all the gas oscillates in phase at a particular frequency. The modes are all independent and therefore represent independent ways in which the gas can oscillate. In addition, the set of modes with $0 \leqslant l, m, n \leqslant \infty$ is a *complete set of orthogonal modes*, so that *any* pressure distribution can be described as a sum over all the orthogonal modes. In the absence of damping or coupling between the modes, the waves oscillate independently and indefinitely with constant amplitude.

We now substitute the expression (9.37) into the wave equation and find the relation between the values of l, m, n and the angular frequency of the wave, ω.

$$\frac{\omega^2}{c^2} = \frac{\pi^2}{L^2}(l^2 + m^2 + n^2) \tag{9.38}$$

REMARKS UPON THE LAW OF COMPLETE RADIATION.

[*Philosophical Magazine*, XLIX. pp. 539, 540, 1900.]

BY complete radiation I mean the radiation from an ideally black body, which according to Stewart* and Kirchhoff is a definite function of the absolute temperature θ and the wave-length λ. Arguments of (in my opinion†) considerable weight have been brought forward by Boltzmann and W. Wien leading to the conclusion that the function is of the form

$$\theta^5 \phi(\theta\lambda)\, d\lambda, \quad \text{.................................(1)}$$

expressive of the energy in that part of the spectrum which lies between λ and $\lambda + d\lambda$. A further specialization by determining the form of the function ϕ was attempted later‡. Wien concludes that the actual law is

$$c_1 \lambda^{-5} e^{-c_2/\lambda\theta}\, d\lambda, \quad \text{..............................(2)}$$

in which c_1 and c_2 are constants, but viewed from the theoretical side the result appears to me to be little more than a conjecture. It is, however, supported upon general thermodynamic grounds by Planck§.

Upon the experimental side, Wien's law (2) has met with important confirmation. Paschen finds that his observations are well represented, if he takes

$$c_2 = 14{,}455,$$

θ being measured in centigrade degrees and λ in thousandths of a millimetre (μ). Nevertheless, the law seems rather difficult of acceptance, especially the implication that as the temperature is raised, the radiation of given wave-length approaches a limit. It is true that for visible rays the limit is out of range. But if we take $\lambda = 60\,\mu$, as (according to the remarkable researches of Rubens) for the rays selected by reflexion at surfaces of Sylvin, we see that for temperatures over 1000° (absolute) there would be but little further increase of radiation.

The question is one to be settled by experiment; but in the meantime I venture to suggest a modification of (2), which appears to me more probable *à priori*. Speculation upon this subject is hampered by the difficulties which attend the Boltzmann-Maxwell doctrine of the partition of energy. According to this doctrine every mode of vibration should be alike favoured;

* Stewart's work appears to be insufficiently recognized upon the Continent. [See *Phil. Mag.* I. p. 98, 1901; p. 494 below.]
† *Phil. Mag.* Vol. XLV. p. 522 (1898).
‡ *Wied. Ann.* Vol. LVIII. p. 662 (1896).
§ *Wied. Ann.* Vol. I. p. 74 (1900).

Rayleigh's original paper on the spectrum of black body radiation.
(From *Scientific papers by John William Strutt* (Baron Rayleigh), Vol. 4, 1892–1901, p. 483–5.)

and although for some reason not yet explained the doctrine fails in general, it seems possible that it may apply to the graver modes. Let us consider in illustration the case of a stretched string vibrating transversely. According to the Boltzmann-Maxwell law the energy should be equally divided among all the modes, whose frequencies are as 1, 2, 3, ... Hence if k be the reciprocal of λ, representing the frequency, the energy between the limits k and $k + dk$ is (when k is large enough) represented by dk simply.

When we pass from one dimension to three dimensions, and consider for example the vibrations of a cubical mass of air, we have (*Theory of Sound*, § 267) as the equation for k^2,

$$k^2 = p^2 + q^2 + r^2,$$

where p, q, r are integers representing the number of subdivisions in the three directions. If we regard p, q, r as the coordinates of points forming a cubic array, k is the distance of any point from the origin. Accordingly the number of points for which k lies between k and $k + dk$, proportional to the volume of the corresponding spherical shell, may be represented by $k^2 dk$, and this expresses the distribution of energy according to the Boltzmann-Maxwell law, so far as regards the wave-length or frequency. If we apply this result to radiation, we shall have, since the energy in each mode is proportional to θ,

$$\theta k^2 dk, \qquad \text{.................................(3)}$$

or, if we prefer it,

$$\theta \lambda^{-4} d\lambda. \qquad \text{.................................(4)}$$

It may be regarded as some confirmation of the suitability of (4) that it is of the prescribed form (1).

The suggestion is that (4) rather than, as according to (2),

$$\lambda^{-5} d\lambda \qquad \text{.................................(5)}$$

may be the proper form when $\lambda\theta$ is great*. If we introduce the exponential factor, the complete expression will be

$$c_1 \theta \lambda^{-4} e^{-c_2/\lambda\theta} d\lambda. \qquad \text{.................................(6)}$$

If, as is probably to be preferred, we make k the independent variable, (6) becomes

$$c_1 \theta k^2 e^{-c_2 k/\theta} dk. \qquad \text{.................................(7)}$$

Whether (6) represents the facts of observation as well as (2) I am not in a position to say. It is to be hoped that the question may soon receive an answer at the hands of the distinguished experimenters who have been occupied with this subject.

* [1902. This is what I intended to emphasize. Very shortly afterwards the anticipation above expressed was confirmed by the important researches of Rubens and Kurlbaum (*Drude Ann.* IV. p. 649, 1901), who operated with exceptionally long waves. The formula of Planck, given about the same time, seems best to meet the observations. According to this modification of Wien's formula, $e^{-c_2/\lambda\theta}$ in (2) is replaced by $1 \div (e^{c_2/\lambda\theta} - 1)$. When $\lambda\theta$ is great, this becomes $\lambda\theta/c_2$, and the complete expression reduces to (4).]

If we write $p^2 = l^2 + m^2 + n^2$, we find that

$$\frac{\omega^2}{c^2} = \frac{\pi^2 p^2}{L^2} \tag{9.39}$$

We thus find a relation between the modes as parameterised by $p^2 = l^2 + m^2 + n^2$ and ω. According to the equipartition theorem, we share out energy equally among each mode and therefore we need to know how many modes there are in the range p to $p + \mathrm{d}p$. We find this by the standard procedure of drawing our three dimensional lattice in l, m, n space and evaluating the number of modes in the octant of a sphere as illustrated in Figure 9.4. If p is large, the number of modes is just

$$n(p)\,\mathrm{d}p = \tfrac{1}{8}4\pi p^2\,\mathrm{d}p \tag{9.40}$$

Expressing this result in terms of ω rather than p,

$$p = \frac{L\omega}{\pi c} \qquad \mathrm{d}p = \frac{L\mathrm{d}\omega}{\pi c} \tag{9.41}$$

and hence

$$n(p)\mathrm{d}p = \frac{L^3\omega^2\mathrm{d}\omega}{2\pi^2 c^3} \tag{9.42}$$

We remember that we are dealing with electromagnetic waves and there are two independent polarisations for a given value of **k**, the wave vector, rather than one. We therefore have twice as many modes as given by the above result. Now all we have to do is to apply the 'Maxwell–Boltzmann doctrine' of the equipartition of energy and give each mode of oscillation an amount kT of energy

Figure 9.4. Illustrating how the number of modes in the interval dl, dm, dn can be replaced by an increment in the phase space volume $\frac{1}{2}\pi p^2\mathrm{d}p$ where $p^2 = l^2 + m^2 + n^2$.

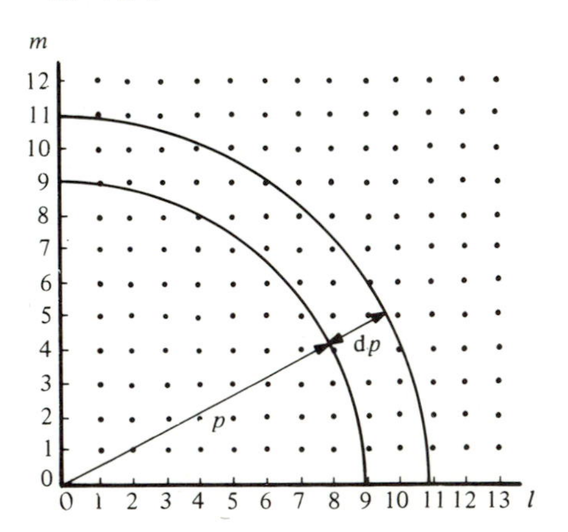

and we obtain the following result for the energy density of electromagnetic radiation in the box:

$$En(p)\mathrm{d}p = u(\nu)\mathrm{d}\nu L^3 = \frac{L^3\omega^2\mathrm{d}\omega}{\pi^2c^3}E \tag{9.43}$$

$$u(\nu) = \frac{8\pi\nu^2}{c^3}E \tag{9.44}$$

or

$$u(\nu) = \frac{8\pi\nu^2}{c^3}kT \tag{9.45}$$

This is exactly the same result as the one which had been derived by Planck from electrodynamics (expression (9.19)) but Rayleigh did not hesitate in setting $E = kT$ according to the Maxwell–Boltzmann doctrine of equipartition. It is intriguing that two such different approaches result in exactly the same answer. A number of comments are in order.

(i) Notice that in Rayleigh's argument, he deals directly with the *waves themselves* rather than the *oscillators* which are the basis of Planck's approach and which are in equilibrium with the waves.

(ii) I do not know if you are happy about the doctrine of equipartition of energy. It always bothered me and it is not until one realises that one is dealing with independent modes of oscillation that it becomes clearer physically. We can decompose any motion of the gas in the box into its normal modes of oscillation and these will be independent of one another. However, what you are not told is that there are processes by which energy *can* be exchanged between these so-called independent modes of oscillation. In fact the modes are not completely independent – for example, the way in which a furnace comes into thermodynamic equilibrium is through Compton scattering – i.e. the small changes in frequency of the photons because the particles are at a finite temperature. No-one ever told me this and it was only when I was working on a related problem in astrophysics that it became obvious. Thus, if one mode gets more energy than another, there are processes by which energy can be redistributed among the modes. What the Maxwell–Boltzmann doctrine states is that if you leave the system long enough, irregularities in the energy distribution among the oscillations will be smoothed out by these energy interchange mechanisms. In many natural phenomena, these equilibrium distributions are set up very quickly and so there is no need to worry, but remember that there is a definite assumption about the ability of the interaction processes to bring about the equipartition distribution of energy.

(iii) You will notice that Rayleigh is aware of 'the difficulties which attend the Boltzmann–Maxwell doctrine of the partition of energy'. Among these is the fact that equation (9.45) must break down at high frequencies because the spectrum of black body radiation does not increase as ν^2 to infinite frequency. However, he proposes that the analysis 'may apply to the graver modes', i.e. at long wavelengths and low frequencies.

Then, right out of the blue, we read in the fifth paragraph 'If we introduce the exponential factor, the complete expression is', in our notation,

$$u(\nu) = \frac{8\pi\nu^2}{c^3} kT\, e^{-\beta\nu/T} \tag{9.46}$$

Obviously, Rayleigh included this factor empirically so that the radiation spectrum would converge at high frequencies and because the exponential in Wien's law provided a good fit to the data.

Rayleigh's analysis is brilliant but as we will see it did not get due credit in 1900. Part of the problem is immediately obvious – the relation (9.46) is not consistent with Wien's displacement law (8.44).

9.6 Comparison of the laws for black body radiation with experiment

On October 25, 1900, Rubens and Kurlbaum compared their precise new measurements of the black body spectrum with five different predictions. These were (i) Planck's formula (9.34), (ii) Wien's relation (9.22), (iii) Rayleigh's result (9.46) and (iv) and (v) two empirical relations which had been proposed by Thiesen and by Lummer and Jahnke. Rubens and Kurlbaum concluded that Planck's formula was superior to all the others and that it gave precise agreement with experiment. Rayleigh's proposal was found to be a poor representation of the experimental data. He was justifiably somewhat upset by the tone of voice in which they discussed his result. When his scientific papers were republished two years later he remarked on his important conclusion that the intensity of radiation should be proportional to temperature at low frequencies. He pointed out 'This is what I intended to emphasise. Very shortly afterwards the anticipation above expressed was confirmed by the important researches of Rubens and Kurlbaum who operated with exceptionally long waves'.[12]

In other words, the essential theoretical point that the radiation can be described accurately by Rayleigh's analysis in the wavelength region in which it is applicable had been missed by the experimenters who had simply compared the relation (9.46) with their experiments. There is a warning to all theorists and experimenters in this tale. Experimenters sometimes use the results of theory without a full appreciation of the range of applicability of the theory. In addition, experimenters like to have lots of theories between which to discriminate experimentally, no matter how outlandish they may be. Thus, in the present example, the theories of Planck and Rayleigh are much more profound than the other three. However, the experimenter is right to maintain an unbiased stance between theories.

It seems inconceivable that Planck was not aware of Rayleigh's work. Rubens had told Planck about his experiments and he must have seen the comparison of the various curves with the experiments. Again, we must assume that it was the statistical basis of the equipartition theorem which put Planck off, plus the facts that Rayleigh's overall law did not account for the experimental data and did not agree with Wien's displacement law. Even so, he ought to have been impressed by the ease with which Rayleigh found the correct low frequency, high temperature relation and also by the method by which he obtained exactly Planck's relation between the energy density of radiation and the mean energy of each mode (relation (9.44)), although Rayleigh's paper does not exactly make this point clear.

Planck had not, however, explained anything. All he had was a formula and it had no firm theoretical basis. He immediately embarked upon this problem. The new formula was produced by October 19, 1900 and on December 14, 1900, he presented another paper to the German Physical Society, 'On the theory of the Energy Distribution Law in the Normal Spectrum'.[13] In his memoirs, he writes 'After a few weeks of the most strenuous work of my life, the darkness lifted and an unexpected vista began to appear'.[14]

10

PLANCK'S THEORY OF BLACK BODY RADIATION

10.1 Introduction

> On the very day when I formulated this law, I began to devote myself to the task of investing it with a true physical meaning. This quest automatically led me to study the interrelation of entropy and probability – in other words, to pursue the line of thought inaugurated by Boltzmann.[1]

Planck recognised that the way forward consisted of adopting a point of view which he had rejected in essentially all his previous work.

We have already written down the basic relation between entropy and probability derived by Boltzmann, $S \propto \ln W$. The constant of proportionality was not known at the time and we will call it C for the moment, i.e. $S = C \ln W$. Planck must have been working at a remarkable pace because he had not specialised in statistical physics. We will show that what he did was not at all what he ought to have done according to classical statistical mechanics and yet he found the correct answer. There are basic flaws in his argument which we will study in a moment. First of all, let us demonstrate how he should have proceeded according to classical statistical mechanics.

10.2 Boltzmann's procedure in statistical mechanics

We are interested in working out the average energy E of an oscillator. We suppose that there are r of these in a black body enclosure and therefore the total energy of the oscillators is $E_{\text{tot}} = rE$. The entropy is also an additive function and so if S is the average entropy of one oscillator, the entropy of the whole system is $S_{\text{tot}} = rS$.

It is one of the tricks of classical statistical mechanics that one begins by considering the molecules as having discrete energies, $0, \epsilon, 2\epsilon, 3\epsilon, \ldots$. This is done in order to work out exact probabilities using finite numbers rather than using a distribution which is continuous and infinite. At the appropriate step in the argument we let the value of ϵ become infinitely small whilst the total energy of the system remains finite so that the energy distribution becomes continuous.

To begin with, this is what Planck does. He says: let us suppose that there exists an energy unit ϵ and that we have a fixed amount of energy to distribute among the oscillators E_N. Then, we can make this up by awarding different amounts of energy to the individual oscillators. As an example, here is the distribution which Planck gives in his paper of 1900.[2] Suppose we have $r = 10$ oscillators and we have to distribute 100ϵ among them; we can do this in the following way.

Number of oscillator i	1	2	3	4	5	6	7	8	9	10	
Energy in units of ϵ, E_i	7	38	11	0	9	2	20	4	4	5	= 100

Now there are obviously many ways of distributing the energy among the oscillators besides this one. If Planck had followed Boltzmann's prescription, this is what he would have done.

Boltzmann noted that each such distribution of energies can be represented by a set of numbers $w_0 w_1 w_2 w_3 \ldots$ which describe the number of molecules or oscillators with energies 0, 1, 2, 3 ... in units of ϵ. We now work out the number of possible ways in which the energy elements can be distributed over the oscillators which will result in the same distribution of energies E_i.

Let us revise some basic elements of permutation theory. The number of different ways in which we can order n different objects is just $n!$ If m of these are identical, the number of different orderings is reduced because we can interchange these m objects and it will make no difference to the distribution. Since the m objects can be ordered in $m!$ ways, the numbers of different arrangements is reduced to $n!/m!$. If a further l objects are identical, the number of different arrangements is reduced to $n!/m!\,l!$ and so on.

Now we ask, in how many ways can we select x objects from n objects? In this case we can divide the set of n objects into two groups of identical objects, the x objects which make up the set selected and the $n-x$ objects which are not selected. From the last paragraph, the number of different ways of making this selection is $n!/(n-x)!\,x!$. This is often written $\binom{n}{x}$ and will be recognised as the set of coefficients in the binomial expansion of $(1+t)^n$, i.e.

$$
\begin{aligned}
(1+t)^n &= 1 + nt + \frac{n(n-1)}{2!}t^2 + \ldots + \frac{n!}{(n-x)!x!}t^x + \ldots + t^n \\
&= \sum_{x=0}^{n} \binom{n}{x} t^x \qquad (10.1)
\end{aligned}
$$

Let us now proceed to distribute the N elements according to $w_0, w_1, w_2, w_3, \ldots w_r$. First of all, we select w_0 elements from N which we can do in $\binom{N}{w_0}$ ways. This leaves $(N - w_0)$ from which we can select w_1 in $\binom{N-w_0}{w_1}$ ways. We then select w_2 from the remaining $(n - w_0 - w_1)$ in $\binom{N-w_0-w_1}{w_2}$ ways and so on until all the elements are used up. The number of possible ways of selecting the distribution $w_0, w_1, w_2, w_3, \ldots w_r$ is therefore the product of all these ways which is

$$p_i(w_0, w_1, w_2 \ldots w_r) = \binom{N}{w_0}\binom{N-w_0}{w_1}\binom{N-w_0-w_1}{\;}\ldots \times \binom{N-w_0-w_1-\ldots w_{r-1}}{w_r}$$

$$= \frac{N!}{w_0!\,w_1!\,w_2!\ldots w_r!} \tag{10.2}$$

Notice that we have used the fact that $N = \Sigma_{i=0}^{r} w_i$. From our previous discussion, $p_i(w_0, w_1 \ldots)$ is just the number of ways of selecting $w_0, w_1, w_2 \ldots$ identical objects from N.

If each state is equally likely (the principle of equal *a priori* probabilities) then the probability of finding a particular state is

$$W_i = \frac{p_i(w_0, w_1, w_2, \ldots w_r)}{\sum_i p_i(w_0, w_1, w_2, \ldots w_r)} \tag{10.3}$$

According to Boltzmann, the equilibrium state is that which has the greatest value of W_i. This is plainly equivalent to maximising the entropy if we introduce the definition

$$S = C \ln W_i$$

i.e. the state of maximum W_i also corresponds to the state of maximum entropy (see Section 7.5). Therefore, taking logarithms of expression (10.2),

$$\ln W_i = \ln N! - \sum_j \ln w_j! \tag{10.4}$$

Now using Stirling's formula,

$$n! \approx (2\pi n)^{\frac{1}{2}} \left(\frac{n}{e}\right)^n; \qquad \ln n! \approx n \ln n - n \tag{10.5}$$

and substituting into (10.4)

$$\ln W_i = N \ln N - N - \sum_j w_j \ln w_j + \sum_j w_j + 0(\ln N)$$

$$= N \ln N - \sum_j w_j \ln w_j \qquad (10.6)$$

Notice that in this analysis we have used the facts that $N = \Sigma_j w_j$ and that $\ln N$ is negligible in comparison with $N \ln N$ because N is very large, typically $\sim 10^{23}$.

To find the state of maximum probability, we have to find the maximum of $\ln W_i$ subject to the constraints that

$$\left.\begin{aligned} &\text{number of oscillators,} \quad N = \sum_j w_j = \text{constant} \\ &\text{total energy of oscillators,} \; E = \sum_j \epsilon_j w_j = \text{constant} \end{aligned}\right\} \qquad (10.7)$$

This is a classic example of the use of the technique of undetermined multipliers, i.e. we find the turning value of the function

$$-\sum_j w_j \ln w_j - A \sum_j w_j - B \sum_j \epsilon_j w_j \qquad (10.8)$$

where A and B are constants to be found from the boundary conditions. Maximising $S(w_j)$, we find that

$$\delta(S(w_j)) = -\delta \sum_j w_j \ln w_j - A\delta \sum_j w_j - B\delta \sum_j \epsilon_j w_j$$

$$= -\sum_j [(\ln w_j \delta w_j + \delta w_j) + A\delta w_j + B\epsilon_j \delta w_j] = 0$$

$$= -\sum_j \delta w_j [\ln w_j + \alpha + \beta\epsilon_j] = 0 \qquad (10.9)$$

This must be true for all j and therefore

$$w_j = e^{-\alpha-\beta\epsilon_j} \qquad (10.10)$$

This is the primitive form of Boltzmann's distribution. The α is a constant in front of the exponential, $w_j \propto e^{-\beta\epsilon_j}$. There is nothing in this analysis to tell us what the constant β should be. We have to appeal to phenomena such as Maxwell's energy distribution to find it (see Section 7.3). From that example, it can be seen that $\beta \propto T^{-1}$ and, in fact, in modern notation,

$$w_j = A_j e^{-\epsilon_j/kT} \qquad (10.11)$$

where k is Boltzmann's constant. Notice that, in this analysis, Boltzmann has to let the energy elements ϵ tend to zero so that he ends up with a continuous distribution.

10.3 Planck's analysis

Planck's analysis begins like a classical piece of statistical mechanics. He makes the assumption that any particular arrangement of the values of w_r is admissible. We have a fixed total energy E_N to be divided among the oscillators and again we introduce the energy elements ϵ. Therefore, we know exactly how many elements we have to spread over the oscillators. We will call this number P and hence $E_N = P\epsilon$. Now what Planck did was simply to work out the total number of ways in which one can distribute the P elements over the N oscillators. Let us work this out using the permutation theory which we outlined in the last section.

We can represent the problem by the diagrams shown in Figures 10.1(*a*) and (*b*). Planck had two fixed quantities – the total number of energy elements P and the number of boxes into which he wished to put them N. In the Figure 10.1(*a*) we show one way in which 20 elements can be put into 10 boxes. Now you will notice that the whole problem is really just to determine the total number of possible ways you can rearrange the elements and the walls between the end stops. Another example is shown in Figure 10.1(*b*). You can see that we are asking in how many possible ways we can redistribute P elements and $N-1$ walls remembering that the P elements are identical and the $N-1$ walls are identical. We deduced the answer in the last section:

$$\frac{(N+P-1)!}{P!(N-1)!} \tag{10.12}$$

This elegant argument was first given by Ehrenfest in 1914. In our present notation, expression (10.12) represents the total number of ways of distributing an energy E_N over N oscillators where E_N is to be composed of P energy elements of energy ϵ. Now the next step is the difficult one in the argument. Planck *defines* this is to be the probability which should be used in the relation

$$S = C \ln W$$

This is what results. N and P are very large and so we can use Stirling's approximation.

$$n! \sim (2\pi n)^{\frac{1}{2}} \left(\frac{n}{\mathrm{e}}\right)^n \left(1 + \frac{1}{12n} + \ldots\right) \tag{10.13}$$

Figure 10.1

× | × × × | | × × × | × × | × × × × × | × | × × × × | | × (*a*)

× × × × | × × | | × | × × × | × × | × | × × × | × × × | × (*b*)

To a good approximation $n! \approx n^n$. Therefore,

$$W = \frac{(N+P-1)!}{P!(N-1)!} \approx \frac{(N+P)!}{P!N!}$$
$$\approx \frac{(N+P)^{N+P}}{P^P N^N} \tag{10.14}$$

We can safely make these approximations since we are about to take logarithms of very large numbers.

$$S_N = C[(N+P)\ln(N+P) - P\ln P - N\ln N]$$
$$P = \frac{E_N}{\epsilon} = \frac{NE}{\epsilon} \tag{10.15}$$

where E is the average energy of the oscillators. Therefore,

$$S_N = C\left[N\left(1+\frac{E}{\epsilon}\right)\ln N\left(1+\frac{E}{\epsilon}\right) - \frac{NE}{\epsilon}\ln\frac{NE}{\epsilon} - N\ln N\right] \tag{10.16}$$

Therefore,

$$S = \frac{S_N}{N} = C\left[\left(1+\frac{E}{\epsilon}\right)\ln\left(1+\frac{E}{\epsilon}\right) - \frac{E}{\epsilon}\ln\frac{E}{\epsilon}\right] \tag{10.17}$$

But this looks familiar. This is exactly the expression which Planck had introduced to account for the spectrum of black body radiation. From expression (9.35), we find that

$$S = a\left[\left(1+\frac{E}{b}\right)\ln\left(1+\frac{E}{b}\right) - \frac{E}{b}\ln\frac{E}{b}\right]$$

with the requirement that $b \propto \nu$.

Thus the energy elements ϵ must be proportional to frequency and Planck wrote this result in the form which has persisted to this day

$$\epsilon = h\nu \tag{10.18}$$

This is where the concept of quanta came from. According to classical statistical mechanics, we ought to let $\epsilon \to 0$, but evidently we cannot obtain agreement with the expression for the entropy of the oscillator unless the energy elements do not disappear and have finite magnitude $\epsilon = h\nu$. In addition, we have now determined the value of a in terms of the universal constant C. So we can now write the complete expression for the Planck distribution:

$$u(\nu) = \frac{8\pi h\nu^3}{c^3}\frac{1}{e^{h\nu/CT}-1} \tag{10.19}$$

Finally, what about C? Planck pointed out that C is a universal constant relating the entropy to the probability that the system be in that state and Boltzmann had implicitly worked out what the constant had to be for a perfect gas. Since

C is a universal constant, any single law such as the perfect gas law which determines its value defines C for all processes. It is, in fact, the ratio $k = R/N_0$ where R is the gas constant and N_0 is Avogadro's number, the number of molecules per kilogram molecule $k = 1.38 \times 10^{-23}$ J K^{-1}, i.e. $C = k$ (see Section 7.5). Therefore, let us write down the Planck distribution once and for all in its final form:

$$u(\nu) = \frac{8\pi h\nu^3}{c^3}\frac{1}{e^{h\nu/kT} - 1} \qquad (10.20)$$

What is one to make of this argument? There are two fundamental criticisms:

(1) Planck does not follow Boltzmann's procedure for finding the equilibrium distribution. Indeed, what he defines as a probability is not really a probability of anything drawn from any parent population. Planck had no illusions about this. In his own words: 'In my opinion, this stipulation basically amounts to a definition of the probability W; for we have absolutely no point of departure, in the assumptions which underlie the electromagnetic theory of radiation, for talking about such a probability with a definite meaning'.[3] Einstein repeatedly pointed out the weak point in Planck's argument – 'The manner in which Mr Planck uses Boltzmann's equation is rather strange to me in that a probability of a state W is introduced without a physical definition of this quantity. If one accepts this, then Boltzmann's equation simply has no physical meaning'.[4]

(2) The second main problem concerns a logical inconsistency in Planck's analysis. On the one hand, the oscillators can only take energies $E = r\epsilon$ and yet a classical result has been used to work out the rate of radiation of the oscillator (Section 9.2.1). Implicit in that analysis is the assumption that the energies of the oscillators can vary continuously rather than take on only discrete values.

These are major stumbling blocks in the theory and it is fair to say that nobody understood quite what Planck had done and the theory did not in any sense gain immediate acceptance. However, we should note the following key points. First of all, whether one likes it or not, the concept of quanta has been introduced by the energy elements without which it is not possible to reproduce the Planck function. It was some time before Planck fully appreciated the significance of his work. In 1906, Einstein showed that if Planck had followed strictly Boltzmann's procedure he would have obtained the same answer and still maintained the essential concept of energy quanta. We will repeat Einstein's analysis in the next chapter.

The second point concerns the contrast between the quantum and classical parts of the derivation of the Planck spectrum. Einstein was not nearly as worried as other physicists were about using classical formulae in the description of quantum phenomena. He regarded equations such as Maxwell's equations for the electromagnetic field as being statements about only the average values of the quantities measured. In the same way, the expression relating $u(\nu)$ and E may be another of these relations which has significance independent of electromagnetic theory, although it can be derived by that means. More important, while the relation might not be precisely true on the microscopic scale, it may still be a good representation of the *average behaviour* of the system which is what one normally measures in laboratory experiments. This was a very advanced point of view for the time and no-one else was thinking along these lines. We will see in a moment, however, that it led to Einstein's most important contributions to quantum theory.

Why was it that Planck took his derivation so seriously? Part of the reason may have been that he saw that the two fundamental constants in the theory, k and h, might have significance far beyond the radiation spectrum. It was he who showed that the proper constant in front of the Boltzmann relation $S = C \ln W$ was in fact the ratio of the gas constant to Avogadro's number – which we have called k. Thus, k could be derived directly from the radiation spectrum and, in combination with a knowledge of R, it gave by far the best estimate of N_0 known at that time. He also noted that k and h, in conjunction with the constant of gravitation and the velocity of light, enabled a set of 'natural' units to be defined in terms of fundamental constants. In addition, the electrical charge carried by a gram equivalent of monovalent ions was known from electrolytic theory – this is called Faraday's constant. Knowing N_0 precisely, one can derive the elementary unit of charge. Again Planck's value was by far the best available at that time.

As for h, he failed to find any physical significance for it beyond its appearance in the radiation formula. Planck spent many years trying to reconcile his theory with classical physics. In his words:

> My futile attempts to fit the elementary quantum of action somehow into the classical theory continued for a number of years and they cost me a great deal of effort. Many of my colleagues saw in this something bordering on tragedy. But I feel differently about it. For the thorough enlightenment I thus received was all the more valuable. I now knew for a fact that the elementary quantum of action played a far more significant part in physics than I had originally been inclined to suspect and this recognition made me see clearly the need for the introduction of totally new methods of analysis and reasoning in the treatment of atomic problems.[5]

Indeed it was not until after 1911 that Planck fully appreciated the absolutely fundamental nature of quantisation which has no counterpart in classical physics. His original view was that the introduction of energy elements was 'a purely formal assumption and I really did not give it much thought except that no matter what the cost, I must bring about a positive result'. In fact, this quotation is from a letter by Planck to R.W. Wood written in 1931, 30 years after the events we are describing in this chapter. I find it a rather moving letter and it is worthwhile reproducing it in full.

> October 7 1931
>
> My dear colleague,
>
> You recently expressed the wish, after our fine dinner in Trinity Hall, that I should describe from a psychological viewpoint the considerations which had led me to propose the hypothesis of energy quanta. I shall attempt herewith to respond to your wish.
>
> Briefly summarised, what I did can be described as simply an act of desperation. By nature I am peacefully inclined and reject all doubtful adventures. But by then I had been wrestling unsuccessfully for six years (since 1894) with the problem of equilibrium between radiation and matter and I knew that this problem was of fundamental importance to physics; I also knew the formula that expresses the energy distribution in the normal spectrum. A theoretical interpretation therefore *had* to be found at any cost, no matter how high. It was clear to me that classical physics could offer no solution to this problem and would have meant that all energy would eventually transfer from matter into radiation. In order to prevent this, a new constant is required to assure that energy does not disintegrate. But the only way to recognise how this can be done is to start from a definite point of view. This approach was opened to me by maintaining the two laws of thermodynamics. The two laws, it seems to me, must be upheld under all circumstances. For the rest, I was ready to sacrifice every one of my previous convictions about physical laws. Boltzmann had explained how thermodynamic equilibrium is established by means of a statistical equilibrium, and if such an approach is applied to the equilibrium between matter and radiation, one finds that the continuous loss of energy into radiation can be prevented by assuming that energy is forced at the outset to remain together in certain quanta. This was purely a formal assumption and I really did not give it much thought except that no matter what the cost, I must bring about a positive result.
>
> I hope that this discussion is a satisfactory response to your inquiry. In addition I am sending you as printed matter the English version of my Nobel lecture on the same topic. I cherish the memory of my pleasant days in Cambridge and the fellowship with our colleagues.

With kind regards
Very truly yours
M. Planck.[6]

It is intriguing to note that he was prepared to give up the whole of physics, *except the two laws of thermodynamics*, in order to understand the radiation spectrum.

One cannot but feel a touch of sadness about Planck's achievements. They are very great and one can sense the titanic intellectual struggle involved in arriving at this completely new level of understanding. Yet, they are in a sense incomplete. The obvious struggle involved contrasts with the elegance of the next giant steps taken by Einstein. Before we move on to this new phase of development, let us conclude the story of Planck's statistical mechanics.

10.4 Why Planck got the right answer

Why was it that Planck found the correct result for the radiation spectrum despite the fact that he had used essentially nonsense statistics? There are two answers, one methodological, the other physical. The first answer is that it seems highly likely that Planck, in fact, worked backwards. It has been suggested by Rosenfeld and by Klein that he started with his answer for the entropy of an oscillator (9.35) and worked backwards to find W from $\exp(S/k)$. This results in the permutation formulae (10.12) which he then regards as a definition. The second answer is that he had stumbled upon the correct method of counting indistinguishable particles which have the same properties as particles of light – what we now call *photons*. This was first shown by the Indian physicist Bose in a manuscript which he sent to Einstein in 1924. Einstein immediately appreciated its significance and arranged for it to be translated and published in German.

Let us show how the correct result is derived. We have to alter the model slightly from the one used by Planck. Consider a particular state k with degeneracy g_k, i.e. up to g_k particles can occupy the same state which has energy ϵ_k. Now suppose that we have n_k particles which we wish to distribute over these g_k states and that they are indistinguishable. Then Planck's result tells us the number of different distinguishable ways the n_k particles can be distributed over these states – i.e.

$$\frac{(n_k + g_k - 1)!}{n_k!\,(g_k - 1)!} \tag{10.21}$$

Notice how this approach differs from Boltzmann's procedure. Boltzmann requires us to evaluate all possible ways of making up a given distribution and this determines W. However, if the particles are identical and indistinguishable, a given distribution has only unit weight, according to the Bose–Einstein prescription.

The result (10.21) refers only to a single energy state. Suppose we now have a large number of states. Then the total number of possible states of all the particles is the product of all these probabilities (or total number of possible arrangements), i.e.

$$P = \prod_k \frac{(n_k + g_k - 1)!}{n_k!\,(g_k - 1)!} \tag{10.22}$$

Now we have not specified anything about how the numbers n_k are distributed among the k states. Therefore, we can ask 'What is the arrangement of n_k over the states which results in the maximum value of P?' This is now following exactly the recommended Boltzmann procedure. We maximise P subject to the constraints $\Sigma n_k = N$; $\Sigma n_k \epsilon_k = E$. We know how to do this. As before,

$$\ln P = \ln \prod_k \frac{(n_k + g_k - 1)!}{n_k!(g_k - 1)!} \approx \sum \ln \frac{(n_k + g_k)^{n_k + g_k}}{n_k^{n_k}(g_k)^{g_k}} \tag{10.23}$$

Now we apply the method of undetermined multipliers to the problem.

$$\delta(\ln P) = 0 = \sum_k \mathrm{d}n_k\{[\ln(g_k + n_k) - \ln n_k] - \alpha - \beta\epsilon_k\}$$

$$\ln[(g_k + n_k)/n_k] = \alpha + \beta\epsilon_k$$

$$n_k = \frac{g_k}{\mathrm{e}^{\alpha + \beta\epsilon_k} - 1} \tag{10.24}$$

This is known as the *Bose–Einstein distribution* and is the correct statistics for indistinguishable particles or *bosons.* Bosons turn out in quantum mechanics to be particles with even spin. For example photons are spin 1 particles, gravitons spin 2 particles, etc.

Now in the case of black body radiation, we do not need to specify the number of photons present. We can see this from the fact that the whole distribution is determined solely by one parameter – the total energy or the temperature of the system. Therefore, when we use the method of undetermined multipliers we can drop the restriction on the total number of particles. It will readjust itself to the total amount of energy present, i.e. $\alpha = 0$. Therefore,

$$n_k = \frac{g_k}{\mathrm{e}^{\beta\epsilon_k} - 1} \tag{10.25}$$

By inspecting the low frequency behaviour of the Planck spectrum, we can show that $\beta = 1/kT$ as in the classical case. Finally, the degeneracy of the state k has already been worked out in our discussion of Rayleigh's approach to the origin of the black body spectrum (Section 9.5).

$$\begin{aligned} \mathrm{d}N &= \frac{8\pi\nu^2}{c^3}\,\mathrm{d}\nu \\ \epsilon_k &= h\nu \end{aligned} \tag{10.26}$$

Therefore,

$$u(\nu)\,\mathrm{d}\nu = \frac{8\pi h\nu^3}{c^3}\,\frac{1}{\mathrm{e}^{h\nu/kT}-1}\,\mathrm{d}\nu \tag{10.27}$$

This is Planck's expression for black body radiation and it has been derived correctly using Bose–Einstein statistics for indistinguishable particles. The statistics are applicable not only to photons but also to integral spin particles of all types.

We have now run far ahead of our story. None of this was known in 1900 and it was in the following years that Einstein made his revolutionary contributions to modern physics.

11

EINSTEIN AND THE QUANTISATION OF LIGHT

11.1 Einstein in 1905

By 1905, Planck's work had made little impression and he was no further forward in understanding the profound implications of what he had done. As we have discussed, he spent a great deal of effort trying to find a classical interpretation for the 'quantum of action' which is what he correctly believed the fundamental constant h referred to. The next great steps were due to Einstein and I do not think that it is an overstatement to claim that it was Einstein who first appreciated the full significance of quanta and showed that they must be a fundamental part of all physical phenomena rather than just a 'formal device' for explaining the Planck distribution. From 1905 onwards, he never deviated from his belief in quanta – it was a long time before any of the great figures of the time conceded that Einstein was correct. He came to this conclusion in a series of brilliant papers of dazzling scientific virtuosity.

Einstein completed what we would now call his undergraduate studies in August 1900. Between 1902 and 1904, he wrote three papers on the foundations of Boltzmann's statistical mechanics. Once again, you will notice how a very solid grounding in thermodynamics proved to be a crucial starting point in the investigation of fundamental problems of theoretical physics. You can appreciate why this should be so because thermodynamics does not deal with specific physical processes which we might not understand very well. It deals with the overall properties of physical systems and gives us broad rules about their expected behaviour. If you have a completely new problem, the thermodynamic approach will often give important clues about how to attack the detailed physics.

In 1905, Einstein was 26 and was employed as 'technical expert third class' at the Swiss patent office in Bern. In that year, he published three papers which are among the greatest classics of all physics. Any one of them would have ensured that his name remained a permanent fixture in scientific history. These papers are

(1) 'On the theory of Brownian motion'[1]
(2) 'On the electrodynamics of moving bodies'[2]
(3) 'On a heuristic viewpoint concerning the production and transformation of light'.[3]

The first paper is the one in which he explained the phenomenon of *Brownian movement*, the random motions of very fine specks of dust in fluids, in terms of the effects of collisions between molecules and the particles. Although each impact is very small, the net result of a large number of them randomly hitting the particle is to give it a 'drunken man's walk'. Einstein quantified this problem by relating the diffusion of the particles to the properties of the molecules responsible for the collisions. In other words, he was able to relate the molecular theory of gases to the observed diffusion of particles. Einstein's predictions were found to be precisely correct by Perrin. Since the agitational motion of the particles is just heat, they must obey the laws governing the chaotic movements of particles. That is, the macroscopic particles were behaving exactly as the molecules must be on a very much smaller scale. This work convinced everyone, including the sceptics, of the reality of the molecular or atomic nature of matter. Notice that, in his analysis, Einstein needed a value for Avogadro's number and he took this from Planck's paper. In other words, there is no question but that he was already familiar with Planck's work on the radiation spectrum.

The second paper is the famous paper on *special relativity* and we will study it in Chapter 13.

The third is the paper which is often referred to as Einstein's paper on the photoelectric effect. This is a gross misrepresentation of a paper of the greatest profundity. In Einstein's own words, the paper is 'very revolutionary'. Well, with all respect, I do not believe that the photoelectric effect has that status. What Einstein is referring to is the theoretical content of the paper. It really is revolutionary. Let us go through this great paper in detail.

11.2 'On a heuristic viewpoint concerning the production and transformation of light'

Let us reproduce the opening paragraphs. I find it revolutionary and startling. It demands attention like the opening of a great symphony.

> There is a profound formal difference between the theoretical ideas which physicists have formed concerning gases and other ponderable bodies and Maxwell's theory of electromagnetic processes in so-called empty space. Thus, while we consider the state of a body to be completely defined by the positions and velocities of a very large but finite number of atoms and electrons, we use continuous three-dimensional

> functions to determine the electromagnetic state existing within some region, so that a finite number of dimensions is not sufficient to determine the electromagnetic state of the region completely . . . .
>
> The undulatory theory of light, which operates with continuous three-dimensional functions, applies extremely well to the explanation of purely optical phenomena and will probably never be replaced by any other theory. However, it should be kept in mind that optical observations refer to values averaged over time and not to instantaneous values. Despite the complete experimental verification of the theory of diffraction, reflection, refraction, dispersion and so on, it is conceivable that a theory of light operating with continuous three-dimensional functions will lead to conflicts with experience if it is applied to the phenomena of light generation and conversion.[4]

In other words, there may well be circumstances when the classical theory of Maxwell cannot explain all electromagnetic phenomena and among these are problems such as the black body spectrum, the photoelectric effect and fluorescence. His proposal is that, for some purposes, it may be more appropriate to consider light as consisting of particles – or light quanta. In his own words '[the proposal] might prove useful to some investigators in their researches'.

Now this is pretty inflammatory stuff. The full implications of Maxwell's theory were still being worked out and Einstein was proposing to replace all of this by light particles. To all physicists, it must have looked like a re-run of the controversy between the wave picture of Huygens and the particle or corpuscular picture of Newton and everyone knew which had won. It must have seemed a particularly inappropriate time to reopen this question when Maxwell's discovery of the electromagnetic nature of light had been so completely vindicated only 15 years previously by Hertz.

Now notice carefully what Einstein is proposing in comparison with Planck's approach. Planck had found that he needed 'energy elements' $\epsilon = h\nu$ which did not vanish, but these are associated with the *oscillators.* Planck had absolutely nothing to say about the radiation emitted by the oscillators and indeed he firmly believed that the waves emitted by them were simply the classical electromagnetic waves of Maxwell. Einstein is proposing that *we should quantise the radiation field as well.*

Like other Einstein papers, the article is beautifully written and very clear. It all looks so very simple and obvious that you forget just how revolutionary its contents are. We will use the same notation which we have used up till now rather than Einstein's notation.

After the introduction, Einstein derives the formula which we have now seen many times relating the average energy of an oscillator to the energy density of black body radiation in thermodynamic equilibrium. However, he writes the classical result in the provocative form

$$u(\nu) = \frac{8\pi\nu^2}{c^3} kT$$

$$\text{total energy} = \int_0^\infty u(\nu)\mathrm{d}\nu = \frac{8\pi kT}{c^3}\int_0^\infty \nu^2 \mathrm{d}\nu = \infty \tag{11.1}$$

This is exactly the same problem which had been pointed out indirectly by Rayleigh in 1900. The phenomenon was later called the 'ultraviolet catastrophe' by Ehrenfest because of the divergence of the formula at short wavelengths.

The next thing Einstein does is to show that the fundamental constant k which Planck derived from his theory of the black body spectrum is, in fact, independent of the details of Planck's theory. Namely, the above arguments which lead to the Rayleigh–Jeans law (11.1) are known to be good for low frequencies and high temperatures and hence, if the temperature of the black body and its spectrum in that region are known, the value of k can be found directly. Einstein showed that this value agreed precisely with the value found by Planck.

Now we come to the heart of the paper. Einstein considers only the observed facts of black body radiation. You will remember that we have emphasised the central role which the entropy plays in the thermodynamics of radiation. Einstein sets out to derive a suitable form for the entropy of the radiation using only thermodynamics and the observed form of the radiation spectrum.

We remember that entropies are additive and since in thermal equilibrium we may consider the radiation of different wavelengths to be independent we can write the entropy of the radiation enclosed in volume V as

$$S = V\int_0^\infty \phi(u(\nu), \nu)\mathrm{d}\nu \tag{11.2}$$

The function ϕ is the entropy of the radiation per unit frequency range per unit volume. Now the aim of the calculation is to find the function ϕ in terms of the spectral energy density $u(\nu)$ and frequency. Clearly, there are no other quantities besides temperature T which can serve to describe the equilibrium spectrum. This problem had already been solved by Wien but Einstein gave an elegant proof of the result as follows.

We know that the function ϕ is such that in thermal equilibrium, the entropy is a maximum for a fixed value of the total energy, i.e. we can write this as a problem in the calculus of variations:

$$\delta S = \delta\int_0^\infty \phi(u(\nu), \nu)\mathrm{d}\nu = 0 \tag{11.3}$$

with the constraint

$$\delta E = \delta \int_0^\infty u(\nu)\,\mathrm{d}\nu = 0$$

where E is the total energy. Using undetermined multipliers, this gives us

$$\int_0^\infty \left(\frac{\partial\phi}{\partial u}\,\mathrm{d}u\,\mathrm{d}\nu - \alpha\,\mathrm{d}u\,\mathrm{d}\nu \right) = 0$$

where the constant α is independent of frequency. The integrand must be zero to ensure that the integral is zero. Therefore,

$$\frac{\partial\phi}{\partial u} = \alpha$$

Now suppose we change the temperature of unit volume of the black body radiation by $\mathrm{d}T$. Then the increase in entropy is

$$\mathrm{d}S = \int_{\nu=0}^{\nu=\infty} \frac{\partial\phi}{\partial u}\,\mathrm{d}u\,\mathrm{d}\nu$$

But $\partial\phi/\partial u$ is independent of frequency and therefore

$$\mathrm{d}S = \frac{\partial\phi}{\partial u}\,\mathrm{d}E \tag{11.4}$$

since

$$\mathrm{d}E = \int_{\nu=0}^{\nu=\infty} \mathrm{d}u\,\mathrm{d}\nu$$

But $\mathrm{d}E$ is just the energy added and we also know that

$$\frac{\mathrm{d}S}{\mathrm{d}E} = \frac{1}{T} \tag{11.5}$$

Therefore,

$$\frac{\partial\phi}{\partial u} = \frac{1}{T} \tag{11.6}$$

This is the equation we have been looking for. You will notice the pleasant symmetry between the relations

$$\left.\begin{aligned} S &= \int_0^\infty \phi\,\mathrm{d}\nu \qquad & E &= \int_0^\infty u(\nu)\,\mathrm{d}\nu \\ \frac{\mathrm{d}S}{\mathrm{d}E} &= \frac{1}{T} \qquad & \frac{\partial\phi}{\partial u} &= \frac{1}{T} \end{aligned}\right\} \tag{11.7}$$

Einstein goes on to use this relation for the spectrum of black body radiation. Now Einstein does not use the Planck formula. He uses the Wien formula for the following reason. He recognises that it is not exact but it is the correct law in the region where the classical theory breaks down and therefore is likely to give most enlightenment into the way in which the theory is inadequate.

First, he writes down the spectrum derived from experiment, which in our notation is

$$u(\nu) = \frac{8\pi\alpha}{c^3}\frac{\nu^3}{e^{\beta\nu/T}} \tag{11.8}$$

From this, we can immediately find an expression for $1/T$.

$$\frac{1}{T} = \frac{1}{\beta\nu}\ln\frac{8\pi\alpha\nu^3}{c^3 u(\nu)} = \frac{\partial\phi}{\partial u} \tag{11.9}$$

Then, by integration, we obtain an expression for ϕ.

$$\begin{aligned}\frac{\partial\phi}{\partial u} &= -\frac{1}{\beta\nu}\left(\ln u + \ln\frac{c^3}{8\pi\alpha\nu^3}\right)\\ \phi &= -\frac{u}{\beta\nu}\left(\ln u - 1 + \ln\frac{c^3}{8\pi\alpha\nu^3}\right)\\ &= -\frac{u}{\beta\nu}\left(\ln\frac{uc^3}{8\pi\alpha\nu^3} - 1\right)\end{aligned} \tag{11.10}$$

Now we come to the very clever step. Let us consider the total radiation in the spectral range ν to $\nu + \Delta\nu$ and let it have energy $\epsilon = Vu\Delta\nu$, where the volume is V. Then the entropy of this amount of radiation is

$$S = V\phi\Delta\nu = -\frac{\epsilon}{\beta\nu}\left(\ln\frac{\epsilon c^3}{8\pi\alpha\nu^3 V\Delta\nu} - 1\right) \tag{11.11}$$

Now suppose we change the volume from, say, V_0 to V, keeping the total energy constant. Then the entropy change is

$$S - S_0 = \frac{\epsilon}{\beta\nu}\ln(V/V_0) \tag{11.12}$$

Now Einstein shows that this entropy change is just exactly the same as one would obtain if one were dealing with an ideal gas composed of particles. Let us repeat the analysis we gave in Section 7.5. Boltzmann's relation can be used to work out the difference of entropy $S - S_0$ between the states: $S - S_0 = k\ln W/W_0$. In this initial state, the system has volume V_0 and the particles move randomly throughout this volume. We then ask what is the probability that all the particles occupy a smaller volume V, i.e. what is the random probability that all the particles end up in volume V by chance. The probability for one particle is V/V_0 and hence the probability that all N end up in the volume V is $(V/V_0)^N$. Therefore, for a gas we can write

$$S - S_0 = kN\ln(V/V_0) \tag{11.13}$$

Now Einstein notes that expressions (11.12) and (11.13) are identical in form. Therefore, the expression for the entropy change of the radiation must equally correspond to the probability that all the radiation is found in a sub-volume V and hence

$$S - S_0 = kN' \ln (V/V_0)$$

where the exponent N' is just equal to $\epsilon/k\beta\nu$. From this, Einstein concludes that 'Monochromatic radiation of low density (within the limits of validity of Wien's radiation formula) behaves thermodynamically as though it consisted of a number of independent energy quanta of magnitude $k\beta\nu$'. Rewriting this in Planck's notation, this means that, since $\beta = h/k$, $\epsilon = h\nu$.

Finally, Einstein works out the average energy of these quanta in terms of the energy that we know is present in the Wien distribution. The energy in the frequency interval ν to $\nu + d\nu$ is ϵ and hence the number of quanta is $\epsilon/k\beta\nu$. Therefore, the average energy is

$$\bar{E} = \frac{\int_0^\infty (8\pi\alpha/c^3)\nu^3 e^{-\beta\nu/T} d\nu}{\int_0^\infty (8\pi\alpha/c^3)(\nu^3/k\beta\nu) e^{-\beta\nu/T} d\nu} = k\beta \frac{\int_0^\infty \nu^3 e^{-\beta\nu/T} d\nu}{\int_0^\infty \nu^2 e^{-\beta\nu/T} d\nu}$$

Integrating the denominator by parts,

$$\int_0^\infty \nu^2 e^{-\beta\nu/T} d\nu = \left[\frac{\nu^3}{3} e^{-\beta\nu/T}\right]_0^\infty + \frac{\beta}{3T}\int_0^\infty \nu^3 e^{-\beta\nu/T} d\nu$$

$$\bar{E} = k\beta \times \frac{3T}{\beta} = 3kT \tag{11.14}$$

Thus, the average energy of the quanta is closely related to the mean kinetic energy per particle in the black body enclosure which is $\frac{3}{2}kT$. This is a highly suggestive result.

So far, Einstein had only said that the radiation 'behaved as though' it consisted of a number of independent particles. Is this meant to be taken seriously or is it just an analogy? The last sentence of section 6 of his paper reads 'the next obvious step is to investigate whether the laws of emission and transformation of light are also of such a nature that they can be interpreted or explained by considering light to consist of such energy quanta'. In other words, 'Yes, let us assume they *are* real particles and see whether or not we can understand other phenomena'.

He then considers three phenomena which cannot be explained by classical electromagnetic theory: (i) Stokes' rule of photoluminescence, (ii) the photoelectric effect and (iii) ionisation of gases by ultraviolet light.

Stokes' rule is the statement that the frequency of photoluminescent emission is less than the frequency of the incident light. This is explained as a consequence of the conservation of energy. The incoming quanta have energy $h\nu_1$ and hence the emitted quantum can have this energy at most. If some of the energy of the quantum is absorbed by the material, the emitted quantum will have less energy. We now know how to interpret this result properly in terms of atomic physics: $h\nu_1 \geqslant h\nu_2$.

The photoelectric effect. This is probably the most famous result of the paper because Einstein makes a definite quantitative prediction on the basis of his theory. The photoelectric effect had been discovered, ironically, by Hertz in the same experiments which demonstrated the validity of Maxwell's equations. Perhaps the most remarkable fact about the process was Lénard's discovery that the energies of the electrons emitted by the metal surface are independent of the intensity of the incident radiation.

Einstein's proposal gave an immediate answer to this problem. Radiation of a given frequency consists of quanta with the same energy. If one of these is absorbed by an atom, the electron receives sufficient energy to remove it from the surface of the material against the forces which bind the electron to the material. If the intensity of the light is increased, more electrons are ejected, but their energies are unchanged. Einstein wrote this result in the following form. The maximum kinetic energy which the ejected electron can have, E_k, is

$$E_k = h\nu - W$$

where W is the amount of work necessary to remove the electron from the surface of the material – this is what we would now call the *work function* of the the material. Experiments to demonstrate this involve putting the apparatus in an opposing potential so that when the potential reaches a particular value V, the electrons can no longer reach the collecting electrode, i.e. the photoelectric current is cut off. This corresponds to the potential at which $E_k = eV$. Therefore, we can write for this experimental arrangement

$$V = \frac{h}{e}\nu - \frac{W}{e} \tag{11.15}$$

In Einstein's words, 'If the formula derived is correct, then V must be a straight line function of the frequency of the incident light, when plotted in Cartesian coordinates, whose slope is independent of the nature of the substance investigated.' i.e. one can find directly the quantity h/e, the ratio of Planck's constant to the electronic charge. These were amazing predictions because nothing at all was known at that time about the dependence of the photoelectric effect upon frequency. In fact, after ten years of very difficult experimentation, all aspects of Einstein's equation were fully confirmed experimentally. In 1916, Millikan was able to summarise the results of his very extensive experiments: 'Einstein's photoelectric equation has been subjected to very searching tests and it appears in every case to predict exactly the observed results'.[5]

Photoionisation of gases. The third piece of evidence in Einstein's paper was that the energy of each photon has to be greater than the ionisation potential of the gas if one is to obtain ionisation. He showed that the smallest energy quanta for the ionisation of air were approximately equal to the ionisation potential determined independently by Stark. Once again, the quantum hypothesis is in agreement with experiment.

At this point, the paper ends. It is one of the great papers in physics and indeed it was the work described in Einstein's Nobel prize citation. I hope you will look at the translation of the original article.[3]

Now it is important to recognise that at this point Planck and Einstein had very different views about what was going on. Planck had quantised the *oscillators* and we find absolutely no mention of this in Einstein's paper. Indeed, it seems that Einstein was not clear at this stage that they were actually talking about the same thing. However, in 1906, he published his second fundamental paper on quanta[6] in which he showed that the two approaches were, in fact, the same. Then, in a paper submitted to *Annalen der Physik* in November 1906 and published in 1907[7], he went on to extend the idea of quantisation to solids.

11.3 The quantum theory of solids

In the first of these papers, Einstein asserts that he and Planck are actually describing the same phenomena of quantisation.

> At that time [when Einstein wrote the 1905 paper] it seemed to me as though Planck's theory of radiation formed a contrast to my work in a certain respect. New considerations which are given in the first section of this paper demonstrate to me, however, that the theoretical foundation on which Planck's radiation theory rests differs from the foundation that would result from Maxwell's theory and the electron theory and indeed differs exactly in that Planck's theory implicitly makes use of the hypothesis of light quanta just mentioned.[6]

These arguments are developed further in the paper of 1907.[7] What Einstein demonstrated is that, if Planck had followed the Boltzmann procedure, he would have obtained the correct formula whilst maintaining the assumption that the oscillator can only take definite energies, $0, \epsilon, 2\epsilon, 3\epsilon, \ldots$. Let us repeat Einstein's argument.

If Planck had worked through the classical analysis of Boltzmann, he would have ended up with the Boltzmann expression for the probability that a state of energy $E = r\epsilon$ is occupied, even though he does not take the limit $\epsilon \to 0$ (see Section 10.2).

$$p(E) \propto \mathrm{e}^{-E/kT}$$

Again, we assume that the energy of the oscillator comes in units of ϵ. Thus, if there are N_0 in the ground state, the number in the $r = 1$ state is $N_0\,\mathrm{e}^{-\epsilon/kT}$, in the $r = 2$ state $N_0\,\mathrm{e}^{-2\epsilon/kT}$ and so on. Therefore, the average energy of the oscillator is given by the usual expression for an average,

$$\begin{aligned}\bar{E} &= \frac{N_0 \cdot 0 + \epsilon N_0\,\mathrm{e}^{-\epsilon/kT} + 2\epsilon\,\mathrm{e}^{-2\epsilon/kT} + \ldots}{N_0 + N_0\,\mathrm{e}^{-\epsilon/kT} + N_0\,\mathrm{e}^{-2\epsilon/kT} + \ldots} \\ &= \frac{N_0\epsilon\,\mathrm{e}^{-\epsilon/kT}[1 + 2(\mathrm{e}^{-\epsilon/kT}) + 3(\mathrm{e}^{-\epsilon/kT})^2 + \ldots]}{N_0[1 + \mathrm{e}^{-\epsilon/kT} + (\mathrm{e}^{-\epsilon/kT})^2 + \ldots]} \end{aligned} \quad (11.16)$$

We recall the following series.

$$\begin{aligned}\frac{1}{(1-x)} &= 1 + x + x^2 + x^3 + \ldots \\ \frac{1}{(1-x)^2} &= 1 + 2x + 3x^2 + \ldots\end{aligned} \quad (11.17)$$

and hence the mean energy of the oscillator is

$$\bar{E} = \frac{\epsilon\,\mathrm{e}^{-\epsilon/kT}}{1 - \mathrm{e}^{-\epsilon/kT}} = \frac{\epsilon}{\mathrm{e}^{\epsilon/kT} - 1} \quad (11.18)$$

Thus, using the proper Boltzmann procedure, one can recover Planck's relation for the mean energy of the oscillator, this only being possible if the size of the energy elements ϵ does not vanish. Einstein's approach has the advantage of indicating clearly where the departure from the classical results occurs. The result of classical statistical mechanics

$$\bar{E} = kT$$

is based upon the assumption that equal volumes of phase space should be given equal weights in the averaging process. This results in the equipartition theorem. Einstein shows that Planck's formula requires that the above assumption is wrong and only those regions of phase space corresponding to energies 0, ϵ, 2ϵ, 3ϵ, . . . should have non-zero weights and that these should all be equal.

Einstein then relates this directly to his own previous paper which concerned only light quanta: 'we must assume that for ions which can vibrate at a definite frequency and which make possible the exchange of energy between radiation and matter, the manifold of possible states must be narrower than it is for the bodies in our direct experience. We must in fact assume that the mechanism of energy transfer is such that the energy can assume only the values 0, ϵ, 2ϵ, 3ϵ . . .'.[8]

But this is only the beginning of the paper. Much more is to follow – Einstein puts it beautifully.

> I now believe that we should not be satisfied with the result. For the following question forces itself upon us. If the elementary oscillators that are used in the theory of the energy exchange between radiation and matter cannot be interpreted in the sense of the present kinetic molecular theory, must we not also modify the theory for the other oscillators that are used in the molecular theory of heat? There is no doubt about the answer, in my opinion. If Planck's theory of radiation strikes to the heart of the matter, then we must also expect to find contradictions between the present kinetic molecular theory and experiment in other areas of the theory of heat, contradictions that can be resolved by the route just traced. In my opinion, this is actually the case, as I try to show in what follows.[9]

This is the paper which is often described as the application of quantum theory to solids but you can see that, like the 'photoelectric paper', these works are very deep and strike right to the fundamentals of the quantum nature of matter and radiation.

The problem discussed by Einstein is that of the specific heat of solids. According to the *Dulong and Petit law*, the heat capacity per mole of a solid is about $3R$ and this law can be derived simply from the equipartition theorem. We build a model of the solid which consists of N_0 atoms per mole and suppose that they all vibrate with three independent directions of vibration. According to the equipartition theorem, the total internal energy of the solid should therefore be $3N_0kT$ since each independent mode of vibration is awarded kT of energy. The heat capacity follows directly by differentiation: $C = \partial U/\partial T = 3N_0k = 3R$.

Now it was known that some materials do not obey the Dulong and Petit law in that they have much smaller heat capacities than $3R$. This was particularly true for the lightest elements such as beryllium, boron and carbon. In addition, by 1900, it was known that the specific heats of some elements vary rapidly with temperature and only attain the Dulong and Petit value at high temperatures.

The problem is readily solved if we adopt Einstein's point of view, which is that, for oscillators, we should not use the classical formula for the average energy of the oscillator kT, but the quantum formula

$$\bar{E} = \frac{h\nu}{e^{h\nu/kT} - 1}$$

Now we know that atoms are complicated things. However, let us suppose for simplicity that all the atomic vibrations are at the same frequency and that they are independent. The vibration frequency is known as the Einstein frequency ω_E (we will use ν_E). Then, giving each atom three degrees of freedom, we find that the internal energy is

$$U = 3N_0 \frac{h\nu_E}{e^{h\nu_E/kT} - 1} \tag{11.19}$$

and the specific heat is

$$\begin{aligned}\frac{dU}{dT} &= 3N_0 h\nu_E (e^{h\nu_E/kT} - 1)^{-2} e^{h\nu_E/kT} \frac{h\nu_E}{kT^2} \\ &= 3R \left(\frac{h\nu_E}{kT}\right)^2 \frac{e^{h\nu_E/kT}}{(e^{h\nu_E/kT} - 1)^2}\end{aligned} \tag{11.20}$$

This turns out to be a remarkably good fit to the actual variation of the specific heat with temperature. Schematically, this variation is shown in Figure (11.1).

We can now explain why it is that the light elements have smaller specific heats than the heavier elements. They presumably have higher vibrational frequencies and hence, at a given temperature, ν_E/T is larger and the specific heat is smaller. In fact, to account for the experimental data, the frequency ν_E must lie in the infrared waveband $\sim 5 \times 10^{12}$ Hz for copper. Thus, we see that all vibrations at higher frequencies can make only a vanishingly small contribution to the specific heat. Indeed, as one would expect, there is strong infrared absorption corresponding to frequencies $\nu \approx \nu_E$.

Perhaps the most remarkable prediction of this theory is that all specific heats should decrease to zero at low temperatures as indicated on the diagram. This was a very important prediction from the point of view of furthering the acceptance to Einstein's ideas, because at about this time Nernst began a series of experiments to measure the specific heats of solids at low temperatures.

Figure 11.1. The variation of the specific heat of solids with temperature accord-to Einstein's quantum theory.

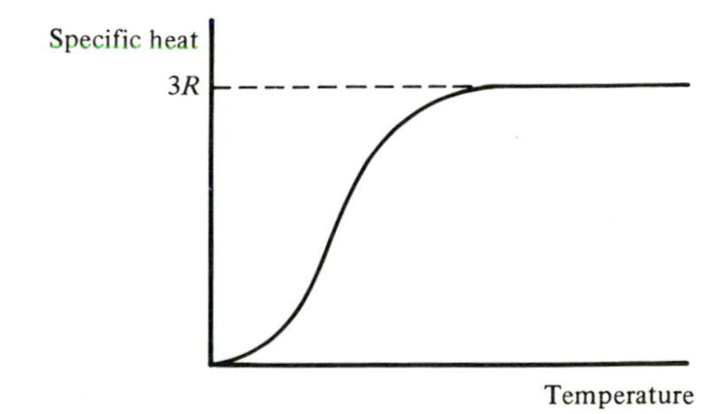

11.4 The specific heats of gases revisited

Einstein's quantum theory also completely resolves the problems of accounting for the specific heats of gases in terms of Maxwell's kinetic theory (see Section 7.3). According to Einstein, all the energies associated with atoms and molecules have to be quantised and then the energy states which can be excited depend upon the temperature of the gas. For example, Einstein has shown that the average energy of an oscillator of frequency ν at temperature T is

$$\bar{E} = \frac{h\nu}{e^{h\nu/kT} - 1}$$

Evidently, if $kT \gg h\nu$, the average kinetic energy is kT as expected from the kinetic theory. If however $kT \ll h\nu$, the average energy tends to $h\nu\, e^{-h\nu/kT}$ which becomes very small if $T \ll h\nu/k$.

Let us therefore look again at the various modes in which energy can be stored in the molecules of a gas. According to Maxwell's theory, the molecules have on average $\frac{3}{2}kT$ of energy associated with their *translational velocities.* Energy can also be stored in the rotational motion of the molecule as a whole. The energy which can be stored in rotational motion can be found from the fundamental result of quantum mechanics that angular momentum J is quantised such that only discrete values are found according to the relation

$$J = [j(j+1)]^{\frac{1}{2}}\hbar$$

where $j = 0, 1, 2, 3, 4, \ldots$ and $\hbar = h/2\pi$. We recall that if the particles have intrinsic angular momentum, or spin, j can take half-integral values. The rotational energy of the molecule is given by the analogue of the classical relation between energy and angular momentum

$$E = \tfrac{1}{2}J^2/I$$

where I is the moment of inertia of the molecule about the rotation axis. Therefore,

$$E = \frac{1}{2I}[j(j+1)]\hbar^2$$

If we write the rotational energy of the molecule $E = \frac{1}{2}I\omega^2$, we find the angular frequency of rotation ω:

$$\omega = [j(j+1)]^{\frac{1}{2}}\hbar/I$$

Therefore, the condition that rotational motion contributes to the storage of energy at temperature T is

$$kT \gtrsim E$$

which is equivalent to $kT \gtrsim \hbar\omega$.

Let us work out the value of E for molecular oxygen. The mass of each atom is $m = 16$ atomic mass units and the distance between the centres of the atoms is

$r_0 = 1.207$ Å. Therefore, the moment of inertia is $I = mr_0^2/2$ and the energy of rotation is

$$E = 1.794j(j+1) \times 10^{-4} \text{ eV}$$

Thus, for the lowest excited state, $j = 1$, the energy is 3.6×10^{-4} eV. This should be compared with kT for, say, room temperature $T = 300$ K. In this case, $kT = 2.6 \times 10^{-2}$ eV and hence $kT \gg E$ and the rotational modes of the molecule are excited. Now we need only ask how many modes are excited per molecule. Taking the three orthogonal axes as shown in Figure 11.2, it is evident that there are two orthogonal modes corresponding to the rotational motion of the molecule worked out above. The moment of inertia about the third axis, parallel to the axis of the molecule, is very small and corresponds to quantisation of the electronic orbitals within the molecule rather than the bulk rotational motion of the molecule as a whole. In fact, to excite this last mode of energy storage requires temperatures such that the molecule would be dissociated. There are therefore only two modes which can be excited and, awarding each of them $\frac{1}{2}kT$ of energy according to Maxwell's principle of equipartition, we find that

$$U = \underbrace{\tfrac{3}{2}kT}_{\text{translational}} + \underbrace{kT}_{\text{rotational}} = \tfrac{5}{2}kT$$

We should note that the above result applies to linear molecules. If the molecules are more complex with no linear axis, they can have significant moments of inertia about all three axes and then rotational energy can be stored in all three independent modes. In this case,

$$U = \underbrace{\tfrac{3}{2}kT}_{\text{translational}} + \underbrace{\tfrac{3}{2}kT}_{\text{rotational}} = 3kT$$

Figure 11.2. Illustrating the modes by which angular momentum can be stored in a linear diatomic molecule.

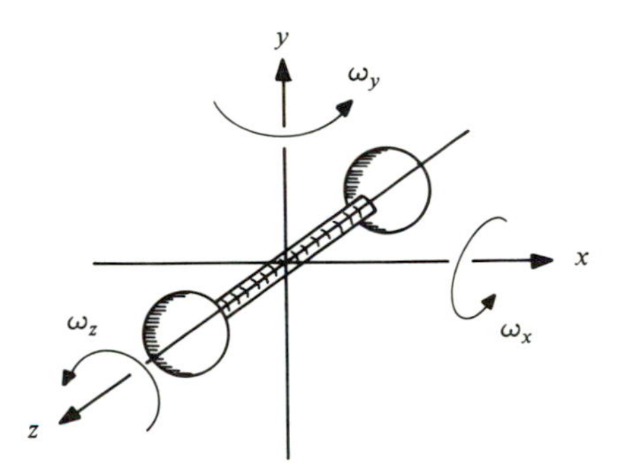

This is exactly the answer Maxwell found for the total internal energy of a gas consisting of rough spheres. It is the value appropriate for polyatomic non-linear molecules.

The next mechanism of energy storage is in the vibrational modes of the molecule. As above, these modes will be excited if $kT > E = h\nu_0$ where ν_0 is the vibrational frequency of the molecule. For oxygen, $E \approx 0.2$ eV but $kT \approx 2.6 \times 10^{-2}$ at $T = 300$ K, therefore $kT \ll E$ and these modes cannot be excited at room temperature. At high temperatures, however, the vibrational mode can be excited and it can store a further kT of energy, i.e.

$$U = \underbrace{\tfrac{3}{2}kT}_{\text{translational}} + \underbrace{kT}_{\text{rotational}} + \underbrace{kT}_{\text{vibrational}} = \tfrac{7}{2}kT$$

The specific heats at constant volume corresponding to these different cases are

Translation and rotation (linear molecule), $C_V = \frac{5}{2}R$

Translation, rotation and vibration, $C_V = \frac{7}{2}R$

The corresponding values for the ratio of specific heats $\gamma = (C_V + R)/C_V$ are

Translation and rotation, $\gamma = \frac{7}{5} = 1.4$

Translation, rotation and vibration, $\gamma = \frac{9}{7} = 1.286$ at high temperatures

The variation of the internal energy of a molecular gas as a function of temperature is shown schematically in Figure 11.3.

Thus, Einstein's quantum theory can account for the properties of real gases and explain the problems which had beset Maxwell and Boltzmann.

Figure 11.3. The variation of the internal energy of a diatomic molecule with temperature.

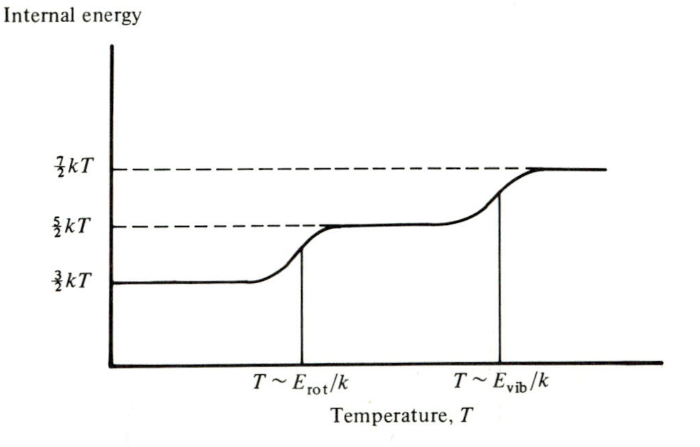

11.5 A word of caution

Einstein's revolutionary idea of quantisation predates the discovery of wave and quantum mechanics by Schrodinger and Heisenberg in the 1920s by about 20 years. Consequently, to today's physicist, Einstein's arguments appear schematic rather than formally precise in the sense of being derived from a complete quantum theory. Obviously, he did not have available Schrodinger's equation which is necessary for the proper description of quantum mechanical systems.

For example, Einstein's theory of specific heats is really a very primitive form of quantisation and the proper quantum theory, the Debye theory of solids, which is derived from Schrodinger's equation, gives much more accurate results. In the same way, we need a proper quantum theory of angular momentum and vibration in order to give a proper treatment of the kinetic theory of gases.

Despite these technical problems, Einstein's fundamental insights proved to be absolutely correct and opened up an entirely new realm of physics – all physical processes are basically quantum in nature, even though classical physics is successful in explaining so much.

12

THE STORY CONCLUDED – A FURTHER CLASSICAL PAPER BY EINSTEIN

12.1 The situation in 1909

In no sense were these startling new quantum ideas of Planck and Einstein immediately accepted by the scientific community at large. Indeed, most of the major figures in physics rejected the idea that light could be considered to be made up of discrete quanta. In a letter to Einstein in 1907, Planck wrote

> I look for the significance of the elementary quantum of action (light quantum) not in vacuo but rather at points of absorption and emission and assume that processes in vacuo are accurately described by Maxwell's equations. At least, I do not yet find a compelling reason for giving up this assumption which for the time being seems to be the simplest.[1]

Planck still rejected the light quantum hypothesis as late as 1913.

In 1909, we find Lorentz writing

> While I no longer doubt that the correct radiation formula can only be reached by way of Planck's hypothesis of energy elements, I consider it highly unlikely that these energy elements should be considered as light quanta which maintain their identity during propagation.[2]

Nevertheless, Einstein continued to work out other ways in which the observed facts of black body radiation lead inevitably to the conclusion that light consists of quanta. One of his most beautiful papers was written in 1909[3] and shows how fluctuations in the intensity of black body radiation provide further evidence on the quantum nature of light. I consider this to be a paper of the very highest quality of inspiration. In my experience, the subject of fluctuations is one which causes problems for students and so let us revise some elementary ideas on statistical fluctuations of particles and waves before looking at Einstein's paper.

12.2 Fluctuations of particles and waves

12.2.1 Particles in a box

Let us deal first of all with the problem of particles in a box. We can imagine dividing the box into a large number of cells N and then counting the numbers of particles in each cell. If the number of particles is very large, the mean number in each cell is roughly the same but there is a finite real scatter about the mean because of statistical fluctuations.

It is worthwhile recalling just how we arrive at the exact expression for these variations. In the simplest case, we begin with the example of tossing coins and ask what is the probability of obtaining x heads (or successes) in n throws. In each throw, the probability of success $p = \frac{1}{2}$ and that of failure $q = \frac{1}{2}$ so that $p + q = 1$. If we throw two coins, we know that the possible results of the experiment are

HH, HT, TH, TT

If we toss three coins, the possible results are

HHH, HHT, HTH, THH, HTT, THT, TTH, TTT

Since order is not important, these correspond to frequencies of

HH	H T	TT
1	2	1

or to probabilities of $\frac{1}{4}, \frac{1}{2}, \frac{1}{4}$ for two coins. Similarly, for three coins, we have

HHH	HH T	H TT	TTT
1	3	3	1

or probabilities $\frac{1}{8}, \frac{3}{8}, \frac{3}{8}, \frac{1}{8}$. It is well known that these probabilities correspond to the terms in t of the binomial expansions of $(p + qt)^2, (p + qt)^3, \ldots$.

There is an alternative way of looking at this problem. Suppose we ask, 'What is the probability of tossing one head and then two tails?' The answer is plainly $pq^2 = \frac{1}{8}$ for this particular ordering. However, if the order does not matter, we need to know how many ways we could have obtained one success and two failures. This is exactly the same problem we discussed in connection with the derivation of the Boltzmann distribution (Section 10.2). The number of different ways of selecting x and y identical objects from n is $n!/x!y!$. In the present case, $y = n - x$ and hence the answer is $n!/x!(n-x)!$. Thus, in this case the number of ways is $3!/1!2! = 3$, i.e. the total probability is $\frac{3}{8}$ in agreement with above.

It immediately follows that, in general, the probability of x successes out of n events is $(n!/x!(n-x)!)p^x q^{n-x}$. It is readily recognised that this is just the coefficient of the term in t^x in the expansion of

$$(q + pt)^n \tag{12.1}$$

If we write the probability of x successes out of n attempts $P_n(x)$ we have

$$P_n(x) = \frac{n!}{(n-x)!\,x!} p^x q^{n-x} \tag{12.2}$$

and

$$(q + pt)^n = P_n(0) + P_n(1)t + P_n(2)t^2 + \ldots + P_n(x)t^x + \ldots + P_n(n)t^n \tag{12.3}$$

Setting $t = 1$, we obtain

$$1 = P_n(0) + P_n(1) + P_n(2) + \ldots + P_n(x) + \ldots + P_n(n) \tag{12.4}$$

showing that the total probability is 1, as it ought to be.

Now let us take derivatives with respect to t. Differentiating (12.3) with respect to t,

$$pn(q + pt)^{n-1} = P_n(1) + 2P_n(2)t + \ldots + xP_n(x)t^{x-1} + \ldots + nP_n(n)t^{n-1} \tag{12.5}$$

Now setting $t = 1$,

$$pn = \sum_{x=0}^{n} xP_n(x) \tag{12.6}$$

But the quantity on the right-hand side is just the average value of x, i.e. the mean value of x is $\bar{x} = pn$ which makes sense completely.

Let us use the same procedure to find the variance of the distribution. Taking the next derivative with respect to t,

$$p^2 n(n-1)(q + pt)^{n-2} = 2P_n(2) + \ldots + x(x-1)P_n(x)t^{x-2} + \ldots + n(n-1)P_n(n)t^{n-2} \tag{12.7}$$

Again, setting $t = 1$, we find that

$$\begin{aligned} p^2 n(n-1) &= \sum_{x=0}^{n} x(x-1)P_n(x) \\ &= \sum_{x=0}^{n} x^2 P_n(x) - \sum_{x=0}^{n} xP_n(x) \\ &= \sum_{x=0}^{n} x^2 P_n(x) - np \end{aligned} \tag{12.8}$$

i.e.

$$\sum_{x=0}^{n} x^2 P_n(x) = np + p^2 n(n-1) \tag{12.9}$$

Now we notice that $\Sigma_{x=0}^{n} x^2 P_n(x)$ is a measure of the variance of the distribution of x but it is measured with respect to the origin rather than the mean. Fortunately, there is a rule which tells us how to measure the variance with respect to mean:

$$\sigma^2 = \sum_{x=0}^{n} x^2 P_n(x) - \bar{x}^2 \tag{12.10}$$

i.e.

$$\begin{aligned}\sigma^2 &= \sum_{x=0}^{n} x^2 P_n(x) - (pn)^2 \\ &= np + p^2 n(n-1) - (pn)^2 \\ &= np(1-p) \\ &= npq \end{aligned} \tag{12.11}$$

Finally, we can go over from a discrete to a continuous distribution. This procedure, which is performed in all the standard text books, results in the answer that the continuous probability distribution is the *normal distribution* $Y(X)\mathrm{d}X$ which can be written

$$Y(X)\mathrm{d}X = \frac{1}{(2\pi)^{\frac{1}{2}}\sigma} \exp\left(-\frac{X^2}{2\sigma^2}\right)\mathrm{d}X \tag{12.12}$$

σ^2 is the variance and again has the value npq; X is expressed with respect to the mean value np.

This is the answer we have been seeking. If we divide our box into N sub-boxes, the probability of a particle being in a single sub-box in one experiment is $p = 1/N$, $q = (1 - 1/N)$. The total number of particles is n. Therefore, the average number of particles per sub-box is n/N and the variance about this value, i.e. the statistical fluctuation about the mean is

$$\sigma^2 = \frac{n}{N}\left(1 - \frac{1}{N}\right) \tag{12.13}$$

For N large, $\sigma^2 = n/N$ = average number of particles in each sub-box, i.e. $\sigma = (n/N)^{\frac{1}{2}}$. Notice that for large values of N the mean is equal to the variance.

This is the origin of the approximate rule that the fractional fluctuation about the average value is $1/N^{\frac{1}{2}}$ where N is the number of discrete objects counted. This is how we would expect particles in a box to behave.

12.2.2 Random superposition of waves

The case of waves is rather different. Let us suppose that the electric field E at any point is the superposition of the electric field from N sources where N is very large. For simplicity, we assume that the amplitude of each wave is ξ. Then the quantity $E^*E = |E|^2$ is proportional to the energy density of the radiation where E^* is the complex conjugate of E. Expanding in terms of the random phases ϕ_i of the waves,

$$E^*E = \xi^2\left(\sum_k e^{i\phi_k}\right)^*\left(\sum_j e^{i\phi_j}\right) = \xi^2\left(\sum_k e^{-i\phi_k}\right)\left(\sum_j e^{i\phi_j}\right) \tag{12.14}$$

$$= \xi^2\left(N + \sum_{jk}{}' e^{i(\phi_j - \phi_k)}\right)$$

$$= \xi^2\left[N + 2\sum_{j>k} \cos(\phi_j - \phi_k)\right] \tag{12.15}$$

The dash means omitting the terms for which $j = k$. The average over the term in $\cos(\phi_j - \phi_k)$ is zero since the phases are random and therefore

$$\langle E^*E\rangle = N\xi^2 \tag{12.16}$$

This is a result with which we are familiar. For incoherent radiation, i.e. waves with random phases, we add together the energy in each wave to find the total energy density.

Let us now look at the fluctuations in the average energy density of the waves. We have to work out the quantity $\langle (E^*E)^2\rangle$ with respect to the mean value. As above (relation (12.10)), we recall that $\langle \Delta n^2\rangle = \langle n^2\rangle - \langle \bar{n}\rangle^2$ and hence

$$\langle \Delta E^2\rangle = \langle (E^*E)^2\rangle - \langle E^*E\rangle^2 \tag{12.17}$$

Now

$$(E^*E)^2 = \xi^4\left(N + \sum_{jk}{}' e^{i(\phi_j - \phi_k)}\right)^2$$

$$= \xi^4\left(N^2 + 2N\sum_{jk}{}' e^{i(\phi_j - \phi_k)} + \sum_{lm}{}' e^{i(\phi_l - \phi_m)}\sum_{jk}{}' e^{i(\phi_j - \phi_k)}\right) \tag{12.18}$$

Once again, the middle part of (12.18) $\Sigma'_{jk}\, e^{i(\phi_j - \phi_k)}$ averages to zero because the phases are random. In the last part, most of the terms average to zero because the phases are random but not all of them do. Those terms for which $l = k$ and $m = j$ do not vanish. Recalling that $l = m$ is excluded from the summation, the matrix of non-zero combinations of l and m is

$m \downarrow$ \ $l \rightarrow$				
–	2,1	3,1	4,1 . . .	N,1
1,2	–	3,2	4,2 . . .	N,2
1,3	2,3	–	4,3 . . .	N,3
1,4	2,4	3,4	– . . .	⋮
⋮	⋮	⋮		
1,N	2,N	3,N	4,N	–

There are clearly $N^2 - N$ terms and therefore we find that

$$(E^*E)^2 = \xi^4 [N^2 + N(N-1)]$$
$$\approx 2N^2\xi^4 \tag{12.19}$$

Therefore,

$$\langle \Delta E^2 \rangle = 2N^2\xi^4 - N^2\xi^4 = N^2\xi^4$$

i.e.

$$\langle \Delta E^2 \rangle = \langle E^*E \rangle^2 \tag{12.20}$$

i.e. the *fluctuations in the field are of the same magnitude as the energy density of the radiation itself.* This is a remarkable property of electromagnetic radiation and is the reason why phenomena like interference and diffraction take place when one is dealing with a superposition of waves of random phase. The physical meaning of this calculation is clear. The matrix of non-vanishing contributions to $(E^*E)^2$ is no more than the sum of all pairs of waves added separately. Every pair of waves of frequency ω interferes to produce fluctuations in intensity of the radiation $\Delta E \approx E$, i.e.

$$\xi^2 \sin(kx - \omega t) \sin(kx - \omega t + \phi) = \frac{\xi^2}{2}\{\cos\phi - \cos[2(kx - \omega t) + \phi]\}$$

Notice that this analysis applies for waves of random phase of a particular angular frequency ω, i.e. what we would refer to as waves corresponding to a particular mode. Let us now look at Einstein's analysis of fluctuations in black body radiation.

12.3 Fluctuations in black body radiation

Einstein began by reversing the Boltzmann relation between entropy and probability:

$$W = e^{S/k} \tag{12.21}$$

Now suppose we consider only the radiation in the interval ν to $\nu + d\nu$. As before, we write $\epsilon = Vu(\nu)d\nu$. Now we divide the volume into a large number of cells and suppose that $\Delta\epsilon_i$ is the fluctuation in the ith cell. Then the entropy of this cell is

$$S_i = S_i(0) + \left(\frac{\partial S}{\partial U}\right)\Delta\epsilon_i + \tfrac{1}{2}\left(\frac{\partial^2 S}{\partial U^2}\right)(\Delta\epsilon_i)^2 + \ldots \tag{12.22}$$

But, over all cells, we know that there is no net fluctuation, $\Sigma\Delta\epsilon_i = 0$, and therefore

$$S = \Sigma S_i = S(0) + \tfrac{1}{2}\left(\frac{\partial^2 S}{\partial U^2}\right)\Sigma(\Delta\epsilon_i)^2 \tag{12.23}$$

Therefore, the probability distribution for the fluctuations is

$$W = (\text{constant}) \exp\left[\tfrac{1}{2}\left(\frac{\partial^2 S}{\partial U^2}\right)\frac{\Sigma(\Delta\epsilon_i)^2}{k}\right] \tag{12.24}$$

We recognise that this is simply the sum of a set of normal distributions which can be written for any individual cell,

$$W_i = (\text{constant}) \exp\left[-\tfrac{1}{2}(\Delta\epsilon_i)^2/\sigma^2\right] \tag{12.25}$$

with

$$\sigma^2 = \frac{k}{(-\partial^2 S/\partial U^2)}$$

Notice that we now have a physical interpretation for the second derivative of the entropy with respect to energy which Planck had used in his original analysis. Let us now derive σ^2 for a black body spectrum:

$$u(\nu) = \frac{8\pi h\nu^3}{c^3}\frac{1}{e^{h\nu/kT}-1}$$

Inverting,

$$\frac{1}{T} = \frac{k}{h\nu}\ln\left(\frac{8\pi h\nu^3}{c^3 u}+1\right) \tag{12.26}$$

Now we express this result in terms of the total energy in the cavity (of volume V) in the frequency interval ν to $\nu + \mathrm{d}\nu$, $\epsilon = Vu\mathrm{d}\nu$. As before, $\mathrm{d}S/\mathrm{d}U = 1/T$ and we may identify ϵ with U and S is just the entropy of the radiation. Therefore,

$$\frac{\partial S}{\partial \epsilon} = \frac{k}{h\nu}\ln\left(\frac{8\pi h\nu^3}{c^3 u}+1\right) = \frac{k}{h\nu}\ln\left(\frac{8\pi h\nu^3 V\mathrm{d}\nu}{c^3 \epsilon}+1\right)$$

$$\frac{\partial^2 S}{\partial \epsilon^2} = -\frac{k}{h\nu}\frac{1}{\left(\dfrac{8\pi h\nu^3 V\mathrm{d}\nu}{c^3\epsilon}+1\right)}\frac{8\pi h\nu^3 V\mathrm{d}\nu}{c^3\epsilon^2}$$

$$\frac{k}{(\partial^2 S/\partial U^2)} = -\left(h\nu\epsilon + \frac{c^3}{8\pi\nu^2 V\mathrm{d}\nu}\epsilon^2\right) = -\sigma^2 \tag{12.27}$$

Or in terms of the fractional fluctuations

$$\frac{\sigma^2}{\epsilon^2} = \left(\frac{h\nu}{\epsilon}+\frac{c^3}{8\pi\nu^2 V\mathrm{d}\nu}\right) \tag{12.28}$$

Einstein noted that the two terms on the right-hand side have quite specific meanings. The first originates from the Wien part of the spectrum and, if we suppose the radiation consists of photons, each of energy $h\nu$, we see that it corresponds to the statement that the fractional fluctuation in the intensity is just $1/N^{\frac{1}{2}}$ where N is the number of photons, i.e.

$$\frac{\Delta N}{N} = \frac{1}{N^{\frac{1}{2}}} \tag{12.29}$$

As we have shown in Section 12.2.1, this is exactly the result we expect if light is considered to consist of discrete particles.

Let us now look more closely at the second term. It originates from the Rayleigh–Jeans part of the spectrum. We have to ask 'How many separate modes are there in the box in the frequency range ν to $\nu + d\nu$?' We have already shown in Section 9.5 that there are $8\pi\nu^2 V d\nu / c^3$ modes. We have shown in Section 12.2.2 that the fluctuations associated with each wave mode are $\Delta\epsilon^2 = \epsilon^2$. When we add together randomly all the independent modes in the range ν to $\nu + d\nu$, we add their variances and hence

$$\frac{\langle \Delta E^2 \rangle}{E^2} = \frac{1}{N_{\text{mode}}} = \frac{c^3}{8\pi\nu^2 V d\nu}$$

which is exactly the same as the second term on the right-hand side of the relation (12.28).

Thus, the two parts of the fluctuation spectrum correspond to particle and wave statistics, the former corresponding to the Wien part of the spectrum and the latter to the Rayleigh–Jeans part. The other amazing aspect of the above formula for the fluctuations is that we remember that we should add the variances due to independent causes and you will notice that the equation

$$\frac{\sigma^2}{\epsilon^2} = \left(\frac{h\nu}{\epsilon} + \frac{c^3}{8\pi\nu^2 V d\nu} \right)$$

says that we should add independently the 'wave' aspects and the 'particle' aspects of the radiation field to find the total value of the fluctuation. I regard this as one of the most miraculous pieces of theoretical physics.

12.4 The story concluded

Einstein had published these results in 1909 but still obtained little support for the idea of light quanta. However, among those who were persuaded of the potential significance of the idea of quanta was Nernst who was at that time measuring the low temperature heat capacities of various materials. By 1910, these experiments showed that indeed Einstein's predictions of the low temperature variation of the specific heat with temperature (equation (11.20)) gave a good description of the experimental results. By 1911, after some degree of hesitation, Nernst was convinced, not only of the validity of Einstein's results but also of the theory underlying it. This was very important because up till then Einstein's ideas had found very little support among his colleagues. Nernst was a friend of the wealthy Belgian industrialist Solvay and he persuaded Solvay to convene a meeting of a select group of physicists to discuss the questions of quanta and radiation. This was the first and, perhaps the most significant, of the famous series of Solvay conferences.

The eighteen official participants met on 29 October 1911 in the Hotel Metropole in Brussels and the meeting took place between the 30th of that month and the 3rd November (Figure 12.1). It is fair to say that, in the years before the meeting, there were very few physicists who took quantum ideas seriously. Indeed, Planck stated that, in 1910, only he, Einstein, Stark, Larmor and J.J. Thomson took the quantum hypothesis seriously. By 1911, things had moved on somewhat so that, on the whole, the participants were supporters of the quantum hypothesis. Here is how they lined up. Two of them were definitely against quanta – Jeans and Poincaré. Rayleigh was invited but did not attend – his views were essentially the same as those of Jeans. Five were initially neutral – Rutherford, Brillouin, Marie Curie, Perrin, Knudsen. The eleven others were basically pro-quanta – Lorentz (chairman), Nernst, Planck, Rubens, Sommerfeld, Wien, Warburg, Langevin, Einstein, Hasenohrl, Onnes. The secretaries were Goldschmidt, de Broglie and Lindemann; Solvay, who paid the bill, was there as well as his collaborators Herzen and Hostelet. Rayleigh and van der Waals were unable to come.

As a result of the meeting, Poincaré was converted to the notion of quanta and the physicists who took a neutral position did so because they were unfamiliar with the arguments. The conference had a profound effect in that it

Figure 12.1. The participants in the first Solvay conference on physics, Brussels 1911. (From *La Théorie du rayonemment et les quanta*, eds. P. Langevin & M. De Broglie, Gautier-Villars, Paris, 1912.)

provided a forum where all the arguments were presented. In addition, all the participants wrote up their lectures beforehand and these were then discussed in detail. These discussions were noted down and the whole proceedings published within a year of the event. Thus, all the important points were available in one volume. The net result was that the next generation of students was brought up to be at least familiar with the arguments. The students of many of the participants set to work immediately to tackle the problems of quanta. In addition, these problems began to be appreciated beyond the central European German-speaking scientific community, and in Great Britain, for example, where atomic studies were flourishing, the concepts began to be applied to atomic structure. In particular, the British phenomenological–empirical tradition was a fruitful ground for the development of these ideas.

However, it would be wrong to believe that suddenly everyone was convinced of the existence of quanta. This was certainly not the case. Let us read what Millikan had to say about his great series of experiments in which he verified the dependence of the photoelectric effect upon frequency, predicted by Einstein in 1905. In 1916, he stated 'We are confronted however by the astonishing situation that these facts were correctly and exactly predicted nine years ago by a form of quantum theory which has now been generally abandoned'.[4] Millikan refers to Einstein's 'bold, not to say reckless, hypothesis of an electromagnetic light corpuscle of energy $h\nu$ which flies in the face of the thoroughly established facts of interference'.[4]

We must remember that there were other explanations of those phenomena which were difficult to accommodate within classical physics and the quantum interpretation was only one of them. The really big problem was the one of reconciling the apparent wave and particle properties of light. This is supposed not to be a problem for the physicists of the 1980s but it certainly was a very major stumbling block for many physicists at that time. Although Einstein's arguments have for us an outstanding clarity and feeling of rightness, many physicists just could not follow them. The statistical mechanical arguments may seem very straightforward now, but they were understood by few physicists at the time. Indeed, outside Germany, Einstein's name was more or less unknown until the theory of relativity caught the public imagination after the first world war.

It was probably only after Compton's beautiful experiments, which showed that photons could undergo collisions in which they behave like particles (the Compton effect or Compton scattering), that the physics community as a whole was convinced. In June 1929, Heisenberg wrote in a review article, entitled 'The development of the quantum theory 1918–1928',

> At this time [1923] experiment came to the aid of theory with a discovery which would later become of great significance for the

development of quantum theory. Compton found that with the scattering of X-rays from free electrons, the wavelength of the scattered rays was measurably longer than that of the incident light. This effect, according to Compton and Debye, could easily be explained by Einstein's light quantum hypothesis; the wave theory of light, on the contrary, failed to explain this experiment. With that result, the problems of radiation theory which had hardly advanced since Einstein's works of 1906, 1909 and 1917 were opened up.[5]

12.5 Reflections

The story which has unfolded over the last five chapters has described one of the most exciting intellectual developments in theoretical physics and indeed in the whole of science. It led directly to the quantum mechanics of Schrodinger and Heisenberg in the 1920s, which has now become the bread and butter of all professional physicists. And yet it all happened so recently.

The contributions of many very great physicists were crucial in reaching this

Figure 12.2. Cartoon of Herblock. (From *Herblock's Here and Now*, Simon and Schuster, 1955.)

new plane of understanding. For me, however, it is Einstein's contribution which stands above all the others. In terms of insight and inspiration, it must rank among the greatest in the history of science. I have been asked by non-scientists whether or not Einstein deserves his reputation as the outstanding figure of modern science as he is portrayed in the popular press (Figure 12.2). The case study of the last five chapters and those to come on special and general relativity should leave the reader in no doubt that Einstein's contributions transcend those of virtually all other physicists, the only physicists who can seriously be mentioned in the same breath being Newton and probably Maxwell.

I am reminded of an occasion a few years ago when I had dinner in Moscow with Academician V.L. Ginzburg. He described to me how Landau, the greatest Russian theoretical physicist of this century, put physicists into leagues, certain physicists belonging to the first division, others to the second division and so on. The first division contained names such as Newton and Maxwell and apparently it was most amusing to learn how he rated other physicists. He rated his own contribution rather modestly. But there was a separate division, the zero division, in which there was only one physicist – that was Einstein. It is an assessment which I believe would be agreed upon by virtually all physicists.

APPENDIX TO CHAPTER 12

JOHNSON NOISE AND THE DETECTION OF FAINT SIGNALS IN THE PRESENCE OF NOISE

We can use the tools we have developed over the last two chapters to study the problem of the detection of faint signals in the presence of noise. In many cases, one is seeking to detect very faint signals indeed and the noise may arise from a number of different sources. For example, the signal itself is of finite magnitude and therefore only determined to a certain statistical precision; the signal may have to be detected in the presence of thermal noise in the detector and there may also be unwanted background radiation incident on the detector. One very useful result, before we consider the general problem, is the expression for the electrical noise in a resistor due to thermal fluctuations. This general result is called Nyquist's theorem.

A12.1 Nyquist's theorem and Johnson noise

Nyquist's theorem is a beautiful application of Einstein's expression for the mean energy per mode in the case of the spontaneous fluctuations of energy present in a resistor at temperature T. This is a very important result for the design of electronic amplifiers and electric circuits used in association with low noise receivers.

Let us derive the expression for the noise power generated in a transmission line which is terminated at either end by matched resistors R, i.e. the wave impedance of the transmission line Z_0 for the propagation of electromagnetic waves is equal to R (Figure A12.1). We suppose that the whole circuit is in thermodynamic equilibrium at temperature T. In this state, we know that energy is shared equally among all the modes present in the system according to Einstein's prescription that, per mode, the average energy is

$$\bar{E} = \frac{h\nu}{e^{h\nu/kT} - 1} \tag{A12.1}$$

where ν is the frequency of the mode (equation (11.18)). Therefore, in thermodynamic equilibrium, we need to know how many modes there are associated with the transmission line and award each of them energy $\bar{E}$ according to Einstein's version of the Maxwell–Boltzmann doctrine of equipartition.

In effect, we are performing the same calculation as Rayleigh carried out for the normal modes of oscillation of waves in a box but now we are dealing with a one dimensional, rather than a three dimensional, problem. The one dimensional case of a line of length L is easily derived from the relation (9.38).

$$\frac{\omega}{c} = \frac{n\pi}{L} \tag{A12.2}$$

where n takes only integral values, $n = 1, 2, 3 \ldots$, and c is the velocity of light. Standard analysis of the nature of these modes starting from Maxwell's equations shows that there is only one polarisation associated with each value of n. In thermodynamic equilibrium, each mode is awarded $\bar{E}$ of energy and hence, to work out the energy per unit frequency range, we only need to know the number of modes in the frequency range $d\nu$. The number is found by differentiating (A12.2):

$$dn = \frac{L}{\pi c}\,d\omega = \frac{2L}{c}\,d\nu \tag{A12.3}$$

and hence the energy per unit frequency range is $2L\bar{E}/c$. Now it is one of the basic properties of standing waves such as those represented by (A12.2) that they correspond precisely to the superposition of waves of equal amplitude propagating at velocity c in opposite directions along the line. These travelling waves correspond to the only two permissible solutions of Maxwell's equations for propagation along the transmission line. There is therefore a certain amount of power being delivered into the matched resistors R at either end of the line and since they are matched to the line, all the energy of the waves is absorbed in them. Equal amounts of energy $L\bar{E}/c$ propagate in either direction and travel along the wire into the resistors in a travel time $t = L/c$. Therefore, the power delivered to each resistor is

Figure A12.1. A transmission line of wave impedance Z_0 terminated at each end with matched resistors R in thermodynamic equilibrium with an enclosure at temperature T.

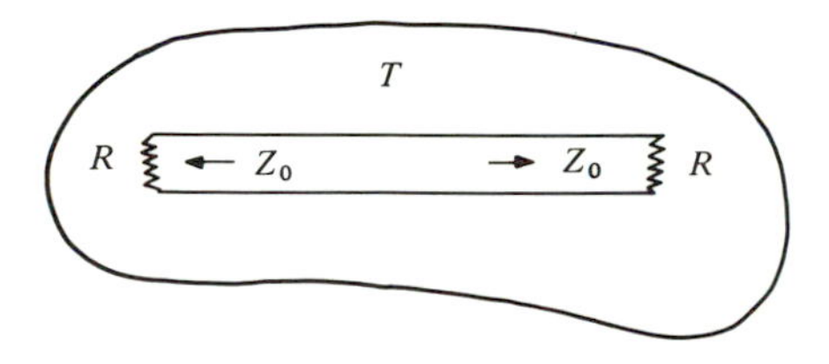

$$P = \frac{L\bar{E}/c}{L/c} = \bar{E}$$

in terms of watts per hertz. However, in thermodynamic equilibrium, the same amount of power must be returned to the transmission line or else the resistor will heat up above temperature T. This proves the fundamental result that the noise power delivered by a resistor at temperature T is

$$P = \frac{h\nu}{e^{h\nu/kT} - 1} \qquad \text{(A12.4)}$$

If we are considering low frequencies $h\nu \ll kT$ which is normally the case for radio and microwave receivers, the expression (A12.4) reduces to

$$P = kT \qquad \text{(A12.5)}$$

This is *Nyquist's theorem* as applied to the thermal noise of a resistor at low frequencies. The power available in the frequency range ν to $\nu + \mathrm{d}\nu$ is $P\mathrm{d}\nu = kT\mathrm{d}\nu$. In the opposite limit $h\nu \gg kT$, the noise power decreases exponentially as $P = h\nu \exp(-h\nu/kT)$.

The result (A12.5) was measured experimentally by J.B. Johnson in 1928. For a variety of resistors, he found exactly the relation predicted by expression (A12.5) and derived an estimate of k within 8% of the correct value.

For radio receivers, expression (A12.5) provides a convenient way of describing the performance of a receiver which delivers a certain noise power P_{n}. We can define an *equivalent noise temperature* T_{n} for the performance of the receiver at frequency ν by the relation

$$T_{\mathrm{n}} = P_{\mathrm{n}}/k \qquad \text{(A12.6)}$$

In the case of low frequencies, we note one other key feature of the result (A12.5). We see that, per unit frequency range per second, the electrical noise power is kT which corresponds exactly to the energy of a single energy mode in thermodynamic equilibrium. Since this energy is in the form of electrical signals, the fluctuations in this wave mode correspond to

$$\Delta E/E = 1 \qquad \text{(A12.7)}$$

as we showed in Section 12.2.2. Therefore, we expect the amplitude of the noise fluctuations per unit frequency interval to be kT. We will use this result in Section A12.3.

A12.2 The detection of photons in the presence of background noise

Let us consider first the case in which $h\nu \gg kT$. The results of Sections 11.2, 12.3 and A12.1 show that in this case we can consider light to consist of independent particles or photons. The statistical properties are then described by the relations developed in Section 12.2.1. If there is no background radiation, the precision with which the signal is measured is

$$\frac{\Delta I}{I} = \frac{1}{N^{\frac{1}{2}}}$$

where N is the number of photons detected from the source.

Often faint sources of radiation are observed in the presence of a much greater background signal. If the number of background photons is $N_b \gg N$, the uncertainty in the measurements is largely determined by the uncertainty with which the background can be determined:

$$\frac{\Delta I}{I} = \frac{1}{(N + N_b)^{\frac{1}{2}}} \approx \frac{1}{N_b^{\frac{1}{2}}}$$

In this case, to make a measure of the intensity of the source, we first measure the source plus background:

$$(N + N_b) \pm (N + N_b)^{\frac{1}{2}} \approx N + N_b \pm N_b^{\frac{1}{2}}$$

We then measure the background on its own:

$$N_b \pm N_b^{\frac{1}{2}}$$

We subtract these measures to produce an estimate of N but in the subtraction we have to add the variances of the error estimates, i.e. the best estimate is

$$N \pm (N + 2N_b)^{\frac{1}{2}} \approx N \pm (2N_b)^{\frac{1}{2}}$$

A12.3 The detection of electromagnetic waves in the presence of noise

In the case $h\nu \ll kT$, we have shown that black body radiation behaves like classical electromagnetic radiation. The signal is detected and a current or voltage induced in the detector which we can model by the resistor considered in Section A12.1. Suppose we detect the signal received within a waveband ν to $\nu + d\nu$. The resultant signal is the vector sum of all the electric fields of the waves as displayed on an Argand diagram (Figure A12.2). This whole diagram rotates at a rate of ν Hz. If all the waves were of exactly the same frequency, the pattern would continue to rotate in this configuration for a long time. However, because there is a spread in frequencies, the vectors change their phase relations so that after a time $T \sim 1/\Delta\nu$, the vector sum of the waves will be different. The time T is called the *coherence time* of the wave, i.e. it is roughly the time during which the vector sum of the waves produces the same result. After this time, we can make an independent estimate of the amplitude and so on after each period of $1/\Delta\nu$.

Thus, whereas when we are dealing with photons, we obtain an independent piece of information every time a photon arrives, in the case of waves we obtain an independent estimate only once per coherence time $1/\Delta\nu$. Thus, if we observe a source for time t, we obtain $t/T = t\Delta\nu$ independent estimates of the intensity of the source. These results tell us how often we need sample the signal.

Often one is interested in measuring very weak signals in the presence of much greater noise present in the receiver. The noise power itself fluctuates with amplitude $\Delta E/E = 1$ per mode per second according to our analysis of Section A12.1. We can therefore reduce the amplitude of the variations by integrating for long periods and by increasing the bandwidth of the receiver. In both cases, we increase the number of independent estimates of the strength of the signal by $\Delta\nu t$ and hence the amplitude of the power fluctuations after time t is reduced to

$$\Delta P = \frac{kT}{(\Delta\nu t)^{\frac{1}{2}}}$$

Thus, if we are prepared to integrate long enough and use large enough bandwidths, very weak sources can be observed.

Figure A12.2. Illustrating the random superposition of electromagnetic waves in the frequency range ν to $\nu + d\nu$ on an Argand diagram.

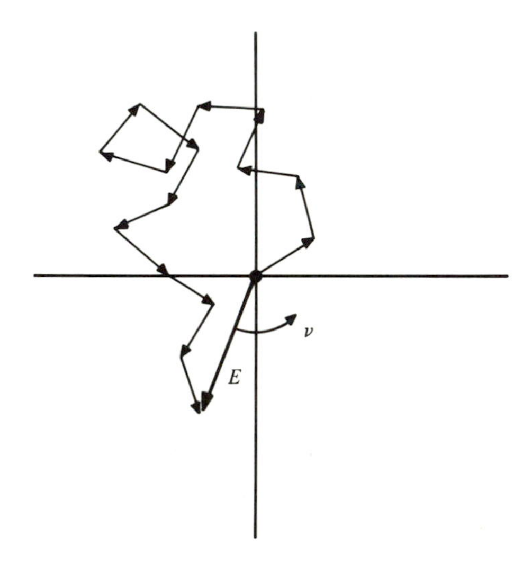

Case Study 6

SPECIAL RELATIVITY

Albert Einstein (1879–1955)
(From *Introduction to Concepts and Theories in Physical Science*, G. Holton & S.G. Brush, p. 439, Addison-Wesley, 1973.)

13

SPECIAL RELATIVITY – A STUDY IN INVARIANCE

13.1 Introduction

Relativity is a subject which is 'difficult' in the sense that the results are not at all intuitively obvious on the basis of our everyday experience. Great care is needed in expounding the basic principles. The number of books on the subject is immense and all add something to one's understanding. However, above all others, I very strongly recommend Einstein's original paper of 1905 'On the electrodynamics of moving bodies'.[1] It is probably as clear as any of the subsequent papers in expounding the basis of the theory. I regard it as one of the miracles of theoretical physics that Einstein should produce such a complete and elegant exposition with such profound implications for our physical understanding of the Universe in this single paper. There has been a degree of controversy over how much of the theory of special relativity was due to Einstein and how much to other theoretical physicists such as Lorentz and Poincaré. Whilst the importance of the transformations between reference frames was clearly indicated in the earlier work, there is no question in my mind but that Einstein's paper went far beyond the work of others in setting the whole subject on a proper formal basis and going on to make predictions which were not even conceived of in other studies.

I want to take a rather different approach from the historical exposition of the last case study on the discovery of quanta. Instead, I want to look at special relativity from the point of view of *invariance*. I do not want to go into many of the formalities but rather to concentrate on deriving the formulae of special relativity with the minimum number of assumptions and exposing clearly where we have to be guided by experiment. The approach will be similar to that of Rindler in his excellent text book *Essential Relativity*.[2] Personally, I find Rindler's exposition one of the most satisfying of all books on relativity.

13.2 Geometry and the Lorentz transformation

We begin as usual with two Cartesian frames of reference moving with velocity V relative to one another. We will call S the laboratory frame of reference or the local rest frame and S' the moving frame which has velocity V in the positive x direction of S. We can choose both frames to be rectangular with their axes parallel. We will refer to the frames of reference in this orientation as 'standard configuration' (Figure 13.1). According to convention, we will refer to frames of reference moving at constant velocity as inertial frames of reference. We will not go into the formalities of showing that it is indeed possible to set up such a standard configuration – this is done by Rindler – but we will assume that it can be proved.

Now the special theory of relativity as formulated by Einstein in his great paper of 1905 contains two basic postulates. Let us quote the words of Einstein himself.

> The same laws of electrodynamics and optics are valid for all frames of reference for which the equations of mechanics hold good. We will raise this conjecture (the purport of which will hereafter be called the 'Principle of Relativity') to the status of a postulate and also introduce another postulate which is only apparently irreconcilable with the former, namely that light is always propagated in empty space with a definite velocity c which is independent of the state of motion of the emitting body.[3]

Rephrasing Einstein's postulates, the first is that the laws of physics are the same in all inertial frames of reference and, second, that the velocity of light should have the same value c in all inertial frames. It is the second postulate which is crucial in leading to the theory of special relativity.

Let us go back to our inertial frames S and S' and suppose that their origins are coincident at $t = 0$ in S and $t' = 0$ in S'. It is always possible to arrange this

Figure 13.1. Inertial frames of reference S and S' in standard configuration.

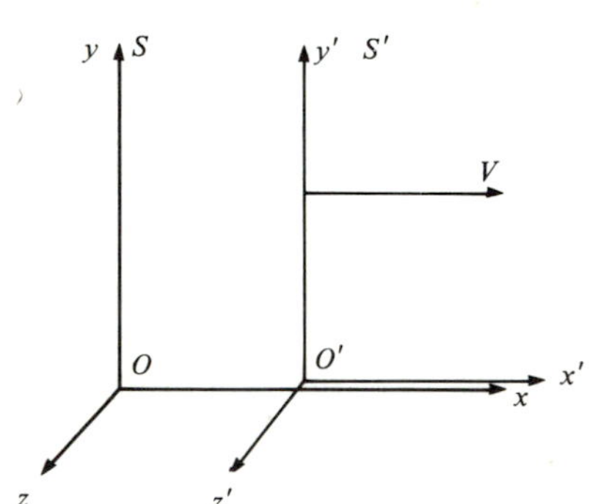

at any particular point in space by resetting clocks S and S' appropriately. Now let us send out an electromagnetic wave from the origin at $t = 0$, $t' = 0$. We can now immediately write down the motion of the wavefront in both frames of reference because of the second postulate of Einstein.

$$\left.\begin{aligned} \text{In } S, \quad x^2 + y^2 + z^2 - c^2 t^2 &= 0 \\ \text{In } S', \quad x'^2 + y'^2 + z'^2 - c^2 t'^2 &= 0 \end{aligned}\right\} \tag{13.1}$$

guaranteeing that in both frames the velocity of light is c. Now the vectors $[x, y, z, t]$ and $[x', y', z', t']$ are simply coordinates in S and S' and therefore what we need is a set of transformations which leaves the expression for the wavefront *form invariant*.

We immediately see that there is a formal analogy between equations (13.1) and the behaviour of the components of a three-vector under rotation. Thus, if we have two frames of reference with the same origin but rotate one with respect to the other, the *norm* (or the square of the magnitude) of the three-vector is the same in both frames, i.e.

$$R^2 = x^2 + y^2 + z^2 = x'^2 + y'^2 + z'^2$$

Let us look at the problem of transforming equation (13.1) in the same way. For simplicity, let us consider the wave propagating along the x axis so that the necessary transform we are seeking must result in $x^2 - c^2t^2 = 0, x'^2 - c^2t'^2 = 0$.

We first convert it into something which we know how to transform by writing $\tau = \mathrm{i}t$, $\tau' = \mathrm{i}t'$. Then

$$x^2 + c^2\tau^2 = x'^2 + c^2\tau'^2 \tag{13.2}$$

This is just the simple rotation formula for two-vectors which is

$$\left.\begin{aligned} x' &= x\cos\theta + c\tau\sin\theta \\ c\tau' &= -x\sin\theta + c\tau\cos\theta \end{aligned}\right\} \tag{13.3}$$

Now τ is imaginary and so the angle θ must be imaginary as well. We get round this by writing $\theta = \mathrm{i}\phi$ where ϕ is real. Because

$$\left.\begin{aligned} \cos(\mathrm{i}\phi) &= \cosh\phi \\ \sin(\mathrm{i}\phi) &= \mathrm{i}\sinh\phi \end{aligned}\right\}$$

we can write

$$\left.\begin{aligned} x' &= x\cosh\phi - ct\sinh\phi \\ ct' &= -x\sinh\phi + ct\cosh\phi \end{aligned}\right\} \tag{13.4}$$

We are thus left with only one constant to determine, ϕ. This we find by noting that, at time t in S, the origin of S' at $x' = 0$, is at $x = Vt$ in S. Substituting into the first relation of (13.4) we find that

$$\begin{aligned} 0 &= x \cosh\phi - ct \sinh\phi \\ \tanh\phi &= x/ct = V/c \end{aligned} \tag{13.5}$$

Then, because $\cosh\phi = (1 - V^2/c^2)^{-\frac{1}{2}}$ and $\sinh\phi = (V/c)/(1 - V^2/c^2)^{\frac{1}{2}}$, we find from (13.4) that

$$\left.\begin{aligned} x' &= \gamma(x - Vt) \\ t' &= \gamma(t - Vx/c^2) \end{aligned}\right\}$$

where $\gamma = (1 - V^2/c^2)^{-\frac{1}{2}}$. In this exposition we also find that $y' = y$, $z' = z$. The term γ is called the *Lorentz factor* and appears in virtually all calculations in special relativity. We have thus derived the complete set of Lorentz transformations by using simply the idea of *form invariance* and Einstein's second postulate which itself is a statement about the *invariance* of the velocity of light between inertial frames of reference. Let us write the Lorentz transformation again in the standard form we will employ throughout this exposition.

$$\left.\begin{aligned} x' &= \gamma(x - Vt) \\ y' &= y \\ z' &= z \\ t' &= \gamma(t - Vx/c^2) \end{aligned}\right\} \gamma = (1 - V^2/c^2)^{-\frac{1}{2}} \tag{13.6}$$

In the standard exposition of special relativity one now goes on to explore the remarkable consequences of this set of transformations and the so-called paradoxes which are supposed to arise. Let me state immediately that there are *no* paradoxes – only some rather non-intuitive features of space–time which result from the Lorentz transformations.

The key to understanding the origin of these phenomena is what Rindler calls the *relativity of simultaneity*. I find this a very helpful concept which elucidates the origin of some of the difficulties which students have in understanding relativity. We have already indicated that, at any particular point in space, we can reset the clocks so that they read the same time *at that point*. In the above example, we set $x' = 0, x = 0; t' = 0, t = 0$. In other words, at that point we can consider the event to be simultaneous in the two frames of reference. However, *at all other points in space*, events are not simultaneous.

From the last equation of (13.6), we see that at $x = 0, t = 0$ implies $t' = 0$, i.e. simultaneity. However, at all other points in space, observers disagree about the time at which events at all other values of x occur, i.e.

$$t' = \gamma(t - Vx/c^2) \tag{13.7}$$

and hence if $x \neq 0, t' \neq t$. In other words, if observers in S and S' can agree on simultaneity at one point in space–time, they will disagree at all other points. This is the origin of the phenomena of time dilation, length contraction, twin paradoxes, etc. It is the fundamental difference between Newtonian relativity and special relativity. In Newtonian relativity, observers in S and S' always agree everywhere about simultaneity. This is clear from the Newtonian limit of relation (13.7). If $V/c \to 0$, $\gamma \to 1$ and $t' = t$ everywhere.

13.3 Three-vectors and four-vectors

You will remember that *three-vectors* have very useful properties – for example, the vector relations remain true however we rotate or translate the axes. Here are some examples.

(i) Vector addition is preserved:

$$\mathbf{a} + \mathbf{b} = \mathbf{c} \equiv \mathbf{a}' + \mathbf{b}' = \mathbf{c}'$$

(ii) Equally,

$$\mathbf{a} \cdot (\mathbf{b} + \mathbf{c}) \equiv \mathbf{a}' \cdot (\mathbf{b}' + \mathbf{c}')$$

(iii) The *norm* or *magnitude* of the three-vector is invariant with respect to rotations and displacements:

$$a^2 = a_1^2 + a_2^2 + a_3^2 = a_1'^2 + a_2'^2 + a_3'^2$$

(iv) Scalar products remain invariant:

$$\mathbf{a} \cdot \mathbf{b} = a_1 b_1 + a_2 b_2 + a_3 b_3 = a_1' b_1' + a_2' b_2' + a_3' b_3'$$

In other words, these vector relations express truths which are independent of the frame of reference in which we perform our calculation.

Now it will plainly be of very great value in our development of relativity if we can find corresponding quantities which remain *form invariant* under *Lorentz transformations.* The aim is to find suitable forms for *four-vectors* which are objects similar to the quantities $[x, y, z, t]$ that can be related to quantities measurable in the laboratory. We will find that only in the simplest cases do the four-vectors resemble the physical quantities which they represent.

There remains the thorny question of what notation to use. I prefer to use the standard Lorentz transformations in the form of relations (13.6) and to define the components of my four-vectors as the quantities equivalent to x, y, z and t. I write the components of the four-vectors in square brackets $[x, y, z, t]$. Some authors use ct for the time component and others ict to get rid of the minus sign which is associated with the time component. I prefer to leave t as it is, t meaning real physical time. I have to remember that to work out the norm of any four-vector, I have to write

$$R^2 = c^2 t^2 - x^2 - y^2 - z^2 \tag{13.8}$$

i.e. multiply t^2 by c^2 and remember the minus sign in front of x^2, y^2 and z^2. This is exactly equivalent to $R^2 = x^2 + y^2 + z^2$ for three-vectors.

In this section we will be concerned only with *kinematic* quantities, i.e. quantities which *describe* motion rather than explain why it happens. We have already introduced our primitive four-vector $[x, y, z, t]$. Let us proceed to the next four-vector which is the *displacement four-vector.*

Displacement four-vector. The four-vector $[x, y, z, t]$ transforms according to the Lorentz transformation and so does the four-vector $[x + \Delta x, y + \Delta y, z + \Delta z, t + \Delta t]$. It is clear from the linearity of the Lorentz transformations that $[\Delta x, \Delta y, \Delta z, \Delta t]$ also transforms like $[x, y, z, t]$. We therefore define the quantity

$$\Delta \mathbf{R} = [\Delta x, \Delta y, \Delta z, \Delta t] \tag{13.9}$$

as a *displacement four-vector.* To express it another way, it must be a four-vector by virtue of being the difference of two four-vectors.

The *proper time* Δt_0 between two events is that time interval between events which takes place at the same spatial coordinates in some frame, i.e. $\Delta x = \Delta y = \Delta z = 0$. We can then easily show that it is the shortest time measured in any frame of reference. Taking the norms of the four-vectors in any two frames of reference,

$$c^2 \Delta t^2 - \Delta x^2 - \Delta y^2 - \Delta z^2 = c^2 \Delta t'^2 - \Delta x'^2 - \Delta y'^2 - \Delta z'^2$$

In the frame in which $\Delta x = \Delta y = \Delta z = 0$, $\Delta t = \Delta t_0$. Therefore,

$$c^2 \Delta t_0^2 = c^2 \Delta t'^2 - \Delta x'^2 - \Delta y'^2 - \Delta z'^2 = \text{constant} \tag{13.10}$$

since it is the norm of a four-vector. Therefore, since $\Delta x'^2$, $\Delta y'^2$, $\Delta z'^2$ are necessarily positive, Δt_0 must be the minimum time.

The other important aspect of Δt_0 is that, in general, it is the only invariant time interval upon which all observers can agree between events separated by $(\Delta x, \Delta y, \Delta z, \Delta t)$. Each observer in different inertial frames of reference can measure $(\Delta x, \Delta y, \Delta z, \Delta t)$ and although they will all measure different Δts for the time interval between the same two events, they will all agree about the value of Δt_0 when they measure Δx, Δy and Δz as well.

The relation (13.10) also provides the explanation for the observation of muons at sea-level. The muons are produced at the top of the atmosphere in collisions between cosmic rays and molecules of the atmosphere. Their half-lives are such that they should decay spontaneously when they have traversed only one-twentieth of the distance to the surface of the Earth. The solution of this apparent paradox is that those particles which we observe at sea-level have Lorentz factors $\gamma \geqslant 20$. Therefore, in their own rest frame, they decay in their own half-life which is measured in units of proper time in the frame of the moving particle. However, this is the shortest time interval ascribed by any observer. The external observer measures Δt and not Δt_0 and, if $\gamma = 20$, this time is 20 times longer than Δt_0.

The velocity four-vector. We want something which is Lorentz invariant and which looks like a velocity. The way to proceed is to look for something which will transform like $\Delta\mathbf{R}$. Therefore, let us begin with $\Delta\mathbf{R}$. The only time we can think of which is Lorentz invariant is the proper time Δt_0, i.e., as we discussed above, any observer no matter what frame of reference he is in can measure Δt, Δx, Δy, Δz for two events and compute for these $\Delta t_0^2 = \Delta t^2 - \Delta x^2 - \Delta y^2 - \Delta z^2$. He knows that any other observer in any other frame would get exactly the same value of Δt_0^2. Therefore, let us *define* a velocity four-vector by

$$\mathbf{U} = \Delta\mathbf{R}/\Delta t_0 \tag{13.11}$$

Let us now rewrite this quantity in such a form that it can be related to observations which we make in any frame of reference. Let us write down the expression for proper time for the interval associated with a particle moving a distance $\Delta\mathbf{r} = (\Delta x, \Delta y, \Delta z)$ in time Δt. Then the proper time is

$$\begin{aligned} c^2\Delta t_0^2 &= c^2\Delta t^2 - \Delta x^2 - \Delta y^2 - \Delta z^2 \\ &= c^2\Delta t^2 - \Delta r^2 \end{aligned} \tag{13.12}$$

But the velocity of the particle V is just $\Delta r/\Delta t$ and hence

$$\Delta t_0^2 = \Delta t^2 \left(1 - \frac{V^2}{c^2}\right)$$

i.e.

$$\Delta t_0 = \frac{\Delta t}{\gamma} \tag{13.13}$$

This is exactly the formula we use in understanding why muons are observable at sea-level. Now we can write $\mathbf{U}$ as

$$\begin{aligned} \mathbf{U} = \frac{\Delta\mathbf{R}}{\Delta t_0} &= \gamma\left[\frac{\Delta x}{\Delta t}, \frac{\Delta y}{\Delta t}, \frac{\Delta z}{\Delta t}, \frac{\Delta t}{\Delta t}\right] \\ &= [\gamma u_x, \gamma u_y, \gamma u_z, \gamma] \\ &= [\gamma\mathbf{u}, \gamma] \end{aligned} \tag{13.14}$$

In this relation $\mathbf{u}$ is the three-velocity of the particle in the frame S and γ the corresponding Lorentz factor; u_x, u_y, u_z are the components of $\mathbf{u}$. We have thus derived a suitable form for the *velocity four-vector.* Notice the procedure we have to use. In some frame of reference, we measure the three-velocity $\mathbf{u}$ and hence γ. We then form the quantities γu_x, γu_y, γu_z and γ and we know they will transform exactly as x, y, z and t.

Let us prove to ourselves that we can use this procedure to work out how to add two velocities relativistically. The relative velocity of the two frames of reference is V in standard configuration and $\mathbf{u}$ is the velocity of the particle in the frame S. What is its velocity in the frame S'? We will call the Lorentz factor associated with the relative motion of S and S' γ_v. First, we write down the velocity four-vectors for the particle in S and S'.

$$\text{In } S,\ [\gamma\mathbf{u}, \gamma] \equiv [\gamma u_x, \gamma u_y, \gamma u_z, \gamma], \gamma = (1 - u^2/c^2)^{-\frac{1}{2}}$$

$$\text{In } S',\ [\gamma'\mathbf{u}', \gamma'] \equiv [\gamma' u_{x'}, \gamma' u_{y'}, \gamma' u_{z'}, \gamma'], \gamma' = (1 - u'^2/c^2)^{-\frac{1}{2}}$$

Our analysis of the velocity four-vector shows that we can relate these components to x, y, z, t and x', y', z', t' as follows.

$$\begin{array}{ll} x' \to \gamma' u_{x'} & x \to \gamma u_x \\ y' \to \gamma' u_{y'} & y \to \gamma u_y \\ z' \to \gamma' u_{z'} & z \to \gamma u_z \\ t' \to \gamma' & t \to \gamma \end{array}$$

Therefore, applying the Lorentz transformation, we find that

$$\left.\begin{aligned} \gamma' u_{x'} &= \gamma_v(\gamma u_x - V\gamma) \\ \gamma' u_{y'} &= \gamma u_y \\ \gamma' u_{z'} &= \gamma u_z \\ \gamma' &= \gamma_v(\gamma - V\gamma u_x/c^2) \end{aligned}\right\} \qquad (13.15)$$

The last relation gives us the relation

$$\gamma\gamma_v/\gamma' = (1 - Vu_x/c^2)^{-1}$$

and therefore from (13.15) we obtain

$$\left.\begin{aligned} u_{x'} &= \frac{u_x - V}{(1 - Vu_x/c^2)} \\ u_{y'} &= \frac{u_y}{\gamma_v(1 - Vu_x/c^2)} \\ u_{z'} &= \frac{u_z}{\gamma_v(1 - Vu_x/c^2)} \end{aligned}\right\} \qquad (13.16)$$

You will recognise these as the standard expressions for the addition of velocities in special relativity. They have many pleasant features. For example, if either u_x, u_y, or u_z are separately equal to c, $u_{x'}$, $u_{y'}$ or $u_{z'}$ are also c, as is required by Einstein's second postulate.

We should also take the norm of the four-velocity:

$$|\mathbf{U}^2| = c^2\gamma^2 - \gamma^2 u_x^2 - \gamma^2 u_y^2 - \gamma^2 u_z^2 = \gamma^2 c^2(1 - u^2/c^2) = c^2 \qquad (13.17)$$

which is an invariant as we would expect.

The acceleration four-vector. We now know exactly how to proceed to generate an acceleration four-vector. First, we form the increment of the four-velocity $\Delta\mathbf{U} \equiv [\Delta(\gamma\mathbf{u}), \Delta\gamma]$ which must itself be a four-vector. Then we define the only invariant acceleration-like quantity we can form by dividing through by the proper time interval $\mathrm{d}t_0$:

$$\mathbf{A} = \frac{\mathrm{d}\mathbf{U}}{\mathrm{d}t_0} = \left[\gamma \frac{\mathrm{d}}{\mathrm{d}t}(\gamma\mathbf{u}), \gamma\frac{\mathrm{d}\gamma}{\mathrm{d}t}\right] \tag{13.18}$$

This is the *acceleration four-vector.* Let us convert this into a more useful form for computational purposes. First, notice how to differentiate $(1-u^2/c^2)^{-\frac{1}{2}}$:

$$\begin{aligned}\frac{\mathrm{d}\gamma}{\mathrm{d}t} &= \frac{\mathrm{d}}{\mathrm{d}t}(1-u^2/c^2)^{-\frac{1}{2}} = \frac{\mathrm{d}}{\mathrm{d}t}(1-\mathbf{u}\cdot\mathbf{u}/c^2)^{-\frac{1}{2}} \\ &= (\mathbf{u}\cdot\mathbf{a}/c^2)(1-u^2/c^2)^{-\frac{3}{2}} = \gamma^3(\mathbf{u}\cdot\mathbf{a}/c^2)\end{aligned} \tag{13.19}$$

Then

$$\frac{\mathrm{d}}{\mathrm{d}t}(\gamma\mathbf{u}) = \gamma\mathbf{a} + \mathbf{u}\,\gamma^3(\mathbf{u}\cdot\mathbf{a}/c^2)$$

$$\mathbf{A} = \frac{\mathrm{d}\mathbf{U}}{\mathrm{d}t_0} = [\gamma^4(\mathbf{u}\cdot\mathbf{a}/c^2)\,\mathbf{u} + \gamma^2\mathbf{a}, \gamma^4(\mathbf{u}\cdot\mathbf{a}/c^2)] \tag{13.20}$$

Notice what this means. We sit in some frame of reference, say S, and we measure at a particular instant the velocity $\mathbf{u}$ and the acceleration $\mathbf{a}$ of a particle. From these quantities we can form the three-vector, $\gamma^4(\mathbf{u}\cdot\mathbf{a}/c^2)\mathbf{u} + \gamma^2\mathbf{a}$ and the scalar $\gamma^4(\mathbf{u}\cdot\mathbf{a}/c^2)$ and we know that these quantities will transform exactly as $\mathbf{r}$ and t between inertial frames of reference.

It is often very useful in relativistic calculations to transform into the frame of reference in which the particle is instantaneously at rest. The particle is accelerated but this does not matter. Notice that there is no dependence upon acceleration in the Lorentz transformations. Plainly this is the frame in which $\mathbf{u} = 0$ and therefore the four-acceleration in the frame is

$$\mathbf{A} = [\mathbf{a}, 0] \equiv [\mathbf{a}_0, 0] \tag{13.21}$$

On the other hand, in the same frame, the four-velocity of the particle is

$$\mathbf{U} = [\gamma\mathbf{u}, \gamma] = [0, 1]$$

Therefore, $\mathbf{A}\cdot\mathbf{U} = 0$. Since four-vector relations are true in any frame of reference, this means that, no matter which frame we care to work in, the scalar product of the velocity and acceleration four-vectors is always zero, i.e. the velocity and acceleration four-vectors are orthogonal. If you are suspicious, then do it the hard way by taking the scalar product of the expressions for **A** and **U**, i.e. the four-vectors (13.14) and (13.20). The term $\mathbf{a}_0$ is called the *proper acceleration.*

13.4 Relativistic dynamics – the momentum and force four-vectors

So far, we have been dealing with *kinematics,* i.e. the description of motion, but now we have to tackle the problem of *dynamics* and this means introducing the concepts of momentum, force, etc.

The momentum four-vector. Let us first of all go through a purely formal exercise of defining suitable momentum and force four-vectors and then asking if they make physical sense. Notice that we really have no *a priori* guides as to what suitable forms we should choose. Let us therefore do the simplest thing we possibly can. Let us introduce the quantity

$$\mathbf{P} = m_0 \mathbf{U} \tag{13.22}$$

where **U** is the velocity four-vector and m_0 is what we will call the *rest mass* of the particle which is taken to be a scalar invariant. We then notice immediately the following consequences.

(i) $m_0 \mathbf{U}$ is certainly a four-vector if m_0 is a constant;

(ii) its space components reduce properly to the Newtonian formula for momentum if $u \ll c$, i.e. $m_0 \gamma \mathbf{u} \to m_0 \mathbf{u}$ as $u \to 0$.

Therefore, *if this agrees with experiment*, we have found a suitable form for the *momentum four-vector.*

$$\mathbf{P} = m_0 \mathbf{U} = [\gamma m_0 \mathbf{u}, \gamma m_0] \tag{13.23}$$

Notice, in addition, that we can define a *relativistic three-momentum* by the spatial components of the four-vector, i.e. $\mathbf{p} = \gamma m_0 \mathbf{u}$ is the relativistic three-momentum; $m = \gamma m_0$ is defined to be the *relativistic inertial mass.* Notice that we have not proved this yet. We will only find this out once we have constructed a proper set of dynamics.

Let us find a relation between the relativistic momentum and the relativistic mass of the particle. We take the norms of the momentum four-vectors in the laboratory frame and in the rest frame of the particle. These must be equal and hence

$$\begin{aligned} m_0^2 c^2 &= \gamma^2 m_0^2 c^2 - \gamma^2 m_0^2 u^2 \\ &= m^2 c^2 - p^2 \end{aligned}$$

or

$$p^2 = m^2 c^2 - m_0^2 c^2 \tag{13.24}$$

Finally, to complete our set of dynamics, we need the four-force vector.

The force four-vector. By analogy with the above procedure, we can define the four-vector generalisation of Newton's second law of motion by

$$\mathbf{F} = \frac{\mathrm{d}\mathbf{P}}{\mathrm{d}t_0} \tag{13.25}$$

where t_0 is proper time. Now, to complete our set of definitions, we need to relate the force which we measure in the laboratory to the four-force. The best way to approach this is through the quantity we have called the relativistic three-momentum. Why did we call it this? Suppose we write down the relation for the collision between two particles which initially have momentum four-vectors $\mathbf{P}_1$ and $\mathbf{P}_2$. After the collision, these have values $\mathbf{P}_3$ and $\mathbf{P}_4$. Then clearly the Lorentz invariant expression

$$\mathbf{P}_1 + \mathbf{P}_2 = \mathbf{P}_3 + \mathbf{P}_4 \tag{13.26}$$

describes the conservation of the momentum four-vectors. In terms of the components, this implies that

$$\begin{aligned} \mathbf{p}_1 + \mathbf{p}_2 &= \mathbf{p}_3 + \mathbf{p}_4 \\ m_1 + m_2 &= m_3 + m_4 \end{aligned} \tag{13.27}$$

i.e. implicit in this formalism is the requirement that (what we have called) the relativistic three-momentum be conserved. Thus, for relativistic particles, $\gamma m_0 \mathbf{u}$ plays the role of momentum. The corresponding force equation is suggested by Newton's second law,

$$\mathbf{f} = \frac{\mathrm{d}\mathbf{p}}{\mathrm{d}t} = \frac{\mathrm{d}}{\mathrm{d}t}(\gamma m_0 \mathbf{u}) \tag{13.28}$$

where $\mathbf{f}$ is the normal three-force.

Now is the definition good enough in relativity? We have to be a bit careful. On the one hand, one could just say 'Let's look at experiment and see if it works' and in many ways this is more or less all that the argument amounts to. We can do a little better. We can note that in point collisions of particles, the relativistic generalisation of Newton's third law should apply, i.e. $\mathbf{f} = -\mathbf{f}$. Notice that we can only use point collisions or else we get into terrible trouble with action at a distance in relativity – you will recall the problems we identified with the relativity of simultaneity in different frames of reference. For a point collision, $\mathbf{f} = -\mathbf{f}$ will be true if we adopt the definition of $\mathbf{f}$ given above because we have already argued that relativistic three-momentum is conserved, i.e.

$$\Delta\mathbf{p}_1 = -\Delta\mathbf{p}_2; \quad \Delta\mathbf{p}_1/\Delta t = -\Delta\mathbf{p}_2/\Delta t; \quad \mathbf{f}_1 = -\mathbf{f}_2$$

However, we should be careful here. We *cannot be strictly sure that we have made the correct choice without appealing to experiment.* We are faced with the same sort of logical problem which we faced when we tried to find out the meaning of Newton's laws of motion (Section 5.2). You will remember that they ended up being a set of definitions which gave results consistent with our experience. Similarly, relativistic dynamics cannot come out of pure thought but it can be put into a logically self-consistent structure which is also consistent with experiment.

We will therefore adopt the definition of **f** as the three-force, in the same sense that it has in Newtonian dynamics, but now the particle may be moving relativistically and the relativistic three-momentum should be used for **p**.

From this framework we can derive a number of very nice results.

(i) $\mathbf{F} = m_0 \mathbf{A}$

This follows directly from the above definition of **F**, i.e.

$$\mathbf{F} = \frac{\mathrm{d}\mathbf{P}}{\mathrm{d}t_0} = m_0 \frac{\mathrm{d}\mathbf{U}}{\mathrm{d}t_0} = m_0 \mathbf{A} \tag{13.29}$$

In addition, because $\mathbf{A} \cdot \mathbf{U} = 0$,

$$\mathbf{F} \cdot \mathbf{U} = 0$$

i.e. the force and velocity four-vectors are orthogonal.

(ii) The relativistic generalisation of $\mathbf{f} = \mathrm{d}\mathbf{p}/\mathrm{d}t$

Let us begin with the force four-vector,

$$\mathbf{F} = [\gamma \mathbf{f}, \gamma f_4] = m_0 \mathbf{A} \tag{13.30}$$

Equating spatial components and using expression (13.20),

$$\mathbf{f} = m_0 \gamma^3 (\mathbf{u} \cdot \mathbf{a}/c^2)\, \mathbf{u} + m_0 \gamma \mathbf{a} \tag{13.31}$$

This is the relativistic generalisation of $\mathbf{f} = m_0 \mathbf{a}$ and is entirely equivalent to $\mathbf{f} = \mathrm{d}\mathbf{p}/\mathrm{d}t$.

Now let us look at the fourth component of the force four-vector, i.e. of

$$\frac{\mathrm{d}\mathbf{P}}{\mathrm{d}t_0} = \left[\gamma \frac{\mathrm{d}}{\mathrm{d}t}(\gamma m_0 \mathbf{u}),\ \gamma \frac{\mathrm{d}(\gamma m_0)}{\mathrm{d}t}\right]$$

From (13.19),

$$\gamma \frac{\mathrm{d}m}{\mathrm{d}t} = \gamma \frac{\mathrm{d}(\gamma m_0)}{\mathrm{d}t} = m_0 \gamma^4 \left(\frac{\mathbf{u} \cdot \mathbf{a}}{c^2}\right)$$

i.e.

$$\frac{\mathrm{d}m}{\mathrm{d}t} = \gamma^3 m_0 \left(\frac{\mathbf{u} \cdot \mathbf{a}}{c^2}\right) \tag{13.32}$$

Let us now inspect the quantity $(\mathbf{f} \cdot \mathbf{u})/c^2$.

$$\begin{aligned}\frac{\mathbf{f} \cdot \mathbf{u}}{c^2} &= m_0 \gamma^3 \left(\frac{\mathbf{u} \cdot \mathbf{a}}{c^2}\right) \frac{u^2}{c^2} + m_0 \gamma \left(\frac{\mathbf{a} \cdot \mathbf{u}}{c^2}\right) \\ &= m_0 \gamma^3 \left(\frac{\mathbf{u} \cdot \mathbf{a}}{c^2}\right) \left(\frac{u^2}{c^2} + \frac{1}{\gamma^2}\right)\end{aligned}$$

$$= m_0\gamma^3\left(\frac{\mathbf{u}\cdot\mathbf{a}}{c^2}\right) \tag{13.33}$$

Therefore, equation (13.32) becomes

$$\frac{\mathrm{d}m}{\mathrm{d}t} = \frac{\mathbf{f}\cdot\mathbf{u}}{c^2}$$

or

$$\frac{\mathrm{d}(mc^2)}{\mathrm{d}t} = \mathbf{f}\cdot\mathbf{u} \tag{13.34}$$

This is one of the amazing results of Einstein's paper of 1905. The quantity $\mathbf{f}\cdot\mathbf{u}$ is just the rate of doing work on the particle, i.e. it is the rate of increase of energy of the particle. Thus mc^2 is identified with the total energy of the particle. This is the formal proof of the equation known to everyone

$$E = mc^2 \tag{13.35}$$

Notice the profound implication of equation (13.34). There is a certain amount of inertial mass associated with the energy produced when work is done. It does not matter what the form of the energy is – electrostatic, magnetic, kinetic, elastic, etc. All energies are the same thing as inertial mass. Likewise, reading the equation backwards, inertial mass is energy. Nuclear power stations and nuclear explosions are vivid demonstrations of the identity of inertial mass and energy.

(iii) A common abuse of the force law

One sometimes sees reference to *longitudinal* and *transverse* masses in text books. These arise from misuse of the relativistic form of Newton's second law of motion. If $\mathbf{u}\|\mathbf{a}$, the force law (13.31) reduces to

$$\mathbf{f} = m_0\gamma^3\left(\frac{u^2\mathbf{a}}{c^2}\right) + \gamma\mathbf{a}m_0 = m_0\gamma^3\left(\frac{u^2}{c^2} + \frac{1}{\gamma^2}\right)\mathbf{a}$$

$$= \gamma^3 m_0\mathbf{a}$$

The quantity $\gamma^3 m_0$ is called the *longitudinal* mass. On the other hand, if $\mathbf{u}\perp\mathbf{a}$, we find that

$$\mathbf{f} = m_0\gamma\mathbf{a}$$

and $m_0\gamma$ is called the *transverse* mass. I think these are horrible names and are only used by people who think Newton's second law of motion is $\mathbf{f} = m_0\mathbf{a}$. It most certainly is *not*. The correct form is $\mathbf{f} = \mathrm{d}\mathbf{p}/\mathrm{d}t$ and once the correct form for $\mathbf{p}$ is included, there is no need to introduce these irrelevant concepts.

(iv) Relativistic particle dynamics in a magnetic field

As a final example of use of the force equation, let us look at particle dynamics in a magnetic field.

$$\mathbf{f} = e(\mathbf{E} + \mathbf{u} \times \mathbf{B})$$

and hence if $\mathbf{E} = 0$,

$$\mathbf{f} = e(\mathbf{u} \times \mathbf{B})$$

Therefore,

$$e(\mathbf{u} \times \mathbf{B}) = m_0\gamma^3(\mathbf{u}\cdot\mathbf{a}/c^2)\mathbf{u} + \gamma\mathbf{a}m_0 \tag{13.36}$$

The left-hand side of this expression and the first term on the right are perpendicular vectors because $\mathbf{u} \perp \mathbf{u} \times \mathbf{B}$ and hence we require simultaneously that

$$\left.\begin{aligned} e(\mathbf{u} \times \mathbf{B}) &= \gamma\mathbf{a}m_0 \\ \text{and} \quad \mathbf{u}\cdot\mathbf{a} &= 0 \end{aligned}\right\} \tag{13.37}$$

i.e. the acceleration impressed by the magnetic field is perpendicular to both $\mathbf{B}$ and the velocity $\mathbf{u}$. This is the origin of the circular or spiral motion of charged particles in magnetic fields.

13.5 The frequency four-vector

Finally, let us look at the way of deriving the frequency four-vector from the scalar product rule: if $\mathbf{A}$ is a four-vector and $\mathbf{A}\cdot\mathbf{B}$ is an invariant, $\mathbf{B}$ must be a four-vector. The simplest way is to look at the phase of a wave. If we write the wave in the form $\exp[\mathrm{i}(\mathbf{k}\cdot\mathbf{r} - \omega t)]$, the quantity $(\mathbf{k}\cdot\mathbf{r} - \omega t)$ is the phase of the wave, i.e. how far through the cycle of the wave from 0 to 360° one happens to be. Now this is an invariant quantity, whatever frame of reference one happens to be in. Thus

$$(\mathbf{k}\cdot\mathbf{r} - \omega t) = \text{constant}$$

is an invariant. But $[\mathbf{r}, t]$ is a four-vector and therefore the quantity

$$\mathbf{K} = [\mathbf{k}, \omega/c^2] \tag{13.38}$$

must also be a four-vector which we call the frequency *four-vector.*

We can also derive this four-vector by considering photons and using the momentum four-vector. Thus

$$\mathbf{P} \equiv [\gamma m_0\mathbf{u}, \gamma m_0] \equiv [h\nu/c, h\nu/c^2]$$

where $h\nu/c$ is the momentum of the photon and $h\nu$ its energy.

$$\frac{h\nu}{c^2} = \frac{\hbar\omega}{c^2}; \quad \frac{h\nu}{c} = \frac{\hbar\omega}{c} = \hbar|\mathbf{k}|$$

and therefore for a photon

$$\mathbf{P} \equiv \hbar[\mathbf{k}, \omega/c^2] \tag{13.39}$$

i.e.

$$\mathbf{P} = \hbar\mathbf{K}$$

13.6 Reflections

I will end this case study here. What I hope you have noticed is how far you can get once you have the appropriate mathematical structure for the theory. I could have gone through the whole thing as a mathematical exercise and it would have hardly been apparent where I had had to look at the real world. In fact, there are assumptions present as we have shown, just as there are in Newtonian theory. It so happens that the mathematical structure is very elegant indeed. It might not have turned out that way.

From the purely practical point of view, the use of four-vectors greatly simplifies all calculations in relativity. You only have to remember a single set of transforms and then remember how to derive the four-vectors. There are great advantages in having a prescription which you can trust for doing relativistic calculations. I am very suspicious of arguments which try to give you simple ways of looking at relativistic problems. Things like length contraction and time dilation should be treated cautiously and the simplest way not to get horribly confused is to write down the four-vectors associated with the events you are considering and then use the Lorentz transforms to relate coordinates.

Case Study 7

GENERAL RELATIVITY AND COSMOLOGY

Albert Einstein (1879–1955)
(from *Einstein: A Centenary Volume*, ed.
A.P. French, p. 26, Heinemann, 1979).

It is often a matter of dispute whether the subjects of general relativity and cosmology should appear in undergraduate syllabuses at all. I firmly believe that they most certainly should be taught for two basic reasons.

First of all, general relativity follows on very naturally from special relativity and results in even more profound changes in our concepts of space and time than those which follow from special relativity. The idea that the geometry of space–time is influenced by the distribution of matter which in turn moves along paths in a curved space–time is a fundamental concept of modern physics. It is, in my view, only right that some flavour of the origins and implications of these remarkable discoveries of Einstein should appear early in a course in physics, once the ideas of special relativity have been introduced. Unfortunately, the subject is technically complex in the sense that the mathematics required goes beyond what is normally introduced at the undergraduate level and it would be wrong to underestimate these technical difficulties. Nonetheless, much can be understood without the use of advanced techniques provided you are prepared to accept certain results which I will quote. It seems a small price to pay for some real feel for the intimate interaction between space–time geometry and gravitation and for some of the more remarkable phenomena expected to occur in strong gravitational fields.

It might be thought even more perverse to introduce the subject of cosmology into this text. My defense is that the large-scale structure and geometry of the Universe are for me just as much an integral part of physics as the structure of the Universe on a microscopic scale, i.e. the structure of atoms and their nuclei. The astrophysical and cosmological sciences are just as much a part of physics as solid state physics, statistical mechanics, etc. The reason I find them attractive for teaching at the undergraduate level is that they bring home the fundamental connection between geometry and gravity on the largest scale, in fact on one of the few scales in which the effects of space–time curvature are strong rather than being minor corrections to Newtonian gravity. Cosmology is also a subject which has flowered over the last 25 years, many more real facts now being known about the Universe. This new understanding has led to a standard view of how the whole Universe came into being, which poses many fundamental questions for physicists and astronomers. In addition, it is a subject which is always ripe for controversy and one can draw a number of important lessons about progress in the observational sciences from some aspects of recent history.

By ending with cosmology, we come right up to date with our studies of how real theoretical physics is done. Many of the issues that we will discuss are still unresolved. It also results in a pleasant symmetry for this series of lectures in which we begin and end with the laws of gravitation.

14

AN INTRODUCTION TO GENERAL RELATIVITY

14.1 Introduction

With the special theory of relativity expounded in 1905, Einstein went on to tackle the much more difficult problem of finding a relativistic theory of gravity. How he did it is one of the greatest stories in theoretical physics and involved the very deepest physical insight, imagination and intuition. Whereas others had come close to the discovery of special relativity, in his discovery of the general theory, Einstein went far beyond all his contemporaries and discovered the relativistic theory of gravity which has proved to be consistent with all tests of the theory so far carried out. It has also led to concepts which could barely have been conceived of, even by a genius like Einstein – the phenomenon of black holes and the possibility of testing theories of gravity in the strong field limit through the observation of relativistic stars.

In this simple introduction, the intention is very modest. We aim to make reasonable the basic ideas about space, time and gravitation as embodied in general relativity. We then look at one specific solution of Einstein's theory, the Schwarzschild solution, and find out how Einstein's relativistic gravity differs from Newtonian gravity. Our exposition leans heavily upon Rindler's exposition in *Essential Relativity*[1] and Berry's introductory text *Principles of Cosmology and Gravitation.*[2] At a more technical level, I can thoroughly recommend Weinberg's authoritative text *Gravitation and Cosmology*[3] and the mammoth volume by Misner, Thorne and Wheeler, *Gravitation.*[4]

14.2 Basic features of the theory of gravity

Newton's law of gravity

$$\mathbf{F} = -G\frac{m_1 m_2}{r^3}\,\mathbf{r} \tag{14.1}$$

has some remarkable properties, especially when written in conjunction with Newton's second law of motion,

$$\mathbf{F} = \frac{\mathrm{d}}{\mathrm{d}t}(m\mathbf{v}) = m\ddot{\mathbf{r}} \tag{14.2}$$

when we are dealing with point masses. There are, in principle, three different types of mass, m, appearing in these two formulae. One of them is quite distinct from the other two and that is the m which appears in equation (14.2). This mass is referred to as the *inertial mass* of the body m_i, by which we mean that it is the constant which appears in the relation between force and acceleration. It simply describes the resistance (or inertia) of the body to changing its state of motion, quite independent of the nature of the impressed force $\mathbf{F}$.

Suppose we ask, 'What is the gravitational force on the mass m_2 due to the presence of the mass m_1 at distance r?' The answer is given by (14.1) but we should note that the sum should be considered in two parts. First, there is a gravitational field at $\mathbf{r}$ due to the mass m_1 which is

$$\mathbf{g} = -\frac{Gm_1}{r^3}\mathbf{r}$$

This equation states that the field is associated with the mass m_1 and since the field originates on m_1, we call this mass the *active gravitational mass.* When we place m_2 at $\mathbf{r}$, the body feels a force which is $m_2\mathbf{g}$. This describes the response of the body to the pre-existing field and we call m_2 the *passive gravitational mass.*

Now the remarkable thing about gravity is that these three masses must be proportional to each other with high precision. The proportionality of the active and passive gravitational masses comes directly from Newton's third law of motion.

$$\mathbf{F}_1 = -\mathbf{F}_2$$

i.e.

$$Gm_{1\mathrm{a}}m_{2\mathrm{p}} = Gm_{2\mathrm{a}}m_{1\mathrm{p}}$$

$$\frac{m_{1\mathrm{a}}}{m_{2\mathrm{a}}} = \frac{m_{1\mathrm{p}}}{m_{2\mathrm{p}}}$$

where the subscripts a and p refer to the active and passive gravitational masses respectively. Since the ratios are the same, m_a and m_p can be made identical by a suitable choice of units.

The equality of the inertial and gravitational masses is much more subtle. We could, in principle, imagine the property of gravitational 'mass' being something totally separate from inertial mass. It is akin to the electric charge in electrostatics and we know that there is a clear distinction between the electric charge of a particle and its inertial mass. The remarkable fact is that the 'gravitational charge' which is included in Newton's law of gravity is identical with the inertial mass.

This feature of Newton's law of gravity was already implicit in his analysis which demonstrated that the planets move in elliptical orbits about the Sun. The inertial mass which is included in the expression for centrifugal force is equated to the gravitational mass present in the inverse square law of gravity. The precise equality of these two masses was subject to increasingly accurate measurement by Eotvos in 1880s, by Dicke in 1964 and by Braginski in 1971. The present limit to the linearity of the ratio between gravitational and inertial mass is about one part in 10^{12}. This is sufficiently accurate to enable rather subtle tests of the identity of the two masses to be made. For example, the inertial mass associated with the kinetic energy of electrons moving at relativistic velocities in atoms, $m = E/c^2$, must behave in exactly the same way as gravitational mass of the same magnitude.

The implications of these amazing properties of gravity led Einstein to the view that there is a very close relation between gravitational phenomena on the one hand and dynamical phenomena as embodied in Newton's second law on the other, i.e. between accelerated motion and gravitational fields. The equivalence of these phenomena was raised to the status of a basic physical principle, known as the *principle of equivalence*, which we will state in Einstein's own words:

> All local, freely falling, non-rotating laboratories are fully equivalent for the performance of all physical experiments.[5]

Stated in this way, the principle appears to be a natural extension of the postulates which form the basis of special relativity. In particular, the postulates are seen to be entirely consistent with the principle of equivalence since two inertial frames in relative motion are 'freely falling' frames in zero gravity. The statement also means that, locally, the interval between events must be given by the standard *metric* of special relativity, known often as the Minkowski metric,

$$ds^2 = dt^2 - \frac{1}{c^2}(d\mathbf{r}^2) \tag{14.3}$$

But the assertion of the principle of equivalence goes far beyond this. By a 'freely falling laboratory', we mean one which is accelerated by the local gravitational field without any restraining forces. In other words, within this laboratory, we feel no gravitational field at all. By transforming into a frame of reference which is freely falling, we completely replace gravity by accelerated frames of reference. The principle of equivalence asserts that the laws of physics should be identical in all such accelerated frames of reference. Thus, Einstein's fundamental insight was to 'abolish' gravity altogether and replace it by appropriate transforms between accelerated frames of reference.

To reinforce the idea of the equivalence of gravity and accelerations locally, let us look at some of the classical pictures of laboratories in free fall, in gravitational fields and accelerated so as to mimic the effects of gravity. In Figure 14.1(*a*) and (*b*), we demonstrate the equivalence of a static gravitational field and an accelerated laboratory with a spring balance. We show for comparison a laboratory in a region of zero gravitational field (Figure 14.1(*c*)) and one in free fall in a gravitational field of field strength **g** (Figure 14.1(*d*)). In Figure 14.2, we show the laboratory accelerated towards a point mass *M*. Whilst we can replace the gravitational field at the centre of the laboratory by an acceleration **g**, this will not give the correct answer for points at the edges of the laboratory. This emphasises the point made in Einstein's statement of the principle, that it should be applicable *locally*. At any point, we can replace **g** by an accelerated frame of reference but, in general, we require different accelerated frames at different points in space.

Figure 14.1. Illustrating the equivalence between gravitational fields and accelerated reference frames. In each drawing the laboratory has a spring suspended from the ceiling with a weight attached. In (*a*) and (*b*) the springs are stretched. In (*c*) and (*d*), there is no extension of the spring.

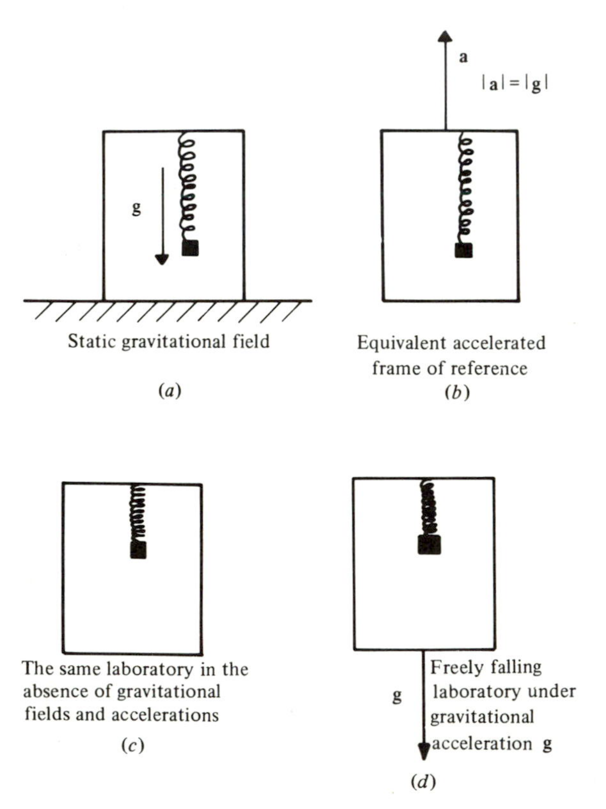

The goal of general relativity becomes clear. We must find transforms between accelerated frames of reference which take account of the influence of the distribution of gravitating matter in the Universe.

14.3 Non-linearity, space curvature and time dilation

Let us note some of the necessary features of a relativistic theory of gravity.

14.3.1 The non-linearity of relativistic gravity

That a relativistic theory of gravity must be non-linear follows from the relation $E = mc^2$ in special relativity. In the simple case of the gravitational field due to some distribution of mass, the gravitational field itself possesses a certain energy density, which, because $E = mc^2$, corresponds to a certain inertial mass density. This inertial mass density in turn is a source of gravitational field. This property contrasts sharply with that of an electrostatic field which produces a certain amount of electromagnetic field energy but this does not generate additional electrostatic charge. Thus, relativistic gravity is intrinsically a non-linear theory which accounts for a great deal of its complexity.

14.3.2 Space curvature

Let us study the propagation of light in a gravitational field **g** and in an equivalent accelerated laboratory as illustrated in Figure 14.3(*a*), (*b*) and (*c*). If we look at the accelerated laboratory from an inertial frame of reference, it is apparent that the light signal follows a curved path according to the observer who is accelerated. If we use Newtonian mechanics, we see that the light signal 'drops' a distance $\frac{1}{2}gt^2$ as it crosses the laboratory, where $t = d/c$. From the

Figure 14.2. Illustrating how it is only possible to interchange accelerated frames and gravitational fields locally. The gravitational field due to the mass *M* may be transformed away at the centre of the box but the same transformation will not eliminate fields at the edge of the box.

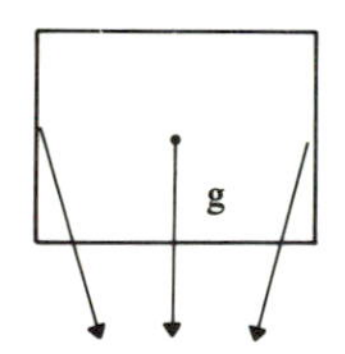

geometry of Figure 14.3(*d*), we can work out the radius of curvature R of the light ray's path, $R\phi = d$. The angle ϕ is also the angle between tangent and chord and hence $\phi = \frac{1}{2}g\,d/c^2$. Therefore, $R = 2c^2/g$. Notice that the radius of curvature of the light ray's path depends only upon **g**, the gravitational acceleration.

This simple argument indicates that, according to the principle of equivalence, the geometry along which light rays propagate in an accelerated laboratory is, in general, not flat but curved space, the spatial curvature depending upon the local value of the gravitational acceleration **g**. Notice that this argument only makes it plausible that the spatial sections through space–time should be curved. In other words, the appropriate geometry for '$d\mathbf{r}^2$' in the metric (14.3) will, in general, refer to curved space. In the full theory, however, the whole of the metric will refer to a curved geometry, i.e. four dimensional space–time should be described by curved four dimensional geometry.

Figure 14.3. Illustrating the gravitational deflection of light according to the principle of equivalence. (*a*) The path of a light beam in a gravitational field g. (*b*) The dynamics of the light beam in the equivalent accelerated frame of reference as viewed by a stationary external observer. The dashed line shows the position of the laboratory after time *t*. (*c*) The path of the light beam as observed in the accelerated frame of reference. (*d*) The geometry of light deflection in an accelerated frame of reference.

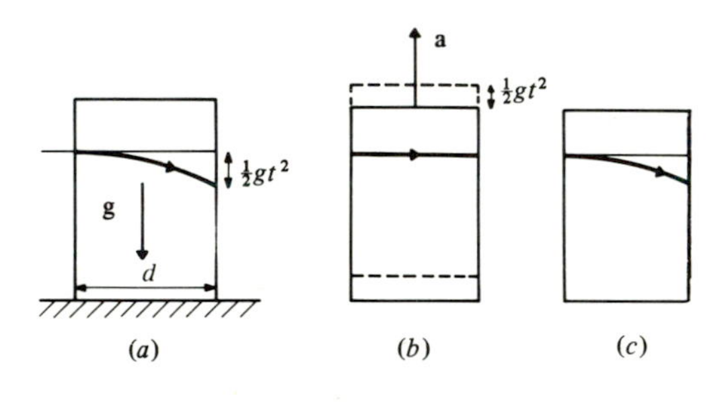

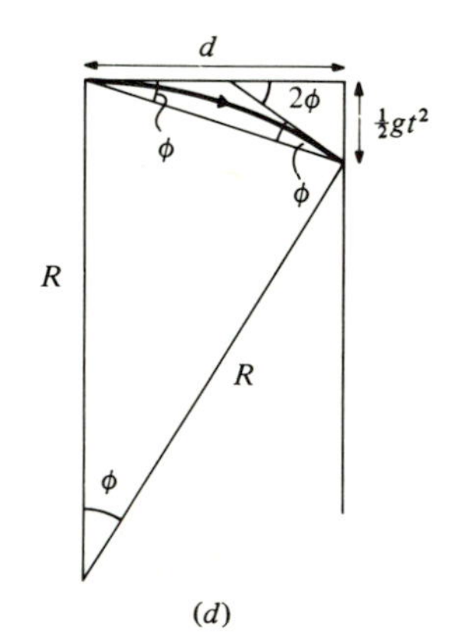

We are therefore led automatically to consideration of four dimensional curved spaces which locally reduce to a metric of the form (14.3). The appropriate choice of the metric, in general, is a *Riemannian metric* which can be written

$$ds^2 = \sum_{\mu,\nu} g_{\mu\nu}\, dx^\mu\, dx^\nu = g_{\mu\nu}\, dx^\mu\, dx^\nu \tag{14.4}$$

i.e. x^μ and x^ν are coordinates defining points in the four dimensional space and ds^2 is given by a homogeneous quadratic differential form in these coordinates. The convention of summing over identical indices is used in the equality of equation (14.4). The indices μ and ν in dx are written as superscripts for the technical reason that in the full theory of four dimensional spaces we have to distinguish between covariant and contravariant quantities. We will not need this distinction in our analysis. These spaces have the property that *locally* they can be converted into Euclidean form, satisfying our requirement that the metric reduces locally to the Minkowski metric (14.3). In this case,

$$g_{\mu\nu} = \begin{bmatrix} 1 & 0 & 0 & 0 \\ 0 & -\dfrac{1}{c^2} & 0 & 0 \\ 0 & 0 & -\dfrac{1}{c^2} & 0 \\ 0 & 0 & 0 & -\dfrac{1}{c^2} \end{bmatrix} \tag{14.5}$$

where $[x^0, x^1, x^2, x^3] = [t, x, y, z]$. The *signature* of the metric is $[+, -, -, -]$ and this property is preserved under general Riemannian transformations.

An important conceptual point is that the coordinates should be considered as 'labels' for the points in the space rather than 'distances' as is conventional in the Euclidean metric. It so happens that x, y and z are real distances in Euclidean space and this is a remarkably convenient labelling. In general, however, the coordinates may not have the same intuitively obvious meaning.

14.3.3 Time dilation in a gravitational field

We can demonstrate why the time-component of the metric should be non-Euclidean by another simple argument which indicates the origin of the gravitational redshift. Suppose a beam of light is propagated from the ceiling of our laboratory to the floor. Then, according to the principle of equivalence, we can replace the laboratory by a frame of reference accelerated in the opposite direction to $\mathbf{g}$ (Figure 14.4). The time it takes the photons to travel from ceiling to floor is $t = h/c$ and in this time the laboratory acquires velocity $v = gt$, i.e.

$v = gh/c$. In the limit of small velocities $v \ll c$, this results in a Doppler shift of the frequency of the light waves to higher frequencies

$$\frac{\Delta\nu}{\nu} = \frac{v}{c} = \frac{gh}{c^2} \tag{14.6}$$

or, if we write the frequencies in terms of ν_0 the frequency at the ceiling of the laboratory and ν_1 the frequency when the wave reaches the floor,

$$\frac{\nu_1}{\nu_0} = 1 + \frac{\Delta\nu}{\nu} = 1 + \frac{gh}{c^2} \tag{14.7}$$

This is the expression for the *gravitational redshift*. Redshift z is defined as

$$z = \frac{\lambda_{obs} - \lambda_{em}}{\lambda_{em}} \tag{14.8}$$

where λ_{obs} is the observed wavelength and λ_{em} the emitted wavelength. If we consider the wave propagating from the floor to the ceiling, we can write $\lambda_{obs} = c/\nu_0$, $\lambda_{em} = c/\nu_1$, and hence

$$z = \frac{\lambda_{obs}}{\lambda_{em}} - 1 = \frac{\nu_1}{\nu_0} - 1$$

$$z = \frac{gh}{c^2} \tag{14.9}$$

It is perhaps more instructive to express the result in terms of the gravitational potential ϕ defined by $\mathbf{g} = -\operatorname{grad} \phi$, i.e.

$$\mathbf{g} \cdot \mathrm{d}\mathbf{l} = -\mathrm{d}\phi \tag{14.10}$$

ϕ becomes more negative as we move to stronger gravitational fields. If we measure distance from the ceiling, we can write $gh = -\mathrm{d}\phi$ and hence

Figure 14.4. Illustrating the gravitational redshift. The laboratory is accelerated to velocity a*t* after time *t*.

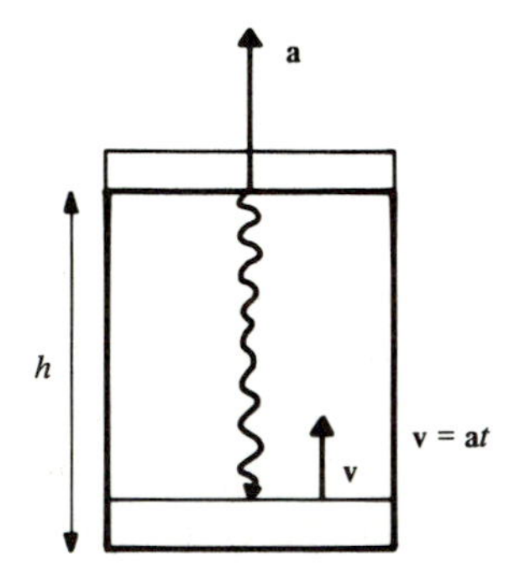

$$z = \frac{d\nu}{\nu} = -\frac{d\phi}{c^2} \tag{14.11}$$

Finally, we can take the integral of (14.11) to give us

$$\int \frac{d\nu}{\nu} = -\int \frac{d\phi}{c^2}$$

i.e.

$$\nu = \nu_0 \exp\left(-\frac{\Delta\phi}{c^2}\right) \tag{14.12}$$

where $\Delta\phi$ is the gravitational potential difference between the points of emission and reception of the radiation.

So far, we have worked entirely in terms of Doppler shifts of accelerated frames of reference, which, according to the principle of equivalence, are exactly the same as the redshift expected when light propagates through regions of different gravitational potential. We note that we can think of this in terms of *time dilation* instead of frequency shifts. If Δt_1 and Δt_0 are the periods of the waves corresponding to ν_1 and ν_0, we know that

$$\Delta t_1 = \nu_1^{-1} \quad \text{and} \quad \Delta t_0 = \nu_0^{-1}$$

and hence

$$\frac{\Delta t_0}{\Delta t_1} = \exp\left(-\frac{\Delta\phi}{c^2}\right) \tag{14.13}$$

This means that identical clocks run at different rates in different gravitational potentials. Notice the sense of the change. If we move to a stronger gravitational field, ϕ becomes more negative and therefore $\Delta t_0 > \Delta t_1$.

Now it is important to appreciate that, in the above analysis, the times Δt_0 and Δt_1 are *proper times.* Thus, if we consider the observer 0 to be located at zero potential, the Δt_0 he measures is proper time on his clock at a fixed point in space, i.e. the interval ds^2 is a pure time interval and hence proper time. In exactly the same way, the observer or source at 1 measures proper time Δt_1. Thus, in general, since $\Delta t_0 \neq \Delta t_1$, we see that there is a major synchronisation problem for our clocks – the rate at which proper time runs depends upon the gravitational potential. Another interesting point is that we have now stumbled across a 'general relativistic' twin paradox problem. If I send my twin into a deeper gravitational potential and back home again, less proper time passes for him than for me. This is no more than time dilation in a gravitational field.

The clock synchronisation problem can be simply resolved if we decide not to use proper time but to use a suitable coordinate time which will take the same value at all points in space. Let us refer all times to the coordinate time measured in zero potential and call the increment of this time dt. We will call the proper time at other points in space $d\tau$ and then we have shown that

$$d\tau = dt \exp\left(\frac{\Delta\phi}{c^2}\right) \tag{14.14}$$

Now a simple Minkowski metric would be written

$$\begin{aligned} ds^2 &= d\tau^2 - \frac{1}{c^2}\, dl^2 \\ &= \left[\exp\left(\frac{\Delta\phi}{c^2}\right)\right]^2 dt^2 - \frac{1}{c^2}\, dl^2 \end{aligned} \tag{14.15}$$

We recall that, whenever we want to find the proper time interval, we have $ds^2 = d\tau^2$ when $dl^2 = 0$ and this is given by the result (14.14) above.

Let us go one step further and consider the case of the gravitational potential of a point mass M. According to Newton's law of gravity, $\Delta\phi = -GM/r$ and hence, expanding for small values of GM/rc^2, we find that the quasi-Minkowski metric becomes

$$ds^2 = \left(1 - \frac{2GM}{rc^2}\right) dt^2 - \frac{1}{c^2}\, dl^2 \tag{14.16}$$

In this *approximate* treatment, we have derived a form which we will encounter again. I must emphasise that, although the above metric looks promising, there are many unsatisfactory steps in the arguments. For example, what exactly do we mean by r in the above metric, what exactly is dl when we know we ought to be using curvilinear coordinates? In fact, we have ended up with a horrid mixture of Newtonian, special relativistic and general relativistic ideas. The aim of this section was not, however, to be precise. We are building up the necessary *physical* components of the theory so that, when we look at a metric derived from general relativity, we will not be confused by Euclidean or even special relativistic concepts of space and time.

One final comment about the gravitational redshift is worth making. Notice that it is no more than an expression of energy conservation in a gravitational field. In our final form of the metric (14.16), this has been transformed into a property of the metric coefficients. Thus, we begin to see bits of real physics being built into the space–time metric.

14.4 Isotropic curved spaces

It is instructive to look a little into isotropic curved spaces because they illuminate some of the problems of interpretation which one encounters in studying the Schwarzschild metric. In flat space, we write the distance between two points separated by dx, dy, dz as

$$dl^2 = dx^2 + dy^2 + dz^2$$

If we write

$$\mathrm{d}l^2 = g_{\mu\nu}\,\mathrm{d}x^\mu\,\mathrm{d}x^\nu$$

$$g_{\mu\nu} = \begin{bmatrix} 1 & 0 & 0 \\ 0 & 1 & 0 \\ 0 & 0 & 1 \end{bmatrix}.$$

Now let us look at a simple *two* dimensional isotropic curved space, the surface of a sphere. This 'two-space' is isotropic because the radius of the sphere is a constant, R. We can set up an orthogonal frame of reference at each point locally on the surface of the sphere. We will choose spherical polar coordinates to describe positions on the surface of the sphere as indicated in Figure 14.5. In this case, the orthogonal coordinates are θ and ϕ, the angular coordinates, and hence

$$\mathrm{d}x^1 = \mathrm{d}\theta$$
$$\mathrm{d}x^2 = \mathrm{d}\phi$$

From the geometry of the sphere, we see that

$$\mathrm{d}l^2 = R^2\,\mathrm{d}\theta^2 + R^2\sin^2\theta\,\mathrm{d}\phi^2 \tag{14.17}$$

Therefore, the elements of the metric tensor are

Figure 14.5. The surface of a sphere as an example of a two dimensional isotropic curved space.

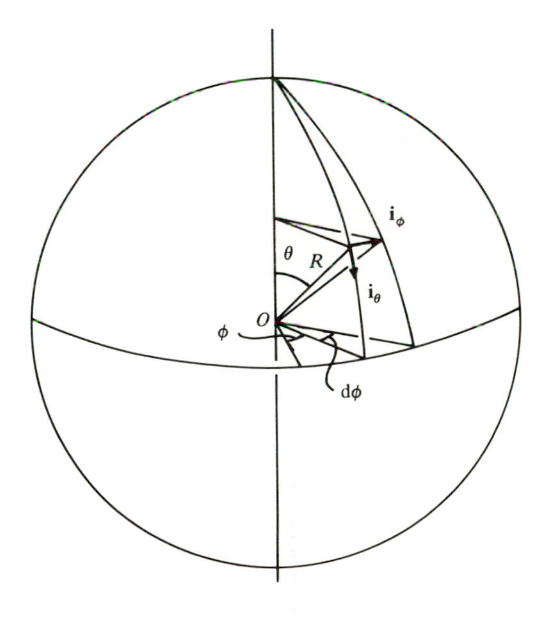

$$g_{\mu\nu} = \begin{bmatrix} R^2 & 0 \\ 0 & R^2 \sin^2\theta \end{bmatrix} \tag{14.18}$$

Now the metric tensor $g_{\mu\nu}$ contains information about the intrinsic geometry of the two-space. We need a prescription which will enable us to find the intrinsic geometry of the surface from the components of the metric tensor. In this simple case, this is hardly necessary, but we could easily have ended up choosing some funny set of coordinates in which the intrinsic geometry of the space would not be at all obvious.

For the case of two dimensional metric tensors which can be reduced to diagonal form with $g_{12} = g_{21} = 0$, Gauss showed that the curvature of the surface is given by the formula

$$K = \frac{1}{2g_{11}g_{22}} \left\{ -\frac{\partial^2 g_{11}}{\partial (x^2)^2} - \frac{\partial^2 g_{22}}{\partial (x^1)^2} + \frac{1}{2g_{11}} \left[\frac{\partial g_{11}}{\partial x^1} \frac{\partial g_{22}}{\partial x^1} + \left(\frac{\partial g_{11}}{\partial x^2} \right)^2 \right] + \frac{1}{2g_{22}} \left[\frac{\partial g_{11}}{\partial x^2} \frac{\partial g_{22}}{\partial x^2} + \left(\frac{\partial g_{22}}{\partial x^1} \right)^2 \right] \right\} \tag{14.19}$$

A proof of this theorem is outlined in Berry's book[6] and the general case for two-spaces is quoted in Weinberg's book.[7] With $g_{11} = R^2, g_{22} = R^2 \sin^2\theta$ and $x^1 = \theta$ and $x^2 = \phi$, it is straightforward to show that $K = 1/R^2$, i.e. the space is one of constant curvature (independent of θ and ϕ). In fact, it can be shown that the only isotropic two-spaces correspond to K being a constant which is positive, negative or zero. This is demonstrated in the Appendix to this chapter. The value $K = 0$ corresponds to *flat Euclidean space* and K negative to *hyperbolic spaces* in which at all points the radii of curvature are in opposite senses. This is sketched in Figure 14.6. In this case, the trigonometric functions such as $\sin\theta$ are replaced by their hyperbolic counterparts such as $\sinh\theta$.

Notice that, in isotropic curved spaces, K is a constant everywhere but in the case of a general two-space, the curvature is a function of the spatial coordinates.

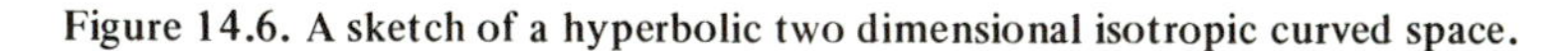

Figure 14.6. A sketch of a hyperbolic two dimensional isotropic curved space.

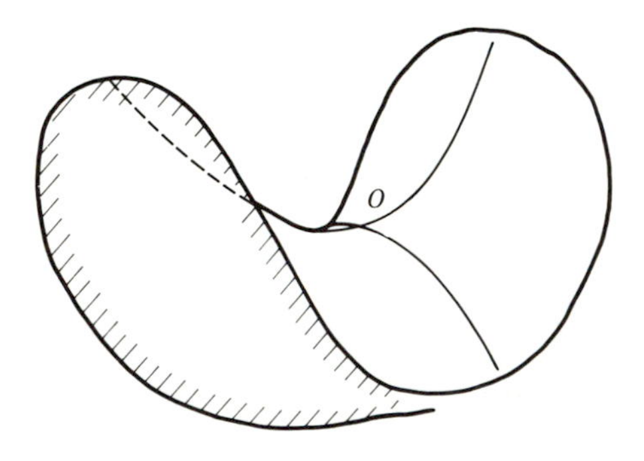

The extension to isotropic three-spaces is, in principle, straightforward but it is no longer possible to envisage the three-space geometrically since we would have to embed it in a four-space. However, we can proceed in a straightforward manner when we realise that a two dimensional section through our three-space must itself be an isotropic two-space for which we have already worked out the metric tensor. Suppose we ask what is the size of a rod L which subtends an angle $d\phi$ at the polar angle θ as illustrated in Figure 14.5. In the notation of that figure, we see that

$$L = R \sin\theta \, d\phi \tag{14.20}$$

If we call the distance round the arc θ, from O to L, r, i.e. a 'geodesic distance' in the surface of the sphere, $\theta = r/R$ and

$$L = R \sin(r/R) \, d\phi \tag{14.21}$$

In this context, by geodesic distance, we mean the shortest distance between two points in the two-space. In this case, this is clearly the great circle joining O and L.

If we want to work out the area of a circular area perpendicular to the 'line of sight' from O, we simply take another perpendicular section through the curved space as illustrated schematically in Figure 14.7. Because of the isotropy of the space, the value of L is the same in the perpendicular direction and hence the area of diameter L is

$$A = \frac{\pi L^2}{4} = \frac{\pi R^2}{4} \sin^2(r/R) \, d\phi^2 \tag{14.22}$$

Notice that, if $R \gg r$, this reduces to $\pi r^2 d\phi^2/4$, the standard Euclidean result.

We can see from this example the problem with non-Euclidean geometries. If we take r to be geodesic distance, the formula for surface area involves something more complex than the Euclidean formula. We can, however, decide to make the formulae for areas look like the Euclidean relation, i.e. $\pi x^2 d\phi^2/4$, and change the definition of the distance coordinate accordingly. In the above example, we would have to write

$$x = R \sin(r/R) \tag{14.23}$$

Figure 14.7. Illustrating how surface areas can be measured in an isotropic three-space by taking orthogonal sections through the space, each of which is an isotropic two-space.

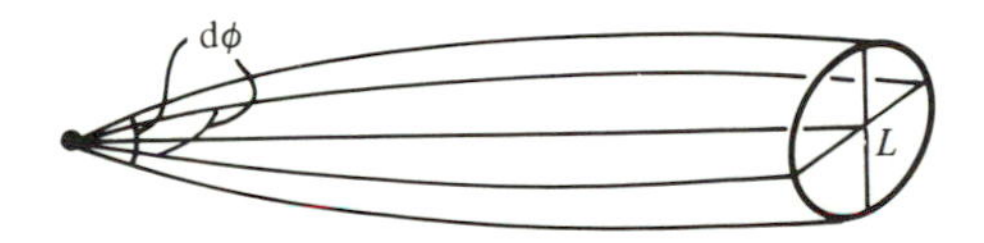

Now the formula for the spatial interval in our curved space is just

$$dl^2 = dr^2 + R^2 \sin^2(r/R)\, d\phi^2 \tag{14.24}$$

Let us rewrite this in terms of x rather than r. Differentiating (14.23),

$$dx = \cos(r/R)\, dr$$

$$dx^2 = \cos^2(r/R)\, dr^2 = [1 - \sin^2(r/R)]\, dr^2$$

$$= \left(1 - \frac{x^2}{R^2}\right) dr^2$$

i.e.

$$dr^2 = \frac{dx^2}{(1 - Kx^2)} \tag{14.25}$$

where $K = 1/R^2$ is the curvature of the two-space. The spatial interval in the metric can therefore be written

$$dl^2 = \frac{dx^2}{(1 - Kx^2)} + x^2\, d\phi^2 \tag{14.26}$$

The formulae (14.24) and (14.26) are exactly equivalent but notice the different meanings of r and x: r is a *geodesic distance* in the three-space whereas x is a distance coordinate which gives the correct answer for distances normal to the geodesic according to the relation

$$L = x\, d\phi$$

Notice how conveniently the metric of form (14.26) takes care of spherical, flat and hyperbolic spaces depending only upon the value of K.

14.5 The route to general relativity

These considerations make it plain why general relativity is a theory of considerable technical complexity. Space–time has to be defined by a general curved metric tensor $g_{\mu\nu}$ which is a function of space–time coordinates. It is apparent from the calculations carried out in the preceding sections that the $g_{\mu\nu}$ are analogous to gravitational potentials. Their variation from point to point in space–time defines the local curvature of space. From the purely geometrical point of view, we have to give mathematical substance to the principle of equivalence, i.e. to devise transformations in Riemannian space which can relate the values of $g_{\mu\nu}$ at different points in space–time. From the dynamical point of view, we have to be able to relate the $g_{\mu\nu}$ to the distribution of matter in the Universe.

The geometrical analysis proceeds from the most general tensor which describes the geometry of space–time, the fourth order Riemann–Christoffel curvature tensor $R_{\mu\nu\lambda\kappa}$. Einstein's analysis showed that the second order tensor which can be directly related to the matter distribution is the Ricci tensor $R_{\mu\nu}$, which is technically a contracted version of the curvature tensor. The tensor which describes the mass–energy content of the Universe is the stress–energy tensor $T_{\mu\nu}$. Einstein's guess was that these two tensors are simply related by the equation

$$R_{\mu\nu} - \tfrac{1}{2} g_{\mu\nu} R = -\frac{8\pi G}{c^2} T_{\mu\nu} \tag{14.27}$$

This looks fairly forbidding (especially when we have not explained what the symbols mean exactly!) but let us indicate how some parts of it can be understood. Consider the simplest sort of gravitating matter – what relativists call 'dust'. All this means is a gas of particles of some sort with zero pressure. Then if the dust has density ρ_0 in its rest frame, the density in a frame of reference moving at velocity v relative to it is $\rho = \gamma^2 \rho_0$, one γ resulting from the fact that the mass of the dust increases by virtue of its motion, the other because of length contraction in the direction of motion. The simplest tensor quantity which we can introduce to form a suitable stress–energy tensor for the dust is

$$T_{\mu\nu} = \rho_0 U_\mu U_\nu \tag{14.28}$$

where U_μ and U_ν are four-velocities. We note that the T_{00} component of the tensor is $\gamma^2 \rho_0$. In fact, the tensor $T_{\mu\nu}$ has some remarkable properties which are elegantly described by Rindler.[1] For example, the single equation

$$\frac{\partial T_{\mu\nu}}{\partial x_\nu} = 0$$

embodies the laws of mass, energy and momentum conservation for dust. Let us show why it is plausible that $T_{\mu\nu}$ and $g_{\mu\nu}$ should be related.

From the metric (14.16), we see that the time-component g_{00} is

$$g_{00} = \left(1 + \frac{2\phi}{c^2}\right) \tag{14.29}$$

Now Poisson's equation for gravity can be written

$$\nabla^2 \phi = 4\pi G \rho$$

and hence, using (14.28) and (14.29), we find that

$$\frac{c^2}{2} \nabla^2 g_{00} = 4\pi G T_{00}$$

i.e.

$$\nabla^2 g_{00} = \frac{8\pi G}{c^2} T_{00} \tag{14.30}$$

The analysis indicates why it is reasonable to expect a close relation between derivatives of $g_{\mu\nu}$ and the stress tensor $T_{\mu\nu}$. Notice that the constant in front of T_{00} is the same as in the full theory. Of course, we cannot expect an exact correspondence between (14.30) and (14.27) because the former results from an analysis in flat space.

In the end, the general theory of relativity results in a set of solutions for the $g_{\mu\nu}$ and consequently a metric of the form

$$ds^2 = g_{\mu\nu}\,dx^\mu\,dx^\nu$$

We still need a rule which tells us how to find the path of a particle in space–time from this metric. In Euclidean three-space, the answer is that we seek that path which minimises the distance ds between the points A and B. We need to find the corresponding result for space–time. If we join the points A and B by a large number of possible routes through space–time, it is evident that the path corresponding to free fall between A and B must be the shortest path. From consideration of the twin paradox in general relativity, the free fall path between A and B must also correspond to the maximum proper time between A and B, i.e. we require $\int_A^B ds$ to be a *maximum* for the shortest path. In terms of variational calculus, we can write this as

$$\delta \int_A^B ds = 0 \tag{14.31}$$

It is interesting to show that, for our naive metric (14.16), this condition is exactly equivalent to *Hamilton's principle* in mechanics and dynamics.

$$\int_A^B ds = \int_{t_1}^{t_2} \frac{ds}{dt}\,dt$$

$$= \int_{t_1}^{t_2} \left[\left(1 + \frac{2\phi}{c^2}\right) - \frac{v^2}{c^2}\right]^{\frac{1}{2}} dt$$

since $dl/dt = v$. For weak fields, $(2\phi/c^2) - (v^2/c^2) \ll 1$, this reduces to

$$\int_{t_1}^{t_2} (U - T)\,dt \tag{14.32}$$

where U is the potential energy of the particle and T its kinetic energy. I find it remarkable that the constraint that ds must be a maximum results exactly in Hamilton's principle in dynamics (see Section 5.4).

14.6 The Schwarzschild metric

The solution of Einstein's equations for the case of a point mass was discovered by Schwarzschild in 1915 only a few months after the first publication of Einstein's theory in its final form. The *Schwarzschild metric* for a point mass M has the form

$$ds^2 = \left(1 - \frac{2GM}{rc^2}\right) dt^2 - \frac{1}{c^2}\left[\frac{dr^2}{\left(1 - \frac{2GM}{rc^2}\right)} + r^2(d\theta^2 + \sin^2\theta\, d\phi^2)\right] \tag{14.33}$$

From our analyses of the previous sections, we can now understand the significance of the various terms in this metric.

(*a*) The *time coordinate* has been written in exactly the form we require for all clocks to be synchronised everywhere, i.e. the interval of proper time is $d\tau = (1 - 2GM/rc^2)^{\frac{1}{2}}\,dt$. Note, however, that whereas the origin of the factor $(1 - 2GM/rc^2)$ in the metric (14.16) was only approximate, being the expansion of $e^{\Delta\phi/c^2}$, in general relativity, the present result is *exact* for all values of the parameter $2GM/rc^2$.

(*b*) The angular coordinates have been written in terms of spherical polar coordinates centred on the point mass. The radial coordinate r is such that metric distances perpendicular to the radial coordinate are correctly given by $r\,(d\theta^2 + \sin^2\theta\, d\phi^2)^{\frac{1}{2}}$; r is often called the 'angular diameter distance' and is distinct from the geodesic distance from the mass to the point labelled by r. The proper distance from 0 to r is given by

$$\int_0^r \frac{dr}{\left(1 - \dfrac{2GM}{rc^2}\right)^{\frac{1}{2}}} \tag{14.34}$$

(*c*) We can see from this formalism how mass influences the curvature of space. We recall that, in isotropic curved spaces, we were able to write the spatial component of the metric:

$$\frac{dr^2}{(1 - Kr^2)} + r^2(d\theta^2 + \sin^2\theta\; d\phi^2) \tag{14.35}$$

We can see that a measure of the local space curvature is given by equating Kr^2 and $2GM/rc^2$:

$$K = \frac{2GM}{c^2 r^3} \tag{14.36}$$

This provides us with a measure of the curvature of space produced by a point mass M at distance r. Notice that this formula has the attractive feature that K tends to zero as r tends to infinity and also that K is proportional to M. If we rewrite the curvature K as $1/R^2$, we see that

$$\frac{r^2}{R^2} = \frac{2GM}{rc^2}$$

This is a measure of the curvature of space relative to the distance from a massive object. Notice that $2GM/rc^2$ is also the 'relativistic' factor present in the time- and radial-components of the metric. Let us look at some typical values of this parameter.

At the surface of the Sun, $2GM_\odot/rc^2 \approx 4 \times 10^{-6}$

At the surface of the Earth, $2GM_E/rc^2 \approx 1.4 \times 10^{-8}$

i.e. within the solar system the effects of space curvature are very small and require measurements of extreme precision in order to be detected at all. The second example indicates the factor by which the sum of the angles of a triangle differ from 180° in measurements made at the surface of the Earth. In fact, Gauss is reputed to have carried out an experiment to determine exactly this in the Harz mountains in the 1820s. Using a theodolite, he found that the angles of the triangle formed by three tall mountains added up to 180° but we now understand how small the effects of space curvature are expected to be. What is remarkable is that he had the insight to realise that Euclid's fifth postulate, which basically asserts that parallel lines only meet at infinity or that the sum of the angles of a triangle is 180°, need not necessarily be correct if we look over large enough distances. However, at the surface of neutron stars, the effects of general relativity are important. For them, $r \approx 10$ km, $M_{NS} \approx 1\,M_{\odot} \approx 2 \times 10^{30}$ kg and $2GM_{NS}/rc^2 \approx 0.3$.

(*d*) Finally, we note that something 'funny' must happen at the radial coordinate $2GM/rc^2 = 1$. The radius defined by this relation, $r_g = 2GM/c^2$, is known as the *Schwarzschild radius* and plays a particularly important role in the study of black holes which we will come to in a moment.

14.7 Orbits about a central point mass

We can learn a lot about the differences between Newtonian gravity and general relativity by studying particle orbits about a central point mass. We will consider orbits in a plane and so the Schwarzschild metric can be written

$$ds^2 = \alpha\,dt^2 - \frac{1}{c^2}(\alpha^{-1}\,dr^2 + r^2\,d\phi^2) \tag{14.37}$$

In this notation, ϕ is the angle in the equatorial plane and $\alpha = (1 - 2GM/rc^2)$. We now need to find the orbits which correspond to maximising ds. Let us introduce a length variable l, whose nature we will not define precisely, with respect to which we vary $\int ds$. We will aim to get rid of l from the formulae as soon as possible. Then

$$ds^2 = \left[\alpha\left(\frac{dt}{dl}\right)^2 - \frac{1}{c^2}\,\alpha^{-1}\left(\frac{dr}{dl}\right)^2 - \frac{r^2}{c^2}\left(\frac{d\phi}{dl}\right)^2\right]^{\frac{1}{2}} dl \tag{14.38}$$

Things will look neater if we write

$$\left(\frac{dt}{dl}\right) = t', \quad \left(\frac{dr}{dl}\right) = l', \quad \left(\frac{d\phi}{dl}\right) = \phi'$$

We must now use the calculus of variations to maximise $\int ds$, i.e.

$$\delta\int_a^b ds = \delta\int_a^b \left(\alpha t'^2 - \alpha^{-1}\frac{r'^2}{c^2} - \frac{r^2}{c^2}\,\phi'^2\right)^{\frac{1}{2}} dl = 0 \tag{14.39}$$

We need Euler's equation for each independent pair of variables, i.e., in the standard notation, if we regard $\int F(y, y', x)\, dx$ as the function to be maximised, we must solve

$$\frac{\partial F}{\partial y} - \frac{d}{dx}\left(\frac{\partial F}{\partial y'}\right) = 0 \tag{14.40}$$

(see Section 5.4). Let us deal with the t and ϕ coordinates first. For the t coordinate,

$$\begin{aligned} F(y, y', x) &= F(t, t', l) \\ &= \left(\alpha t'^2 - \alpha^{-1}\frac{r'^2}{c^2} - \frac{r^2}{c^2}\phi'^2\right)^{\frac{1}{2}} = \frac{ds}{dl} \end{aligned} \tag{14.41}$$

Substituting this form into (14.40), we find, with a little algebra, that

$$\frac{d}{dl}\left[\frac{\alpha t'}{(ds/dl)}\right] = 0$$

i.e.

$$\frac{\alpha t'}{(ds/dl)} = \text{constant} = A \tag{14.42}$$

Notice that, in this maximisation, we are finding that property of the metric which is invariant with respect to the time variable. According to the analysis of Section 5.6, this relation is the general relativistic equivalent of the equation of *energy conservation.*

Let us perform the same analysis for the angular coordinate ϕ. We guess immediately that this will result in the equivalent of *conservation of angular momentum.* In this case, $F(y, y', x) = F(\phi, \phi', l)$ and is the same as (14.41). Substituting again into (14.40), we find that

$$\frac{d}{dl}\left[\frac{r^2\phi'}{(ds/dl)}\right] = 0$$

i.e.

$$\frac{r^2\phi'}{(ds/dl)} = \text{constant} = h \tag{14.43}$$

The relation with conservation of angular momentum is clear if we look at the dimensions of $\phi'/(ds/dl)$.

$$[\phi'/(ds/dl)] = [(d\phi/dl)/(ds/dl)] = [d\phi/d\tau]$$

where τ is proper time, i.e. there is a clear relation between $\phi'/(ds/dl)$ and something which is dimensionally an angular velocity. Now rather than solve for r by the same method, let us substitute the results (14.42) and (14.43) back into the metric in the form of (14.41).

$$\left(\frac{ds}{dl}\right)^2 = \frac{A^2}{\alpha}\left(\frac{ds}{dl}\right)^2 - \frac{1}{\alpha c^2}\left(\frac{dr}{dl}\right)^2 - \frac{h^2}{c^2 r^2}\left(\frac{ds}{dl}\right)^2 \tag{14.44}$$

Dividing through by $(\mathrm{d}s/\mathrm{d}l)^2$, we find that

$$\left(\frac{\mathrm{d}r}{\mathrm{d}s}\right)^2 + \frac{\alpha h^2}{r^2} = c^2(A^2 - \alpha) \tag{14.45}$$

This is beginning to look somewhat familiar. Let us define $\mathrm{d}r/\mathrm{d}s$ to be $\dot{r}$, i.e. something that looks like a time derivative but is actually with respect to s.

$$\dot{r}^2 + \frac{\alpha h^2}{r^2} = c^2(A - \alpha) \tag{14.46}$$

The Newtonian equivalent of this expression is the conservation of energy in a gravitational field, i.e.

$$\text{Newton} \quad \frac{\dot{r}^2}{c^2} + \frac{h^2}{r^2c^2} - \frac{2GM}{rc^2} = \frac{\dot{r}_\infty^2}{c^2} \tag{14.47}$$

In this equation $\dot{r}$ means $\mathrm{d}r/\mathrm{d}t$, h is the specific angular momentum, i.e. $h = r^2\,\mathrm{d}\phi/\mathrm{d}t$, and $\dot{r}_\infty$ is the radial component of velocity at infinity which is just a constant. For comparison, (14.46) can be written

$$\text{GR} \quad \frac{\dot{r}^2}{c^2} + \frac{h^2}{r^2c^2} - \frac{2GM}{rc^2} - \frac{2GMh^2}{r^3c^4} = (A^2 - 1) \tag{14.48}$$

Although this looks very similar to (14.47), we must recall that the definitions of all the variables in (14.48) are different from those of (14.47). Only in the weak field limit will they be the same. We see that, besides the different meanings of the coordinates, the main difference is the appearance of the term $2GMh^2/r^3c^4$ on the left-hand side of equation (14.48). Let us now investigate the influence of this term on particle dynamics.

In Newtonian dynamics, we can represent the radial component of the velocity of the particle in its orbit about the point mass in terms of the variation of the different terms in equation (14.47) with radius. Let us write the equation in terms of $r_g = 2GM/c^2$ and a dimensionless specific angular momentum $\eta = h^2/r_g^2c^2$. Then

$$\text{Newton} \quad \frac{\dot{r}^2}{c^2} = -\left[\frac{\eta}{(r/r_g)^2} - \frac{1}{(r/r_g)}\right] + \frac{\dot{r}_\infty^2}{c^2} \tag{14.49}$$

The term in brackets on the right-hand side acts as a potential Φ:

$$\Phi = \frac{\eta}{(r/r_g)^2} - \frac{1}{(r/r_g)} \tag{14.50}$$

where the first term acts as a 'centrifugal' potential and the second the gravitational potential. In Figure 14.8(a), $-\Phi$ against r/r_g has been drawn. If a particle starts from infinity with zero radial velocity, i.e. from A', the radial component of its velocity can be found from the curve. Obviously, when the particle reaches the radius marked A, the radial component of its velocity is zero and solutions

Figure 14.8. Comparison of the gravitational potential Φ in (*a*) Newtonian theory and (*b*) general relativity. In Newtonian gravity, there are two contributions to the potential whereas in general relativity, there is an additional attractive potential which dominates close to the origin.

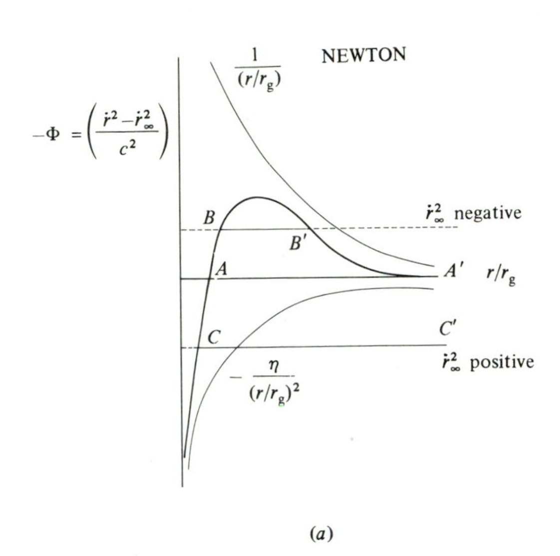

(*a*)

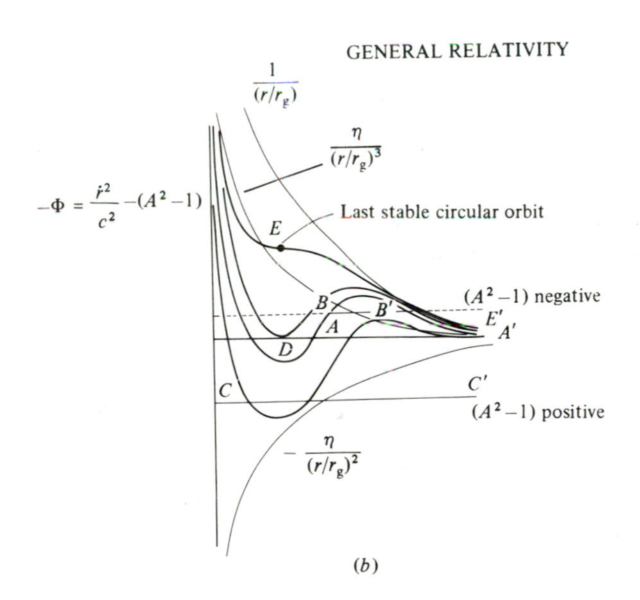

(*b*)

do not exist for smaller values of r because $\dot{r}$ would then be imaginary. In physical terms, what has happened is that, because of conservation of energy and angular momentum, all the kinetic energy is in the ϕ component of the particle's orbit. A is the distance of closest approach of the particle in a parabolic orbit.

Bound orbits are found by taking $\dot{r}_\infty^2$ negative. Then the particle moves in an elliptical orbit between the radii B and B'. Hyperbolic orbits of unbound particles correspond to positive values of $\dot{r}_\infty^2$ and are represented by loci such as CC'. C is the distance of closest approach and is the solution of (14.49) with $\dot{r} = 0$. Notice that only if $\eta = 0$ is it possible for the particle to reach $r = 0$. In other words, even if the angular momentum is very small, this is still sufficient to prevent particles reaching the origin.

We can repeat the analysis in the case of general relativity. Again, we define $r_g = 2GM/c^2$ and $\eta = h^2/r_g^2c^2$ noting that h has a different meaning from that of the Newtonian case. Then equation (14.48) becomes

$$\text{GR} \quad \frac{\dot{r}^2}{c^2} = -\left[\frac{\eta}{(r/r_g)^2} - \frac{1}{(r/r_g)} - \frac{\eta}{(r/r_g)^3}\right] + (A^2 - 1) \tag{14.51}$$

We can draw the same curves as in Figure 14.8(a) but now we must include the specifically general relativistic term $\eta/(r/r_g)^3$ which acts as a negative potential and which increases in magnitude even more rapidly than the centrifugal potential as $r \to 0$. We display the value of $\dot{r}^2/c^2$ in Figure 14.8(b) for different values of η. The effect of the presence of the term in $(r/r_g)^3$ is to introduce a strong attractive potential which becomes dominant if particles can penetrate to small values of r/r_g. The behaviour of the particle depends upon the magnitude of η. Suppose we consider a particle with $A^2 - 1 = 0$, i.e. equivalent to the case of a particle with zero radial velocity at infinity. If η is large, the particle behaves as before and will attain zero radial velocity at the point A. However, if η becomes smaller, the behaviour illustrated by DA is found. At the point D, the negative general relativistic potential term is sufficient to counterbalance the usual centrifugal potential and the particle can fall in to $r = 0$. If $A^2 - 1$ is negative, bound orbits such as BB' are possible, but eventually even these become impossible. So long as the chord BB' is of finite length, a range of elliptical and circular bound orbits is possible, but eventually as η decreases there is a certain value of η when B and B' coincide, as illustrated by E. This corresponds to the last stable circular orbit for particles about the mass M.

We can work out readily the relevant ranges of η appropriate for these different types of behaviour. The limiting case D corresponds to the case in which bound orbits are possible but any particle which falls in from infinity, i.e. $(A^2 - 1) \geqslant 0$, will fall in to $r = 0$. The condition for this to occur is that the roots of $\Phi = 0$ should be equal to zero.

$$\frac{1}{(r/r_g)^3}\left[\eta\left(\frac{r}{r_g}\right)-\left(\frac{r}{r_g}\right)^2-\eta\right]=0$$

One root is clearly $r = \infty$. The others are at the value $\eta = 4$, $r/r_g = 2$.

The other limiting case is that corresponding to the last stable orbit E. In this case, the turning points of the potential function coincide and $d^2\Phi/dr^2 = 0$. The roots coincide when $\eta = 3$ and the corresponding value of r is $3r_g$. This is an important result for the dynamics of orbits about point masses and, in particular, in the study of accretion discs about black holes.

This is the origin of the phenomenon of *black holes*. The potential barrier which is created by the centrifugal potential in Newtonian theory is modified in general relativity so that particles can fall in to $r = 0$ despite the fact that they have a finite amount of angular momentum. We will discover other remarkable phenomena in a moment.

We have been able to deduce these results from equation (14.48) which is an energy integral. We have yet to take the final step of integrating that equation to find the equation of the particle's orbit. The standard procedure is to make the substitution $u = 1/r$ and then to note that, since $r^2 d\phi/ds = h$, we can replace differentials with respect to s as follows.

$$\frac{d}{ds}=\frac{h}{r^2}\frac{d}{d\phi}$$

Then, in a straightforward way, we find that

$$\frac{dr}{ds}=-h\frac{du}{d\phi}$$

Equation (14.48) then becomes

$$\left(\frac{du}{d\phi}\right)^2+u^2=\frac{c^2}{h^2}(A^2-1)+\frac{c^2r_g}{h^2}u+r_gu^3 \tag{14.52}$$

Differentiating again with respect to ϕ and dividing through by $2du/d\phi$, we obtain

$$\text{GR}\quad \frac{d^2u}{d\phi^2}+u=\frac{GM}{h^2}+\frac{3GM}{c^2}u^2 \tag{14.53}$$

The corresponding Newtonian equation can be derived from the Newtonian version of the problem, equation (14.47), and is

$$\text{Newton}\quad \frac{d^2u}{d\phi^2}+u=\frac{GM}{h^2} \tag{14.54}$$

The specifically general relativistic term in equation (14.53) is the term $(3GM/c^2)u^2$. Normally, this is only a very small correction term unless we are dealing with distances of the order of r_g.

The classical example of the use of this equation of motion is in the study of the precession of planetary orbits. The position of closest approach of a planet to the Sun in an elliptical orbit is known as the *perihelion* of the planet's orbit. Most of the planetary orbits are more or less circular but Mercury's orbit has eccentricity $e = 0.2$ and hence precession of its orbit is measurable with high precision. For simplicity, we will consider only the precession of circular orbits by looking at how the phase of the planet's position in its orbit is influenced by the general relativistic term in equation (14.53).

For a circular orbit, in Newtonian theory, $\mathrm{d}^2u/\mathrm{d}\phi^2 = 0$ and hence

$$u = \frac{GM}{h^2} \tag{14.55}$$

Let us therefore look at perturbations about this relation due to the 'correction' term in (14.53). We write

$$u = \frac{GM}{h^2} + g(\phi) \tag{14.56}$$

and substitute into (14.53) preserving terms to first order in $g(\phi)$. We find that

$$\frac{\mathrm{d}^2g}{\mathrm{d}\phi^2} + g\left[1 - \left(\frac{3GM}{c^2}\right)\left(\frac{2GM}{h^2}\right)\right] = \frac{3GM}{c^2}\left(\frac{GM}{h^2}\right)^2 \tag{14.57}$$

It can be seen that this is a harmonic equation for g as a function of ϕ. If there were no term $(3GM/c^2)u^2$ present in (14.53), the equation would read

$$\frac{\mathrm{d}^2g}{\mathrm{d}\phi^2} + \omega^2 g = 0$$

i.e. g is harmonic with period 2π. However, because of the perturbation term, the phase of the particle in its orbit is slightly different from 2π per revolution. If we write

$$\frac{\mathrm{d}^2g}{\mathrm{d}\phi^2} + \omega^2 g = \text{constant}$$

$$\omega^2 = \left[1 - \left(\frac{3GM}{c^2}\right)\left(\frac{2GM}{h^2}\right)\right] \tag{14.58}$$

or the period of the orbit $T = 2\pi/\omega$ is

$$T = \frac{2\pi}{\left[1 - \left(\frac{3GM}{c^2}\right)\left(\frac{2GM}{h^2}\right)\right]^{\frac{1}{2}}}$$

$$= 2\pi\left(1 + \frac{3G^2M^2}{c^2h^2}\right) \tag{14.59}$$

i.e. the orbit closes up slightly later per orbit. The fractional phase change per orbit is

$$\frac{d\phi}{2\pi} = \frac{3G^2M^2}{c^2h^2} = \frac{3}{4}\left(\frac{2GM}{hc}\right)^2 \tag{14.60}$$

For circular orbits, $h = rv$ and hence

$$\frac{d\phi}{2\pi} = \frac{3}{4}\left(\frac{c}{v}\right)^2\left(\frac{r_g}{r}\right)^2 \tag{14.61}$$

For elliptical orbits, the exact result is

$$\frac{d\phi}{2\pi} = \frac{3}{4}\left(\frac{c}{v}\right)^2\left(\frac{r_g}{r}\right)^2\frac{1}{(1-e^2)} \tag{14.62}$$

For the planet Mercury, $r = 5.8 \times 10^{12}$ cm, $T = 88$ days, r_g (Sun) $= 3$ km and $e = 0.2$. It is one of the great triumphs of general relativity that, on putting these values into the above relations, we predict a precession of about 43 arcsec per century, in exact agreement with the residual precession present in Mercury's orbital precession.

14.8 Light rays in Schwarzschild space–time

The basic postulate of relativity is that light propagates along null-geodesics, i.e. $ds^2 = 0$. The dynamical solutions we have found, (14.42) and (14.43), are

$$\frac{\alpha(dt/dl)}{ds/dl} = A \quad \text{and} \quad \frac{r^2(d\phi/dl)}{ds/dl} = h$$

Since $ds = 0$, both A and h must be infinite for light rays, although their ratio A/h has a finite value. Therefore, we find the propagation equations for photons about a point mass from equation (14.53) with $h = \infty$, i.e.

$$\frac{d^2u}{d\phi^2} + u = \frac{3GM}{c^2}u^2 \tag{14.63}$$

It is the term on the right-hand side which describes the influence of space–time curvature on the propagation of light. To estimate the amount by which light is bent, let us work out the deflection of a light ray passing by the limb of the Sun. In the absence of the term $(3GM/c^2)u^2$, the propagation equation is

$$\frac{d^2u}{d\phi^2} + u = 0$$

A suitable solution of this equation is $u_0 = \sin\phi/R$ which is indicated in Figure 14.9. It corresponds to a straight line, tangent to a sphere of radius R. To find the solution for the path of the light ray in the next order of approximation, let us seek solutions of equation (14.63) with $u = u_0 + u_1$. Then

$$\frac{d^2u_1}{d\phi^2} + u_1 = \frac{3GM}{c^2R^2}\sin^2\phi \tag{14.64}$$

By inspection, we find that a suitable trial solution for u_1 is $u_1 = A + B\cos 2\phi$. Differentiating and equating coefficients gives the following solution for u:

$$u = u_0 + u_1 = \frac{\sin\phi}{R} + \frac{3}{2}\frac{GM}{c^2R^2}(1 + \tfrac{1}{3}\cos 2\phi) \tag{14.65}$$

We need only study the asymptotic solutions at large values of r and so $\sin\phi \approx \phi$, $\cos 2\phi \approx 1$. In the limit $r \to \infty$, $u \to 0$, we find that

$$u = \frac{\phi_\infty}{R} + \frac{3}{2}\frac{GM}{c^2R^2}(1 + \tfrac{1}{3}) = 0$$

i.e.

$$\phi_\infty = -\frac{2GM}{Rc^2} \tag{14.66}$$

From the geometry of Figure 14.9, it is clear that the total deflection is twice this value of ϕ_∞, i.e.

$$\Delta\phi = \frac{4GM}{Rc^2} \tag{14.67}$$

This calculation agrees precisely with observations of the deflection of light and radio waves which just graze the Sun's limb, the theoretical prediction amounting to 1.75 arcsec. A Newtonian calculation of the deflection can be performed using the procedures outlined in Section 14.3.2 and this results in a deflection of half that predicted by general relativity. We can show this rather neatly using the formulae for Rutherford scattering which we developed in the appendix to Chapter 2, Section A2.3. To change from electrostatic forces to gravitational forces we replace $(Zze^2/4\pi\epsilon_0)$ by $-(GmM)$ in equation (A2.22), where M is the mass of the Sun and m the mass of the orbiting particle. The 'classical' result for the deflection of photons by the Sun is found by taking the impact parameter $p_0 = R$, the radius of the Sun, and v_0, the velocity of the particle, to be the velocity of light. Conveniently, m cancels out and, in the limit of very small scattering angles, we find that

Figure 14.9. The coordinate system for working out the deflection of light waves by the Sun.

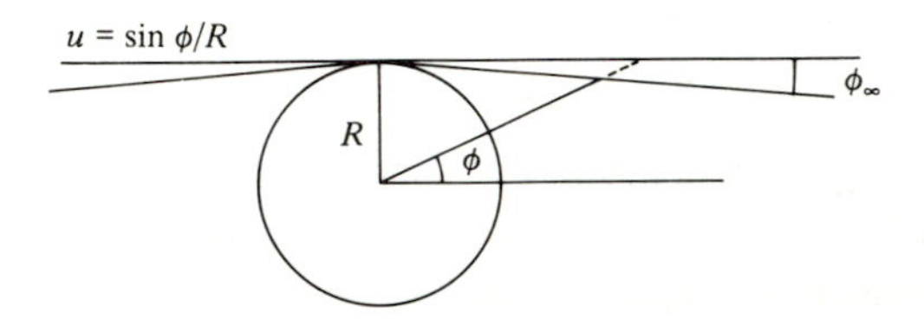

$$\Delta\phi = \frac{2GM}{Rc^2}$$

14.9 Particles and light rays near black holes

Let us consider first a particle falling radially to $r = 0$. In this case, we substitute $h = 0$ into (14.48) and then

$$\left(\frac{\mathrm{d}r}{\mathrm{d}s}\right)^2 - \frac{2GM}{r} = (A^2 - 1)c^2 \tag{14.68}$$

If the particle falls from rest at distance R, the constant on the right-hand side can be rewritten as $-2GM/R$ and hence

$$\left(\frac{\mathrm{d}r}{\mathrm{d}s}\right)^2 = 2GM\left(\frac{1}{r} - \frac{1}{R}\right) \tag{14.69}$$

We can find a simple parametric solution for this equation,

$$\left.\begin{aligned} \int_0^\tau \mathrm{d}s &= R\left(\frac{R}{2GM}\right)^{\frac{1}{2}}\left(\eta - \frac{\sin 2\eta}{2}\right) \\ \frac{r}{R} &= \sin^2\eta \end{aligned}\right\} \tag{14.70}$$

It is probably more instructive to consider the simplest case of infall from rest at infinity, i.e. $R \to \infty$, since this solution contains all the essential features we wish to bring out in this analysis.

$$\left(\frac{\mathrm{d}r}{\mathrm{d}s}\right)^2 = \frac{2GM}{r} \tag{14.71}$$

Now ds is just an interval of proper time and therefore we can work out the proper time it takes a particle to fall in from some arbitrary point on its trajectory to the point at $r = 0$.

$$\int_0^{r_1} r^{\frac{1}{2}}\,\mathrm{d}r = \int_{\tau_1}^{\tau_2} (2GM)^{\frac{1}{2}}\,\mathrm{d}s$$

$$(\tau_2 - \tau_1) = \left(\frac{2}{9GM}\right)^{\frac{1}{2}} r_1^{\frac{3}{2}} \tag{14.72}$$

Thus, particles fall in to $r = 0$ in a *finite proper time*. Notice that nothing funny happens at r_g. Now let us repeat the calculation for the time measured by an observer at infinity, i.e. someone who measures dt rather than proper time at the particle. The metric (14.37) reduces to

$$\mathrm{d}t^2 = \alpha^{-1}\,\mathrm{d}s^2 + \frac{1}{c^2}\,\alpha^{-2}\,\mathrm{d}r^2$$

and, substituting for ds using equation (14.71), we find that

$$\mathrm{d}t^2 = \alpha^{-1}\,\mathrm{d}r^2\left(\frac{r}{2GM} + \frac{1}{c^2}\,\alpha^{-1}\right)$$

After some simple algebra, this reduces to

$$\mathrm{d}t = -\frac{\mathrm{d}r}{c\left(\frac{2GM}{rc^2}\right)^{\frac{1}{2}}\left(1-\frac{2GM}{rc^2}\right)} \tag{14.73}$$

Performing the same integral for collapse from r_1 to r_2,

$$\int_{t_1}^{t_2}\mathrm{d}t = -\int_{r_1}^{r_2}\frac{r^{\frac{3}{2}}\,\mathrm{d}r}{c\left(r-\frac{2GM}{c^2}\right)\left(\frac{2GM}{c^2}\right)^{\frac{1}{2}}}$$

$$= -\frac{1}{c\,r_g^{\frac{1}{2}}}\int_{r_1}^{r_2}\frac{r^{\frac{3}{2}}\,\mathrm{d}r}{(r-r_g)} \tag{14.74}$$

We now see that the coordinate time t as measured by a distant observer diverges as r_2 tends to r_g. Thus, although collapse to the point $r = 0$ occurs in a finite proper time, to the external observer, it takes an infinite time to reach the distance r_g, the Schwarzschild radius, from the origin. This behaviour is indicated schematically in Figure 14.10.

In the process, however, the light signals seen by the observer at infinity become progressively redshifted as $r_2 \to r_g$. The relation between proper and coordinate time is given by

$$\mathrm{d}s = \mathrm{d}\tau = \left(1-\frac{2GM}{rc^2}\right)^{\frac{1}{2}}\mathrm{d}t \tag{14.75}$$

and hence, in terms of frequencies,

Figure 14.10. Comparison of the proper time τ and time observed at infinity for an object falling radially into the origin at $r = 0$.

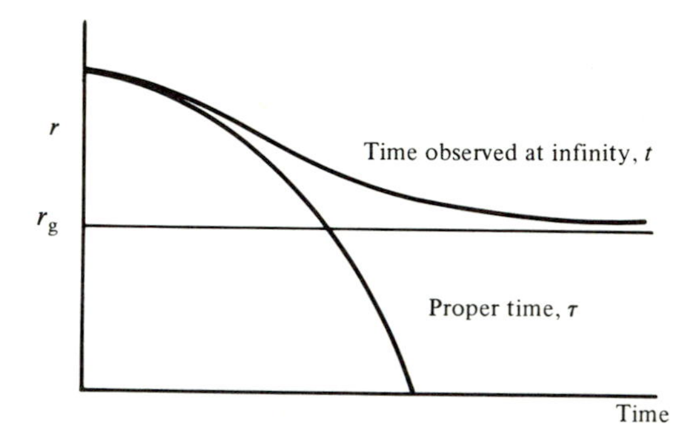

$$\nu_0 = \left(1 - \frac{r_g}{r}\right)^{\frac{1}{2}} \nu_1 \tag{14.76}$$

where ν_1 is the emitted frequency and ν_0 the frequency measured at infinity. In terms of redshift z,

$$z = \left(1 - \frac{r_g}{r}\right)^{-\frac{1}{2}} - 1 \tag{14.77}$$

i.e. as $r \to r_g$, $z \to \infty$. A further consequence of this analysis is that we cannot observe any light signals originating from radii less than r_g. In other words, light signals can certainly travel in to $r = 0$ but those within r_g cannot propagate beyond r_g. We may think of this phenomenon in terms of gravity being so strong that the light rays are bent back on themselves and cannot penetrate the 'surface' at r_g.

We have thus exposed most of the basic properties of *black holes*. They are defined in the case of the Schwarzschild metric only by their *mass* and they possess an effective radius $r_g = 2GM/c^2$. Particles and light rays falling within r_g inevitably collapse to the singularity at $r = 0$. Nothing can come out of the black hole beyond r_g according to our classical analysis. There are stable orbits about the black hole but the closest stable orbit is circular and has a radius $r = 3r_g$.

The nature of the singularity at $r = r_g$ has been the subject of much study and it is now clear that it is not a real singularity but one associated with our particular choice of coordinates. We have already seen that particles can collapse into the origin in a finite proper time and these particles are hardly aware of passing through $r = r_g$. Alternative coordinate systems can be used, for example Kruskal coordinates, in which this coordinate singularity is eliminated. More details of this phenomenon are given in Chapter 8 of Rindler's book.[8] Thus, it appears that for $r < r_g$ the space and time components in the metric (14.37) are interchanged, but this can be shown to be as a result of the particular set of coordinates in which that metric is written. On the other hand, the singularity at $r = 0$ seems to be a *real physical singularity* and has been the subject of intense study. Further discussion of this basic problem is beyond the scope of what I want to cover in this chapter.

Two further recent advances are worth noting. First, in the early 1960s, Kerr discovered a solution of Einstein's equations corresponding to a *rotating black hole*. It turns out that, in addition to mass, black holes can also possess angular momentum. This gives rise to some remarkable properties, including the dragging of inertial frames close to the black hole. Rotating black holes are defined by their mass M and angular momentum J. They possess similar properties to those of spherically symmetric black holes but now there are two significant radii to consider rather than one. The equivalent of the Schwarzschild radius is the radius

$$r_{\mathrm{g}} = \frac{GM}{c^2} + \left[\left(\frac{GM}{c^2}\right)^2 - \left(\frac{J}{Mc}\right)^2\right]^{\frac{1}{2}}$$

If $J > GM^2/c$, no black hole can form. There is, however, another radius

$$r_{\mathrm{stat}} = \frac{GM}{c^2} + \left[\left(\frac{GM}{c^2}\right)^2 - \left(\frac{J}{Mc}\right)^2 \cos^2\theta\right]^{\frac{1}{2}}$$

where θ is the polar angle with respect to the rotation axis of the black hole. This radius lies outside r_{g} and the region between r_{g} and r_{stat} is referred to as the *ergosphere* of the black hole. Within the ergosphere no particle can remain static and there exist bound orbits within it. Thus, the last stable orbits can lie much closer to $r = 0$ in the case of rotating black holes than in the case of the Schwarzschild metric.

The second point is that one of the most remarkable discoveries of recent years has the demonstration of Hawking[9] that, when we consider quantum phenomena in the vicinity of the black hole, there is a finite probability of particles or photons escaping from within r_{g}. We can associate a temperature with a black hole of a particular mass and the black hole will behave as if it were a black body at that temperature. Hawking has shown that the temperature of the black hole is

$$T = \frac{hc^3}{16\pi^2 GMk} \approx \frac{10^{-8}}{(M/M_{\odot})} \text{ K}$$

where $M_{\odot} = 2 \times 10^{30}$ kg is the mass of the Sun. Thus, if we wait long enough, black holes will eventually evaporate away by Hawking radiation.

These last two paragraphs are quite clear illustrations of the excitement of research in contemporary relativity. However, the enthusiastic student must be warned that this is a subject of the greatest technical complexity and should not be embarked upon without a thorough grounding in theoretical physics and general relativity.

APPENDIX TO CHAPTER 14

ISOTROPIC CURVED SPACES

We can give a simple proof of the result that the only isotropic two-spaces are those which have constant curvature K where K can be positive, negative or zero. This proof was shown to me long ago by Dr Peter Scheuer. Suppose we draw on an arbitrary two-space the paths of neighbouring light rays from an origin at O (Figure A14.1). In order to measure the curvature of the space, we perform the equivalent operation to measuring the angles of a triangle. In the case of an arbitrary curved space, we work out the sum of the angles enclosed by our figure by taking a little vector around a closed area by *parallel displacement.* By this we mean that the vector moves round the figure in the two-space so that, at every point on the curve, it makes the same angle with the direction of the light ray. This procedure is very important in general relativity and requires considerable mathematical apparatus to set it up for general four-spaces. However, in the case of two-spaces which are isotropic, the analysis is simple and straightforward. In our picture, Figure A14.1, it is clear that the non-Euclidean aspects of the two-space become important when the vector has to turn the corners.

We define the small area $ABCD$ by two light rays AB and DC which are chosen such that they are perpendicular to the light ray OAD. Let us suppose that our vector starts off parallel to AD at A and is parallelly transported along AB to B. In general, the vector will not be parallel to BC at B and we have to

Figure A14.1

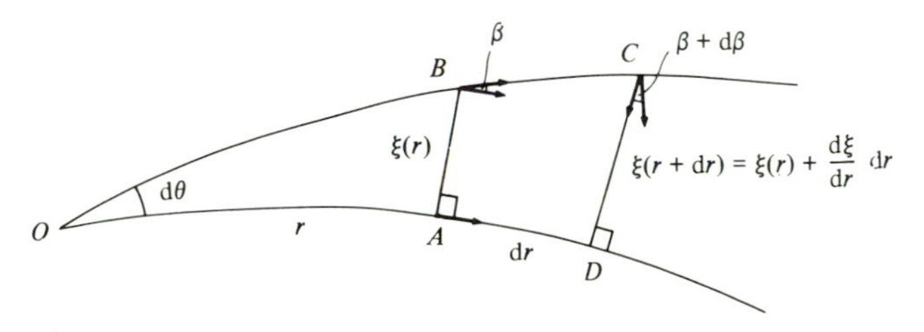

rotate the vector by a small angle β to make it so. The vector is then rotated through $\pi/2$ radians and transported along *BC* to *C*. At *C*, it has to rotate through an angle $\beta + d\beta$ in order to lie parallel to *CD*. Notice that this rotation is in the opposite sense to the rotation which took place at *B*. The vector is then rotated through $\pi/2$ radians and parallelly transported to *D*. It is now parallel to *AD*. Therefore, the vector is rotated by $\pi/2$ and transported back to *A* where it makes one final $\pi/2$ rotation to end up in the same direction in which it started. Thus, in going round the loop, the vector has rotated through an angle $2\pi + d\beta$ since the rotations at *B* and *C* to bring the vector parallel to the light ray are in opposite senses. Notice that only in flat Euclidean space is $d\beta = 0$.

Now the rotation of the vector $d\beta$ must depend upon the area enclosed by the loop. If we divide the area *ABCD* into a number of sub-loops, the separate small rotations for each sub-loop must add up linearly to the total rotation $d\beta$ (Figure A14.2). This suggests that the value of $d\beta$ should be proportional to the area of the loop. Furthermore, because the space is isotropic, we should obtain the same rotation $d\beta$ wherever we locate our loop within the two-space. We conclude that, in isotropic two-spaces, the constant of proportionality relating $d\beta$ to the area of the loop must be constant everywhere in the space. Let us now see what this means mathematically.

We let the distance *AB* be $\xi(r)$ and the distance *DC* be $\xi(r + dr) = \xi(r) + (d\xi/dr)\,dr$. We see by inspection of Figure A14.1 that the angle β is given by

$$\beta = \frac{d\xi}{dr} \tag{A14.1}$$

Therefore,

$$\beta + d\beta = \frac{d\xi}{dr} + \frac{d}{dr}\left(\frac{d\xi}{dr}\right) dr$$

Subtracting,

$$d\beta = \frac{d^2\xi}{dr^2}\, dr \tag{A14.2}$$

Figure A14.2

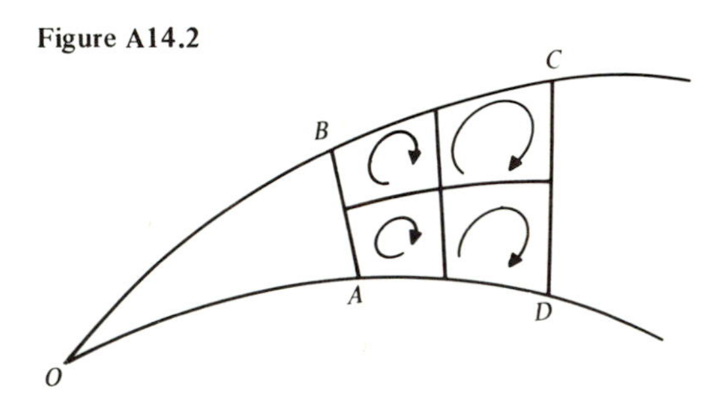

This must be proportional to the area of the loop, $A = \xi(r)\mathrm{d}r$. We will set the constant of proportionality equal to $-K$ where K is an arbitrary constant. Therefore,

$$\frac{\mathrm{d}^2\xi}{\mathrm{d}r^2} = -K\xi \qquad \text{(A14.3)}$$

But this is just an equation of simple harmonic motion for which the solution is

$$\xi = \xi_0 \sin K^{\frac{1}{2}} r \qquad \text{(A14.4)}$$

To find the constant ξ_0, we look at the value of ξ for very small values of r. For r small enough, the geometry must approximate to Euclidean space and then $\mathrm{d}\theta = \mathrm{d}\xi/\mathrm{d}r = K^{\frac{1}{2}}\xi_0 \cos K^{\frac{1}{2}} r \approx K^{\frac{1}{2}}\xi_0$.

Therefore, $\xi_0 = \mathrm{d}\theta/K^{\frac{1}{2}}$. The relation between ξ and r in the isotropic two-space is therefore

$$\xi = \frac{\mathrm{d}\theta}{K^{\frac{1}{2}}} \sin K^{\frac{1}{2}} r \qquad \text{(A14.5)}$$

This result applies equally for negative values of K. If we write $K = -K'$ where K' is positive,

$$\xi = \frac{\mathrm{d}\theta}{K'^{\frac{1}{2}}} \sinh K'^{\frac{1}{2}} r \qquad \text{(A14.6)}$$

We recognise that we have rederived the equations for the size of a rod of length ξ in an isotropic curved space and that K is exactly the same as the curvature $K = 1/R^2$ introduced in Section 14.4. In terms of R, the expressions for ξ are

$$\xi = \mathrm{d}\theta\, R \sin(r/R); \quad \xi = \mathrm{d}\theta\, R \sinh(r/R) \qquad \text{(A14.7)}$$

Only in the case $K = 0$, $R \to \infty$, do we recover the Euclidean result

$$\xi = r\theta \qquad \text{(A14.8)}$$

15

COSMOLOGY

15.1 Cosmology and physics

Cosmology is the application of the laws of physics on the very largest scale in the Universe. By definition, this means that the validation of our theories depends upon *observation* rather than on *experiment*. Obviously, this puts us one stage further from our 'apparatus' than in the case of a laboratory experiment and yet, if we look at the history of astronomical discovery, we find that, in many cases, astronomical observations have provided essential new pieces of physics which have been rapidly incorporated into the established structure of science. In Chapter 2, we illustrated how Tycho's observations of the motions of the planets led to Newton's laws of gravitation. Among the first determinations of the velocity of light was the observation of the time it takes light to propagate across the Earth's orbit about the Sun from observation of the eclipses of the satellites of Jupiter. The orbits of double stars indicated that the velocity of light is independent of the motion of the stars, direct evidence for Einstein's second postulate which forms the basis of special relativity.

A striking recent discovery which could only have been made astronomically was that of gravitational lenses. The quasars 0957+561 A and B are separated by only 6 arcsec on the sky and their properties are identical. It has been convincingly demonstrated that they are actually two separate images of a single object which is being focused by an intervening galaxy because of the gravitational deflection of light (see Section 14.8).

Most of the detailed tests of the general theory of relativity involve the use of astronomical objects. Perhaps the most exciting recent discovery of one such object is that of the pulsar PSR 1913+16 in a binary system. A pulsar is a magnetised rotating neutron star which has mass $M \approx 1\ M_{\odot}$, radius $r = 10$ km and general relativistic parameter $2GM/rc^2 \approx 0.3$. By processes which are poorly understood, this pulsar emits a sharp pulse of radio emission once per rotation (i.e. once every 0.059 seconds) and thus it is an ideal 'clock' in a rotating frame of reference. Since the binary system is rather close, subtle general relativistic effects can be tested by very precise timing of the arrival of the pulses at Earth. For example, the slow-down rate of the binary orbit is exactly that expected from the emission of gravitational waves. These tests have confirmed with high accuracy the validity of the general theory of relativity. Another nice example is the physics of the interiors of neutron stars where matter at nuclear density $\approx 10^{15}$ g cm^{-3} is found in bulk. Observable phenomena can be related to the properties of this nuclear superfluid.

Thus, astronomy has played and continues to play a major role in fundamental physics. The amazing fact is that the very best of laboratory physics enables us to understand in a convincing way the properties of celestial objects where physical conditions are present which are unattainable in the laboratory. However, when we come to the subject of cosmology, we are trying to find out about phenomena on the very largest scale and we have to make *assumptions* about which physical processes are dominant on this scale. A second key point is that we have only *one* universe to study and physicists have to be cautious about wholly accepting the results of a one-off experiment. However, in cosmology, there is no prospect of doing any better.

In this position, there are two possible approaches. On the one hand, one can adopt the very best of contemporary physics and astrophysics and see how much one can explain when this is applied to the Universe as a whole. This is the route which leads to the hot big bang model of the Universe which we describe in Section 15.6.3. On the other hand, one could turn the problem on its head and seek to derive new laws of physics by studying the Universe on the largest scales, which are not accessible in laboratories within the Solar System. This second approach is subject to much less stringent constraints than the first and, indeed, a number of astronomers have argued that this is the way in which one should proceed. They can cite some singular successes which have been achieved through this approach, Kelper's laws of planetary motion and Newton's law of gravitation being classical examples. The value of this second approach can only be judged from how convincing a case can be made from careful analysis of reliable observations for the new laws. We recall that Kepler's and Newton's analyses were wholly convincing because of the reliability of the observations and their interpretation. We should subject cosmological observations to the same degree of scrutiny that we would give to an experiment in the laboratory. We will look at some aspects of this approach to cosmology in Section 15.5.

It turns out that, using conventional physics, we can understand many of the new facts which are now known about the Universe. Indeed, I believe that the conventional approach has turned out to be very much more successful than any cosmologist had any right to expect. What has become the *standard hot big bang model of the Universe* leads to new uncharted areas of physics. It may well turn out that these areas of physics will only be understandable through cosmological observations and theory. Let us therefore proceed to the route which leads us to these exciting possibilities.

15.2 Basic observations

We have to identify the basic large-scale properties of the Universe as a whole which should provide the boundary conditions for our models of the Universe. It is traditional to develop models for our Universe starting from a number of basic facts and observations.

15.2.1 Olbers' paradox

We now know that the paradox was known to astronomers before Olbers but his name has been traditionally associated with it. The paradox arises as follows. Suppose we consider an infinite, static, universe uniformly filled with stars of luminosity L and of space density n. The apparent intensity (or flux density) S of a star at the Earth is given by the inverse law $S = L/4\pi r^2$. The number of stars in the distance range r to $r + \mathrm{d}r$ is $4\pi r^2 n\,\mathrm{d}r$ and consequently the total intensity of radiation arriving at the Earth is

$$I = \int_0^\infty \frac{L}{4\pi r^2}\, 4\pi r^2 n\, \mathrm{d}r$$

$$= Ln \int_0^\infty \mathrm{d}r \to \infty$$

i.e. in a uniform, infinite, static universe, the background light diverges. In fact, because of the finite surface area of stars, the surface brightness of the sky should be as bright as the surface of the stars, plainly in contradiction with our experience. Thus, the Universe cannot be simultaneously infinite, uniform and static. This result can be shown to be independent of the geometry of the Universe.

15.2.2 Hubble's law

In fact, Hubble showed in 1929 that the Universe is not static. He discovered that the observed velocities of distant galaxies are proportional to their distances: $\mathbf{v} = H_0 \mathbf{r}$, where H_0 is a constant, known appropriately as Hubble's constant. This is precisely what is expected in a uniformly expanding medium. By definition, in uniform expansion $\Delta r/r =$ constant in any given time interval Δt, i.e. $\Delta r/\Delta t \propto r$, $v \propto r$. The fact that the system of galaxies expands can resolve the paradox of the divergence of the background sky intensity as we will show below.

15.2.3 The isotropy of the Universe

Any picture of the distribution of galaxies shows immediately that the Universe is highly clumpy and inhomogeneous with galaxies clustered into groups. However, when we look at the distribution of galaxies on a very large scale, it looks rather more uniform. Looking on a large scale at the unobscured central part of the Figure 15.1, we see that the picture is fairly uniform.

Figure 15.1. The distribution of galaxies on the celestial sphere. This image was generated from counts of galaxies over the whole northern sky and as far south as declination −20° by the Lick astronomers, C.D. Shane, C.A. Wirtanen and their colleagues. This picture is an equal area projection of the northern galactic hemisphere, i.e. the centre of the diagram represents looking vertically out of the plane of our galaxy and the circumference corresponds to looking through the galactic equator. Over one million galaxies were counted in this survey. The black segment to the lower right of the diagram is due to the lower limit in declination of the Lick survey. Towards the edges of the diagram the numbers of galaxies decrease because of obscuration by interstellar dust in the plane of our Galaxy. Towards the centre of the diagram, however, the picture is unobscured and represents the true distribution of galaxies in the Universe. It is apparent that the distribution of galaxies is not smooth on a fine scale but, if averages are taken over a large angular scale, the distribution is much more uniform. (Diagram taken from M. Seldner, B. Siebars, E.J. Groth & P.J.E. Peebles (1977), *Astron. J.*, **82**, 249.)

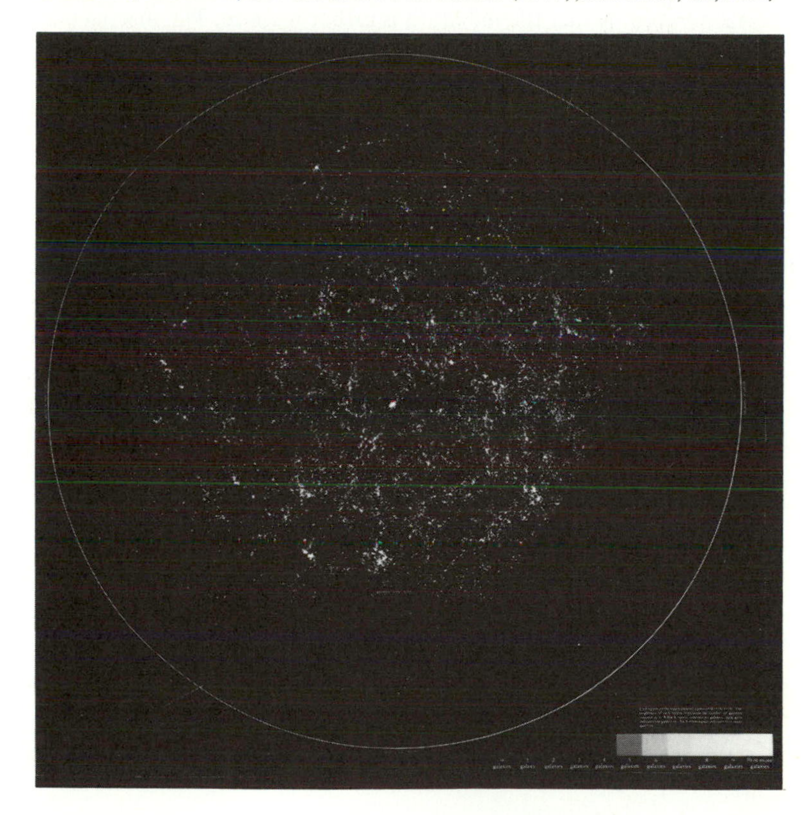

However, on a smaller scale, there is considerable fine structure which is real and which must be explained astrophysically.

Much more powerful evidence on the overall isotropy of the Universe comes from the *microwave background radiation* which was discovered by Penzias and Wilson in 1965. This radiation turns out to be extremely isotropic, i.e. it has the same intensity in all directions to a precision of better than one part in a thousand on all scales from about 1 arcmin to 360°. This is quite remarkable accuracy for any cosmological observation. In fact, a small 360° anisotropy in the radiation has been discovered at just below this level but this is associated with the velocity of the Earth relative to the frame of reference in which the radiation would be isotropic. No matter what the origin of the radiation, this is compelling evidence that our starting point should be the study of isotropic models of the Universe.

Two other features of the microwave background radiation are important. First, its spectrum is Planckian to a high degree of precision with a radiation temperature of about 2.9 K (Figure 15.2). This implies that at some stage the radiation must have been in equilibrium with matter at a particular temperature. The second point is that the radiation is rather massive. The inertial mass density of isotropic radiation at 2.9 K is aT^4/c^2 g cm^{-3} $= 6 \times 10^{-34}$ g cm^{-3} where

Figure 15.2. The spectrum of the microwave background radiation in the millimetre waveband. The hatched areas indicate the uncertainty in the balloon-borne observations made by the Berkeley group. The gaps in the spectrum are due to strong atmospheric emission lines. The solid line shows a black body spectrum at 2.96 K. At long wavelengths ($\nu < 5$ cm^{-1}), the observations can be made from the ground and are all consistent with the radiation temperature of about 2.9 K, even at frequencies as small as 0.01 cm^{-1}. (From D.P. Woody & P.L. Richards (1981), *Astrophys. J.*, **248**, 18.)

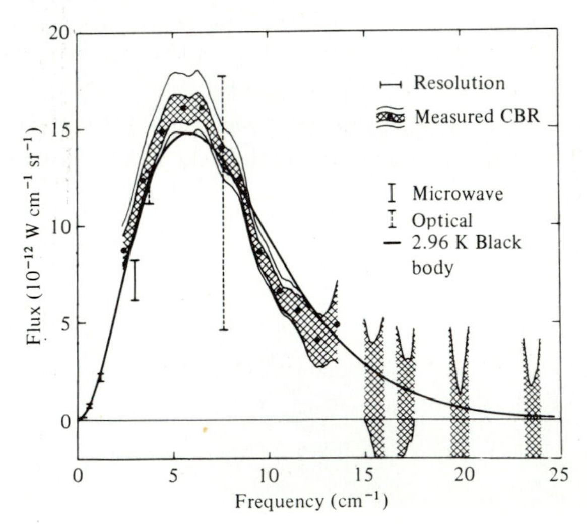

$a = 4\sigma/c$, σ being the Stefan–Boltzmann constant. This is by far the most 'massive' component of diffuse radiation in the Universe. The average density of ordinary matter in the Universe is about 10^{-30}–10^{-29} g cm^{-3} and consequently there is much more inertial mass in the matter than in the radiation at the present time. As we will see, however, the situation must have been very different in the distant past.

15.3 The Robertson–Walker metric

The data of Section 15.2 clearly refer to large-scale features of the Universe which must be a part of any acceptable cosmological model. The simplest models to begin with are therefore isotropic, homogeneous, expanding universes. We can set stringent limits on the possible form of the metric for such universes from rather general considerations, namely the isotropy and homogeneity assumptions and the metric of special relativity.

Formally, we must begin by making an assumption which is called the *cosmological principle* – this is the statement that we are not located at any special place in the Universe. The corollary is that we are at a typical location in the Universe and that a suitably chosen observer located at any other point would observe the same large-scale features that we observe. Isotropic, homogeneous, expanding universes clearly satisfy this requirement since every observer who partakes in the uniform expansion of the Universe will observe the Universe to expand uniformly.

We now introduce a set of *fundamental observers*, defined as observers who move in such a way that the Universe appears to be isotropic to them. Each of them has a clock and proper time measured by that clock is called *cosmic time*. There are no problems of synchronisation of these clocks carried by fundamental observers because, for example, they could be told to set their clocks to the same time when the Universe has a certain density.

We can now write down expressions for the spatial increment in the metric since we have already shown (Section 14.4) that the only possible isotropic curved spaces are given by

$$\mathrm{d}l^2 = \frac{\mathrm{d}r^2}{(1 - Kr^2)} + r^2(\mathrm{d}\theta^2 + \sin^2\theta\ \mathrm{d}\phi^2)$$

where the spatial coordinates are measured with respect to the frame of reference of a fundamental observer. For reasons which will become apparent in a moment, we prefer to write this in the form

$$\mathrm{d}l^2 = \mathrm{d}x^2 + R_c^2 \sin^2(x/R_c)(\mathrm{d}\theta^2 + \sin^2\theta\,\mathrm{d}\phi^2)$$

Then the metric of the isotropic curved space is

$$ds^2 = dt^2 - \frac{1}{c^2}[dx^2 + R_c^2 \sin^2(x/R_c)(d\theta^2 + \sin^2\theta\, d\phi^2)] \qquad (15.1)$$

Notice that, in this form, t is cosmic time and dx is an increment of proper distance; R_c is the radius of curvature of the curved space where $K = 1/R_c^2$ is the spatial curvature.

Now there is a problem in putting this simple metric into a form which is useful for comparing observables at different distances, and consequently different epochs, in the Universe. This is illustrated by the simple space–time diagram shown in Figure 15.3. Since light travels at a finite velocity, we observe all astronomical objects along a *past light cone* centred on the Earth at the present epoch t_0. Therefore, when we observe distant objects, we do not observe them at the present epoch but rather at an earlier epoch when the Universe was still homogeneous and isotropic but all distances between fundamental observers were smaller and the spatial curvature different.

To resolve this problem, we perform the following thought experiment. To measure a proper distance which can be included in the metric (15.1), we line up a set of fundamental observers between the Earth and the galaxy whose distance we wish to measure. The observers are all instructed to measure the

Figure 15.3. A simple space–time diagram showing the past light cone along which we observe all objects in the Universe.

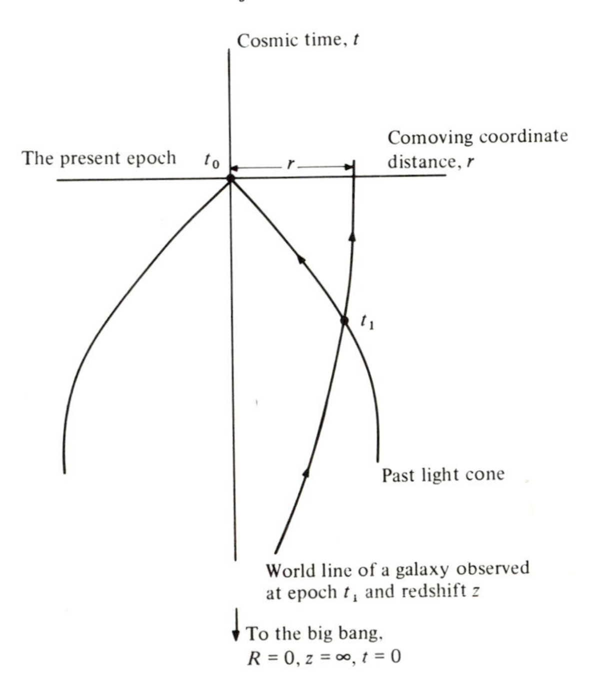

distance dx to the next fundamental observer at a particular cosmic time t which they read on their own clock. In this way, by adding together all the dxs, we can find a proper distance x which is measured at a single epoch and which can be used immediately in the metric (15.1). Notice that this is really a fictitious distance in that we cannot actually measure distances in this way. In particular, it is not a distance measurable by light signals because, between the emission of the light ray and its arrival at the observer, the Universe has expanded. We will work out in a moment how to relate x to measurable quantities.

Let us now look at what happens to the x coordinates in a uniformly expanding Universe. The definition of a uniform expansion is that between two cosmic epochs, t_1 and t_2, the distances of any two fundamental observers, i and j, change such that

$$\frac{x_i(t_1)}{x_j(t_1)} = \frac{x_i(t_2)}{x_j(t_2)} = \text{constant}$$

i.e.

$$\frac{x_i(t_1)}{x_i(t_2)} = \frac{x_j(t_1)}{x_j(t_2)} = \ldots = \text{constant} = \frac{R(t_1)}{R(t_2)} \tag{15.2}$$

$R(t)$ is a universal function which describes how the relative distance between any two fundamental observers changes with cosmic time t. Let us therefore adopt the following definitions. We will set $R(t)$, which is called the *scale factor*, equal to 1 at the present epoch t_0 and let the value of x at the present epoch be called r, i.e. we can rewrite the relation (15.2) as

$$x(t) = R(t)r \tag{15.3}$$

r thus becomes a *distance label* which is attached to a galaxy or fundamental observer for all time, the variation in proper distance in the expanding Universe being taken care of by the scale factor $R(t)$; r is called the *comoving radial distance coordinate*.

Now let us apply the same condition (15.2) to proper distances which expand perpendicularly to the line of sight between the epochs t and t_0.

$$\frac{\Delta l(t)}{\Delta l(t_0)} = R(t)$$

and hence, from the metric (15.1),

$$R(t) = \frac{R_c(t)\sin(x/R_c(t))\,d\theta}{R_c(t_0)\sin(r/R_c(t_0))\,d\theta} \tag{15.4}$$

Reorganising this equation and using (15.3), we see that

$$\frac{R_c(t)}{R(t)}\sin(R(t)r/R_c(t)) = R_c(t_0)\sin(r/R_c(t_0)) \tag{15.5}$$

This is only true if

$$R_c(t) = R_c(t_0)\,R(t) \tag{15.6}$$

i.e. the radius of curvature of the spatial sections is just proportional to $R(t)$. Let us call the value of $R_c(t_0)$, i.e. the radius of curvature at the present epoch, $\mathscr{R}$. Then

$$R_c(t) = \mathscr{R}R(t) \tag{15.7}$$

Substituting relations (15.3) and (15.7) into the metric (15.1), we obtain

$$ds^2 = dt^2 - \frac{R^2(t)}{c^2}\,[dr^2 + \mathscr{R}^2 \sin^2(r/\mathscr{R})\,(d\theta^2 + \sin^2\theta\, d\phi^2)] \tag{15.8}$$

This is the *Robertson–Walker metric* in the form which we will use in all our future analyses. Notice that it contains one unknown function $R(t)$, the scale factor, which describes the dynamics of the Universe and an unknown constant $\mathscr{R}$ which describes the spatial curvature of the Universe at the present epoch. We emphasise that the r appearing in the metric (15.8) is the proper distance the object would have if it were possible to measure it at the present epoch. It is thus a special use of r and different from its use in, say, the Schwarzschild metric in Chapter 14.

Finally, we note that it is possible to rewrite the metric in different ways. For example, if we use a comoving 'angular diameter distance' $r_1 = \mathscr{R}\sin(r/\mathscr{R})$, the metric becomes

$$ds^2 = dt^2 - \frac{R^2(t)}{c^2}\left[\frac{dr_1^2}{1 - Kr_1^2} + r_1^2\,(d\theta^2 + \sin^2\theta\, d\phi^2)\right] \tag{15.9}$$

where $K = 1/\mathscr{R}^2$. Evidently, by a suitable rescaling of the r_1 coordinate $Kr_1^2 = r_2^2$, the metric could equally well be written

$$ds^2 = dt^2 - \frac{R_1^2(t)}{c^2}\left[\frac{dr_2^2}{1 - kr_2^2} + r_2^2\,(d\theta^2 + \sin^2\theta\, d\phi^2)\right] \tag{15.10}$$

with $k = +1, 0$ and -1 for universes with spherical, flat and hyperbolic geometries respectively. This is a rather popular form for the metric but we will use (15.8) because the r coordinate has an important (and clear) physical meaning.

The importance of the metric (15.8) is that it enables us to define the invariant interval ds^2 between events at any epoch or location in the expanding Universe. We now need to invest it with some more meaning. Notice that we have so far specified nothing about the physics of the expanding Universe. All of this has been absorbed in the function $R(t)$.

15.4 Observations in cosmology†

It is useful to produce a catalogue of results which are independent of the particular form of $R(t)$. First of all, let us elucidate the real meaning of redshift in cosmology.

15.4.1 Redshift

By redshift, we mean the shift of spectral lines to longer wavelength. If λ_e is the wavelength of the line as emitted and λ_o the observed wavelength, the redshift is defined by

$$z = \frac{\lambda_o - \lambda_e}{\lambda_e} \tag{15.11}$$

In the case of galaxies receding from the observer, the velocity inferred from the redshift according to special relativity is given by

$$1 + z = \left(\frac{1 + v/c}{1 - v/c}\right)^{\frac{1}{2}} \tag{15.12}$$

You should show this for yourselves using the results of Chapter 13. In the small redshift limit, $v/c \ll 1$, the relation (15.12) reduces to

$$v = cz \tag{15.13}$$

It is this type of velocity which Hubble used in deriving the velocity–distance relation, $v = H_0 r$.

Let us now consider a wave packet of frequency ν_1 emitted between cosmic times t_1 and $t_1 + \Delta t_1$ from a distant galaxy. This wave packet is received by the observer at the present epoch in the cosmic time interval t_0 to $t_0 + \Delta t_0$. The signal propagates along null-cones, i.e. $ds^2 = 0$ and so, considering radial propagation from source to observer, $d\theta = 0$ and $d\phi = 0$, the metric (15.8) gives us

$$dt = -\frac{R(t)}{c}\,dr$$

i.e.

$$\frac{c\,dt}{R(t)} = -dr \tag{15.14}$$

The minus sign appears because the origin of the r coordinate is the observer. Therefore, considering first the leading edge of the wave packet, the integral of (15.14) is

$$\int_{t_1}^{t_0} \frac{c\,dt}{R(t)} = -\int_r^0 dr \tag{15.15}$$

† On first reading, the reader may concentrate upon Sections 15.4.1 and 15.4.2 which are central to the further development of the theory.

The end of the wave packet must travel the same distance in units of comoving distance coordinate since the r coordinate is fixed to the source for all time. Therefore,

$$\int_{t_1+\Delta t_1}^{t_0+\Delta t_0} \frac{c\,\mathrm{d}t}{R(t)} = -\int_r^0 \mathrm{d}r$$

i.e.

$$\int_{t_1}^{t_0} \frac{c\,\mathrm{d}t}{R(t)} + \frac{c\,\Delta t_0}{R(t_0)} - \frac{c\,\Delta t_1}{R(t_1)} = \int_{t_1}^{t_0} \frac{c\,\mathrm{d}t}{R(t)}$$

Since $R(t_0) = 1$, we find that

$$\Delta t_0 = \frac{\Delta t_1}{R(t_1)} \tag{15.16}$$

This is the cosmological expression for the phenomenon of *time dilation.* Since $R(t_1) < 1$ when we observe distant galaxies, phenomena are observed to take longer in our frame of reference than in that of the source. This is exactly the phenomenon which is observed for relativistic muons propagating through the atmosphere (Section 13.3).

One very useful result of this calculation is that it enables us to derive an expression for the comoving radial distance coordinate r. Equation (15.15) can be rewritten

$$r = \int_{t_1}^{t_0} \frac{c\,\mathrm{d}t}{R(t)} \tag{15.17}$$

Thus, once we know $R(t)$ we can immediately find r by integration.

The result (15.16) also provides us with an expression for redshift. If we take $\Delta t_1 = \nu_1^{-1}$ to be the period of the emitted waves and $\Delta t_0 = \nu_0^{-1}$ to be the observed period, we find that

$$\nu_0 = \nu_1 R(t_1) \tag{15.18}$$

Rewriting this result in terms of redshift z, we find that

$$z = \frac{\lambda_o - \lambda_e}{\lambda_e} = \frac{\lambda_0}{\lambda_1} - 1 = \frac{\nu_1}{\nu_0} - 1$$

i.e.

$$1 + z = \frac{1}{R(t_1)} \tag{15.19}$$

This is the basic meaning of redshift in cosmology. Redshift is simply a measure of the scale factor of the Universe when the source emitted its radiation. Thus, when we observe a galaxy with redshift $z = 1$, the scale factor of the Universe when the light was emitted was $R(t) = 0.5$, i.e. the Universe was half its present size. Note, however, that we obtain no information about *when* the light was emitted. If we did, we could determine directly from observation, the function

$R(t)$. We do not understand the physics of galaxies and quasars well enough to be able to estimate times or ages from observation. In this circumstance, we have to have some theory of the dynamics of the Universe in order to determine $R(t)$.

15.4.2 Hubble's law

Using our original prescription of proper distances, we can write Hubble's Law, $v = H_0 x$ as

$$\frac{dx}{dt} = H_0 x$$

Now, substituting $x = R(t)r$, we find that

$$r\,\frac{dR(t)}{dt} = H_0 R(t) r$$

i.e.

$$H_0 = \dot{R}/R$$

Since we measure Hubble's constant at the present epoch, $t = t_0$, $R = 1$, we find that

$$H_0 = (\dot{R})_{t=t_0} \tag{15.20}$$

Notice, however, that we can define a Hubble constant at any epoch through the more general relation

$$H(t) = \dot{R}/R \tag{15.21}$$

Thus, Hubble's constant H_0 defines the present expansion rate of the Universe.

15.4.3 Angular diameters

The great simplification which results from the use of a metric of the form (15.8) is apparent in working out the angular size of an object of proper length d perpendicular to the radial coordinate at redshift z. The relevant spatial component of the metric (15.8) is the term in $d\theta$ and hence the proper length must be

$$\begin{aligned} d &= R(t) D \Delta\theta \\ &= \frac{D \Delta\theta}{(1+z)} \end{aligned} \tag{15.22}$$

where we have introduced an *effective distance* $D = \mathscr{R} \sin(r/\mathscr{R})$. Clearly, for small redshifts $z \ll 1$ and $r \ll \mathscr{R}$, expression (15.22) reduces to the Euclidean relation $d = r\,\Delta\theta$.

15.4.4 Apparent intensities

Suppose the source has luminosity $L(\nu_1)\,\mathrm{W\,Hz^{-1}}$, i.e. the total energy emitted over 4π steradians per unit time per unit frequency interval. Let us suppose that $N(\nu_1)$ photons of energy $h\nu_1$ are emitted by the source in the bandwidth $\Delta\nu_1$ in the proper time interval Δt_1 and that it has redshift z. Then the luminosity of the source is

$$L(\nu_1) = \frac{N(\nu_1)\, h\nu_1}{\Delta\nu_1\, \Delta t_1} \qquad (15.23)$$

These photons are distributed over a 'sphere' centred on the source at epoch t_1 and when the 'shell' of photons arrives at the observer at the epoch t_0, he intersects a certain fraction of them with his telescope. We now need to known how the photons spread out over a sphere between the epochs t_1 and t_0. The photons are observed at t_0 with frequency $\nu_0 = R(t_1)\nu_1$, in a proper time interval $\Delta t_0 = \Delta t_1\, R^{-1}(t_1)$ and in waveband $\Delta\nu_0 = R(t_1)\Delta\nu_1$. The only complication is that we must relate the diameter of our telescope Δl to the angular diameter $\Delta\theta$ which it subtends at the source at epoch t_1. Again, the metric (15.8) provides an elegant answer. The proper distance Δl refers to the present epoch at which $R(t) = 1$ and hence

$$\Delta l = D\,\Delta\theta \qquad (15.24)$$

where $\Delta\theta$ is the angle measured by the appropriate fundamental observer located at the source. Notice the difference between relations (15.22) and (15.24). They correspond to angular diameter measures in opposite directions along the light cone. In fact, the difference between them of a factor $(1 + z)$ is part of a more general relation about angular diameter measures along light cones which is known as the *reciprocity theorem.*

Therefore, the surface area of our telescope is $\pi\Delta l^2/4$ and the solid angle subtended by this area at the source is $\Delta\Omega = \pi\,\Delta\theta^2/4$. The number of photons incident upon the telescope in time Δt_0 is

$$N(\nu_1)\ \Delta\Omega/4\pi$$

but they are now observed with frequency ν_0, i.e. $N(\nu_1) = N(\nu_0)$. Therefore, the flux density of the source, i.e. the energy received per unit time, per unit area and per unit bandwidth ($\mathrm{W\,m^{-2}\,Hz^{-1}}$) is

$$S(\nu_0) = \frac{N(\nu_0)\, h\nu_0 \Delta\Omega}{4\pi\Delta t_0 \Delta\nu_0 (\pi/4)\,\Delta l^2} \qquad (15.25)$$

We can now relate the quantities in the expression (15.25) to properties of the source, using the above relations (15.23) and (15.24).

$$S(\nu_0) = \frac{L(\nu_1) R(t_1)}{4\pi D^2} = \frac{L(\nu_1)}{4\pi D^2 (1 + z)} \qquad (15.26)$$

If the spectra of sources are of power law form $L(\nu) \propto \nu^{-\alpha}$, this relation becomes

$$S(\nu_0) = \frac{L(\nu_0)}{4\pi D^2 (1+z)^{1+\alpha}} \tag{15.27}$$

We can repeat the analysis for *bolometric* luminosities and flux densities in which case we consider the emitted and received radiation integrated over all wavelengths.

$$L_{\text{bol}} = \frac{\Sigma N(\nu_1) h\nu_1}{\Delta t_1}$$

and the relation with S_{bol} becomes

$$S_{\text{bol}} = \frac{L_{\text{bol}}}{4\pi D^2 (1+z)^2} \tag{15.28}$$

The quantity $D_L = D(1+z)$ is often called luminosity distance since this definition makes the relation between S_{bol} and L_{bol} look like an inverse square law. A key point about relations such as (15.26) and (15.27) is the presence of time dilation factors $(1+z)$ which reduce the flux density by more than the inverse square law. However, this is a somewhat misleading way of expressing the result because D itself does not behave like a Euclidean distance with increasing redshift.

15.4.5 Number densities

We often need to know the number of sources in a particular redshift range, z to $z + \mathrm{d}z$. Since there is a one-to-one relation between r and z, which is still undefined at this stage in the development, the problem is very simple because, by definition, r is a radial proper distance defined at the present epoch and hence the number of objects in the range r to $r + \mathrm{d}r$ is given by results already obtained in Section 14.4. The volume of a shell of thickness $\mathrm{d}r$ at comoving distance coordinate r is

$$\mathrm{d}V = 4\pi\mathscr{R}^2 \sin^2(r/\mathscr{R})\,\mathrm{d}r = 4\pi D^2\,\mathrm{d}r \tag{15.29}$$

Therefore, if N_0 is the present space density of objects and their number is conserved in the expanding Universe,

$$\mathrm{d}N = 4\pi N_0 D^2\,\mathrm{d}r \tag{15.30}$$

The definition of comoving coordinates automatically takes care of the expansion of the Universe.

15.4.6 The background radiation in the Universe

We can now work out the brightness of the night sky using the results of Sections 15.4.4 and 15.4.5. The intensity of the background radiation, i.e. the flux density from a steradian of sky, is given by the sum of the flux densities of the sources in the volume $\mathrm{d}V$ at distance r integrated from $r = 0$ to the maximum value of r. For 1 steradian, $\mathrm{d}V = D^2\,\mathrm{d}r$ and using expression (15.27) we find that

$$
\begin{aligned}
I(\nu_0) &= \int_V S(\nu_0)\,N_0\,\mathrm{d}V \\
&= \frac{L(\nu_0)N_0}{4\pi}\int_0^{r_{\max}} \frac{D^2\,\mathrm{d}r}{D^2(1+z)^{1+\alpha}} \\
&= \frac{L(\nu_0)N_0}{4\pi}\int_0^{r_{\max}} \frac{\mathrm{d}r}{(1+z)^{1+\alpha}}
\end{aligned}
\tag{15.31}
$$

$r_{\max}$ is the upper limit to the comoving distance coordinate corresponding to $R = 0$, $z = \infty$ which is the origin of the expansion. Comparison with the analysis of Section 15.2.1 shows that the background light converges more rapidly than in the Euclidean case because of the redshift factor $(1 + z)^{-(1+\alpha)}$ but we cannot provide a complete answer until we know the dependence of r on the redshift z.

15.4.7 The age of the Universe

Finally, let us work out an expression for the age of the Universe. We can do this from a rearranged version of equation (15.17). The basic differential relation is

$$-\frac{c\,\mathrm{d}t}{R(t)} = \mathrm{d}r$$

and hence

$$\int_0^{t_0} \mathrm{d}t = \int_0^{r_{\max}} \frac{R(t)\,\mathrm{d}r}{c} \tag{15.32}$$

Again $r_{\max}$ is the comoving distance coordinate corresponding to $R = 0$, $z = \infty$.

15.5 Approaches to the determination of the function $R(t)$

So far, there has been little to argue about concerning the basic structure of the theory of isotropic cosmological models. However, to determine the function $R(t)$, the variation of the scale factor with cosmic epoch, we have to decide upon the physics which describe the large-scale dynamics of the Universe. In the introduction, we described two basic approaches. One was to use the best of conventional physics and assume that it is applicable on a global scale. We will develop that story in Section 15.6. In this section, we discuss briefly a few examples of the alternative approach of adopting non-conventional physics. We will learn of some intriguing facets of the different approaches which can be taken to cosmological problems.

15.5.1 The steady state theory

This theory caused a great deal of popular interest in the 1950s and 1960s following the original proposal by Bondi, Gold and Hoyle.[1] The basic motivation for the theory was the fact that, in the 1940s, the value of Hubble's constant H_0 was thought to be very much larger than it is now. We will show that all the conventional world models of Section 15.6 have ages which are less than H_0^{-1}. Since H_0 was thought to be about 500 km s^{-1} Mpc^{-1}†, the age of the Universe was calculated to be less than about 2×10^9 years in these models. Since this age is less than the age of the Earth, which is about 4.6×10^9 years, there is an obvious contradiction.

Bondi, Gold and Hoyle came up with an ingenious solution to this problem by replacing the cosmological principle by what they termed the *perfect cosmological principle.* This is the statement that the large-scale properties of the Universe observed today shall be the same *for all observers at all epochs.* In other words, the Universe should present an unchanging appearance for all observers for all time. This postulate immediately determines the dynamics of the Universe since Hubble's constant becomes a fundamental constant of physics, i.e.

$$\dot{R}/R = H_0$$

and hence

$$R = R_0\, e^{H_0 t} \tag{15.33}$$

This means that diffuse matter is continuously dispersing and therefore, for the Universe to preserve the same appearance for all time, matter must continuously appear out of nothing. The theory is sometimes called the theory of *continuous creation.* This is the price one has to pay for a rather elegant theory. The proponents of the theory showed that the creation rate needed to replace the dispersing matter was unobservably small in any terrestrial experiment.

Let us derive some of the other properties of the steady state model. The model must preserve the same global properties for all time and the spatial curvature is such a property. Relation (15.7) tells us that $R_c(t) = \mathscr{R}\, R(t)$ and hence the radius of curvature of the geometry will vary with time unless $\mathscr{R} = \infty$, i.e. $K = 0$. Therefore, the geometry must be flat. Equally simple is the relation between r and z which results from substituting the relation (15.33) into the definition of r, equation (15.17). Recalling that $R = (1 + z)^{-1}$, we find that

$$r = \int_{t_1}^{t_0} \frac{c\,dt}{R(t)} = \int_{t_1}^{t_0} \frac{c\,dt}{R_0\, e^{H_0 t}}$$

† 1 Mpc = 1 Megaparsec ≈ 3×10^{22} m.

$$= \frac{c}{H_0}\left(\frac{1}{R(t)} - 1\right)$$

$$= \frac{cz}{H_0} \qquad (15.34)$$

We recognise that this is identical to Hubble's law. However, whereas in the conventional models the law is only true for small redshifts, in the steady state theory it is true for all redshifts. The metric is thus

$$ds^2 = dt^2 - \frac{R^2(t)}{c^2}\,[dr^2 + r^2\,(d\theta^2 + \sin^2\theta\, d\phi^2)]$$

with $R(t) = e^{H_0(t-t_0)}$ and $r = cz/H_0$. Finally, the Universe is clearly of infinite age, solving the time scale problem, and there is no initial singularity as in the conventional hot big bang.

In Great Britain, the theory caused a great deal of excitement which reached the popular newspapers and a public debate developed between proponents of the steady state and evolutionary pictures of the Universe. In the meantime, it was discovered by Baade that Hubble's constant had been significantly overestimated and a revised value of $H_0 = 180$ km s^{-1} Mpc^{-1} was derived, giving $T = H_0^{-1} = 5.6 \times 10^9$ years, comfortably greater than the age of the Earth. Subsequent studies have reduced the value even further so that the best estimates place H_0 in the range of about 50 to 100 km s^{-1} Mpc^{-1}.

Despite this, the simplicity and elegance of the theory was a profound attraction for a number of theorists who saw in it a unique cosmological model as opposed to the conventional models which, we will show, depend upon the density of the Universe, and possess a singularity at $t = 0$. Of course, the uniqueness of the theory also makes it highly testable. One of the first pieces of evidence against the theory was the number counts of faint radio sources which indicated that there were many more radio sources per unit volume in the past than there are at the present day. Such a result violates the basic postulate of the theory that the overall properties of the Universe should be unchanging with time. The really fatal blow to the theory was, however, the microwave background radiation. There is no natural origin for this radiation in the steady state picture and there do not exist sources which could produce the Planckian spectrum of the radiation and its great energy density. On the other hand, these properties find a natural explanation in the hot big bang picture as we will see in a moment.

The story of the steady state theory is an intriguing one from the point of view of methodology in cosmology. The theory is mathematically much more elegant than the big bang model and it results in a unique model for the Universe. This elegance is, however, achieved at a certain price and that is the introduction of totally new physics – the continuous creation of matter. For many scientists, this step is so objectionable that they rule out the theory immediately. Even after the time scale problem was resolved, many others were prepared to investigate whether or not the theory could survive observational validation despite the introduction of new physics.

Hoyle's approach to the steady state theory was somewhat different from that of Bondi and Gold in that he believed the basic concept of creation of matter to be the fundamental aspect of the theory and sought to give this aspect of the theory a proper theoretical basis. When the simplest model ran into conflict with the observations, various alternatives were proposed whereby a Universe with continuous creation could mimic an evolutionary Universe. In other words, the perfect cosmological principle was dropped but the concept of the creation of matter was maintained with the possibility that the rate of creation varies with time or location in the Universe. In my view, this is getting onto very much more speculative ground. Logically, this modified picture is quite feasible but it has lost much of the attraction of the original picture which was one of simplicity and uniqueness. There comes a point when even an open-minded physicist has to make the value judgement as to whether or not he is straying too far from conventional physics for the ideas to maintain credibility as a physical theory. The problem is a particularly acute one in cosmology because we only have one Universe to observe and there are obvious logical problems with theories that can only be tested on the cosmological scale.

15.5.2 Cosmologies with a variable gravitational constant

Another example of a modification to the laws of physics that would have profound cosmological implications is to be found in those theories in which the gravitational constant G varies with cosmic epoch. Such theories have been proposed by Dirac[2] and by Brans and Dicke.[3] In these theories, gravity was stronger in the past but otherwise the cosmological models are not too dissimilar from the hot big bang models. At small redshifts, the differences would be small but, in the early stages of evolution, the expansion must have been much faster in order to escape the increased gravitational forces. Fortunately, we can set strong limits on such models from arguments concerning primordial nucleosynthesis which depend critically upon the expansion rate at early epochs.

15.5.3 Non-conventional models

To complete this discussion of non-standard models, we should note that we can conceive of observations which would require us to seek new interpretations of cosmological data. For example Arp has claimed to have found associations of galaxies and quasars with widely different redshifts and, if these associations are real, this cannot be readily accommodated within conventional physics. However, this claim requires the most careful scrutiny by independent observers before it poses a serious challenge to conventional physics. I have not been persuaded by any of the advocates of radical new physics for cosmology on the basis of the evidence presented so far. But, we must not exclude the possibility that cosmological observations could tell us something profoundly original and unexpected about fundamental physics.

15.6 The hot big bang model of the Universe

15.6.1 The dynamics of the standard hot big bang

We can derive most of the essential features of the standard big bang models by appealing to Newtonian gravitation. In the case of isotropic models, this works because global physics must be the same as local physics, the one unknown in the problem being the geometry of the spatial components of the metric.

Consider the expanding spherical piece of the Universe shown in Figure 15.4 with a fundamental observer O at the centre. We suppose that the sphere is composed of a pressureless fluid (commonly known as 'dust') of density $\rho(t)$. We ask 'What is the deceleration of the galaxy G of mass m_g at distance x from O due to the matter within distance x?' We use Gauss's law to replace the spherical mass distribution by a point mass at O of the same mass and then Newton's laws of motion and gravitation enable us to write

$$m_g \frac{d^2x}{dt^2} = -\frac{4\pi}{3} \frac{x^3\rho(t)Gm_g}{x^2} = -\frac{4\pi}{3} G\rho(t)m_g x \tag{15.35}$$

Now let us replace x by the comoving coordinate $x = R(t)r$ as in equation (15.3) and normalise the density to its value at the present epoch, ρ_0, by $\rho(t) = \rho_0 R^{-3}(t)$. Then, substituting into equation (15.35), we obtain

Figure 15.4. How to work out the dynamics of Newtonian universes.

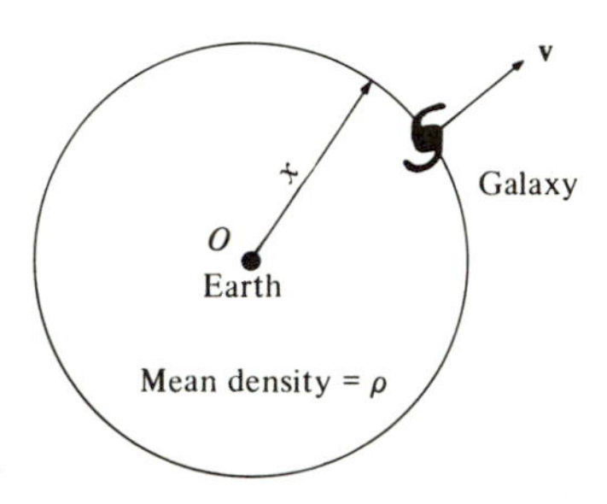

$$\frac{d^2R}{dt^2} = -\frac{4\pi G\rho_0}{3}\frac{1}{R^2} \tag{15.36}$$

Notice that the galaxy has disappeared from the calculation. What we are left with is an expression for the dynamics of the Universe under Newtonian gravity. Equation (15.36) is identical with that which comes out of the theory of general relativity for the case of a dust-filled Universe.

We can now take the first integral of equation (15.36) by multiplying by $\dot{R}$ and integrating with respect to time.

$$\int \dot{R}\,\ddot{R}\,dt = -\frac{4\pi G\rho_0}{3}\int \frac{\dot{R}}{R^2}\,dt$$

Integrating,

$$\dot{R}^2 = \frac{8\pi G\rho_0}{3}\frac{1}{R} + \text{constant} \tag{15.37}$$

Again, this answer is the same as that found in general relativity but, in that theory, an explicit value is given for the 'constant'. We can guess that it must contain some information about the geometry of the Universe and indeed it does. In our notation, the constant is $-Kc^2 = -c^2/\mathscr{R}^2$. For a gas with pressure p, the full equations of general relativity are:

$$\ddot{R} = -\frac{4\pi GR}{3}\left(\rho + \frac{3p}{c^2}\right) + [\tfrac{1}{3}\Lambda R] \tag{15.38}$$

$$\dot{R}^2 = \frac{8\pi G\rho}{3}R^2 - Kc^2 + [\tfrac{1}{3}\Lambda R^2] \tag{15.39}$$

The pressure term in equation (15.38) is a relativistic correction to the inertial mass density. The terms in square brackets are the so-called Λ-terms associated with the cosmological constant Λ which Einstein introduced in the first application of his gravitational field equations to cosmology. Einstein's aim was to produce a static universe in which there are no solutions in the absence of matter. However, de Sitter and Friedmann showed that there exist non-stationary solutions for the Universe even if $\Lambda = 0$ and this realisation was triumphantly vindicated by Hubble's discovery of the expansion of the Universe. Einstein subsequently abandoned the cosmological constant which he regarded as 'theoretically unsatisfactory anyway'.[4] The solutions of the field equations with non-zero Λ were revived in the 1930s and 1940s because of the time-scale problem discussed in Section 15.5.1. Models with positive values of Λ can have ages much greater than H_0^{-1}. With the revision of Hubble's constant in the 1950s, these models again lost much of their appeal. In their most recent reincarnation, they may find application in the very earliest stages of the Universe, but this is not the place to enter into these ideas which must still be regarded as speculative. In what follows, we assume that $\Lambda = 0$.

Let us put the values of R and $\dot{R}$ at the present day, $R = 1$, $\dot{R} = H_0$, into equation (15.39). Then

$$H_0^2 = \frac{8\pi G\rho_0}{3} - Kc^2 \tag{15.40}$$

Let us introduce a *critical density* $\rho_c = 3H_0^2/8\pi G$, whose significance will be apparent in a moment, and also a *density parameter* $\Omega = \rho_0/\rho_c$, which is the mean density of the Universe at the present day relative to the critical value ρ_c. Then relation (15.40) becomes

$$K = \frac{(\Omega - 1)}{(c^2/H_0^2)} \tag{15.41}$$

i.e. there is a one-to-one relation between the mean density of the Universe and the curvature of its spatial sections. If $\Omega > 1$, the geometry is spherical, if $\Omega < 1$, it is hyperbolic and, if $\Omega = 1$, the geometry is flat.

We can also introduce a dimensionless deceleration parameter which tells us how rapidly the Universe is being decelerated at the present epoch, $t = t_0, R = 1$.

$$\ddot{R}(t) = -\frac{4\pi G\rho_0}{3}$$

and hence

$$\frac{\ddot{R}(t_0)}{\dot{R}^2(t_0)} = -\frac{4\pi G\rho_0}{3H_0^2} = -\tfrac{1}{2}\Omega$$

Therefore, we define the *deceleration parameter* q_0 by

$$q_0 = -\frac{\ddot{R}(t_0)}{\dot{R}^2(t_0)} = -\frac{\ddot{R}(t_0)}{H_0^2} = \tfrac{1}{2}\Omega \tag{15.42}$$

Thus, the deceleration of the Universe is wholly determined by the amount of matter present. Notice a crucial test of the theory exhibited by equation (15.42). We are able, in principle, to measure independently q_0, the rate of deceleration of the Universe, and its mean density Ω. If the general relativistic models of the Universe are correct, $q_0 = \frac{1}{2}\Omega$. If not, many of us will regard this as a great step backwards and proceed to other pursuits like viticulture and music criticism! Unfortunately, it is technically very difficult to measure either q_0 or Ω precisely. We know they are probably equal within a factor of about 10 but we need to know this very much more precisely.

We can investigate the dynamics of the Universe by examination of equation (15.39) which we can rewrite, using relation (15.41), as

$$\dot{R}^2 = H_0^2\left(\frac{\Omega}{R} + 1 - \Omega\right) \tag{15.43}$$

Let us look at the dynamics when R becomes very large ($R \to \infty$). Then

$$\dot{R}^2 = H_0^2\,(1 - \Omega)$$

If $\Omega < 1$, the Universe ends up at infinity with a finite velocity. If $\Omega > 1$, the Universe never reaches infinity since $\dot{R}$ goes to zero before R reaches infinity and hence it collapses back to a singularity at $R = 0$. If $\Omega = 1$, the Universe just expands to infinity and at that point has zero velocity. There is a clear analogy with the *escape velocity*. If $\Omega < 1$, the Universe has more kinetic than gravitational potential energy and escapes to infinity. If $\Omega > 1$, the gravitational energy dominates and the Universe does not exceed its own escape velocity. In the case $\Omega = 1$, we find a particularly simple solution, known as the *Einstein–de Sitter model*,

$$K = 0, \quad \dot{R}^2 = \frac{\Omega H_0^2}{R}$$

Integrating,

$$R = (\tfrac{3}{2}H_0 t)^{\frac{2}{3}} \tag{15.44}$$

The dynamics of these three types of behaviour are summarised in Figure 15.5.

The next thing to derive is the relation between r and z. To do this, we insert $R = (1 + z)^{-1}$ into equation (15.43) and then

$$\frac{\mathrm{d}}{\mathrm{d}t}(1 + z)^{-1} = H_0\,[\Omega(1 + z) + 1 - \Omega]^{\frac{1}{2}}$$

$$\frac{\mathrm{d}z}{\mathrm{d}t} = -H_0(1 + z)^2(\Omega z + 1)^{\frac{1}{2}} \tag{15.45}$$

We also recall that $\mathrm{d}r = -c\,\mathrm{d}t/R$ (equation (15.14)). Hence

Figure 15.5. The variation of the scale factor R as a function of time for classical world models having different density parameters Ω.

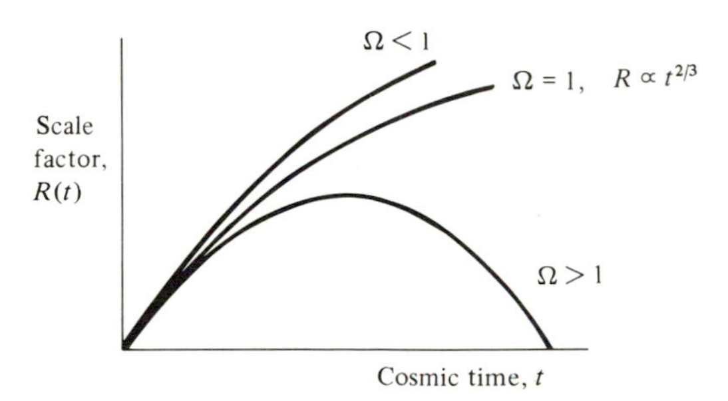

$$\frac{\mathrm{d}r}{\mathrm{d}t} = -c(1+z)$$

and

$$\frac{\mathrm{d}r}{\mathrm{d}z} = \frac{\mathrm{d}r}{\mathrm{d}t}\bigg/\frac{\mathrm{d}t}{\mathrm{d}z} = \frac{c}{H_0}\,\frac{1}{(1+z)(\Omega z+1)^{\frac{1}{2}}} \tag{15.46}$$

Equations (15.45) and (15.46) are particularly useful. From (15.45) we can find directly the age of the Universe.

$$T = \int_0^{t_0} \mathrm{d}t = \frac{1}{H_0}\int_0^{\infty} \frac{\mathrm{d}z}{(1+z)^2(\Omega z+1)^{\frac{1}{2}}} \tag{15.47}$$

It is a useful exercise to show that if $\Omega = 0, T = H_0^{-1}$ and if $\Omega = 1, T = \frac{2}{3}H_0^{-1}$.

The final steps are to work out r, i.e.

$$r = \frac{c}{H_0}\int_0^{z} \frac{\mathrm{d}z}{(1+z)(\Omega z+1)^{\frac{1}{2}}} \tag{15.48}$$

and then to form the quantity

$$D = \mathscr{R}\sin(r/\mathscr{R})$$

where, from the relation (15.41), $\mathscr{R} = (c/H_0)/(\Omega - 1)^{\frac{1}{2}}$. With a bit of algebra, you should be able to show that

$$D = \frac{2c}{H_0\Omega^2(1+z)}\,\{\Omega z + (\Omega - 2)\,[(\Omega z+1)^{\frac{1}{2}} - 1]\} \tag{15.49}$$

This result is true for all values of K and is the expression we need to use in all the results of Section 15.4.

15.6.2 Classical cosmology

For many years, observational cosmology meant the study of distant galaxies with a view to determining the value of Hubble's constant H_0 and the deceleration parameter q_0 (or Ω). I well remember a review article a number of years ago entitled 'Cosmology – the search for two numbers'.[5] It can be seen from the analyses of Section 15.4 and the relations (15.46) and (15.49), that the properties of distant objects depend upon the density parameter Ω (or q_0). For example, Figure 15.6 shows the dependence of angular size and flux density upon redshift for an object of fixed intrinsic properties in cosmological models with different values of Ω. It was hoped that, if one could find objects with the same intrinsic properties at small and large redshifts, by comparing their observed properties, the value of q_0 could be found.

This programme has met with only limited success. From observations of the most luminous galaxies in clusters, we know that q_0 probably lies in the range $0 < q_0 < 2$, but there is certainly not sufficient understanding of the evolutionary effects which could influence the properties of galaxies at large redshifts to measure a reliable value of q_0 to say 25% accuracy at this stage. However, progress is encouraging and it may be that we will find a reasonable value of q_0 over the next 10 years with the advent of facilities such as the Space Telescope.

The value of H_0 probably lies in the range 50 to 100 km s^{-1} Mpc^{-1}, corresponding to a range of $T = H_0^{-1}$ of 10 x 10^9 to 20 x 10^9 years. The main problem in these studies is in measuring accurate distances of galaxies which are

Figure 15.6. ***(a)*** **The variation of the angular size of a standard length with redshift for different world models.** ***(b)*** **The variation of observed intensity of a standard source of luminosity** ***L*** **with redshift for different world models. The source is assumed to have a power law spectrum $L \propto \nu^{-\alpha}$.**

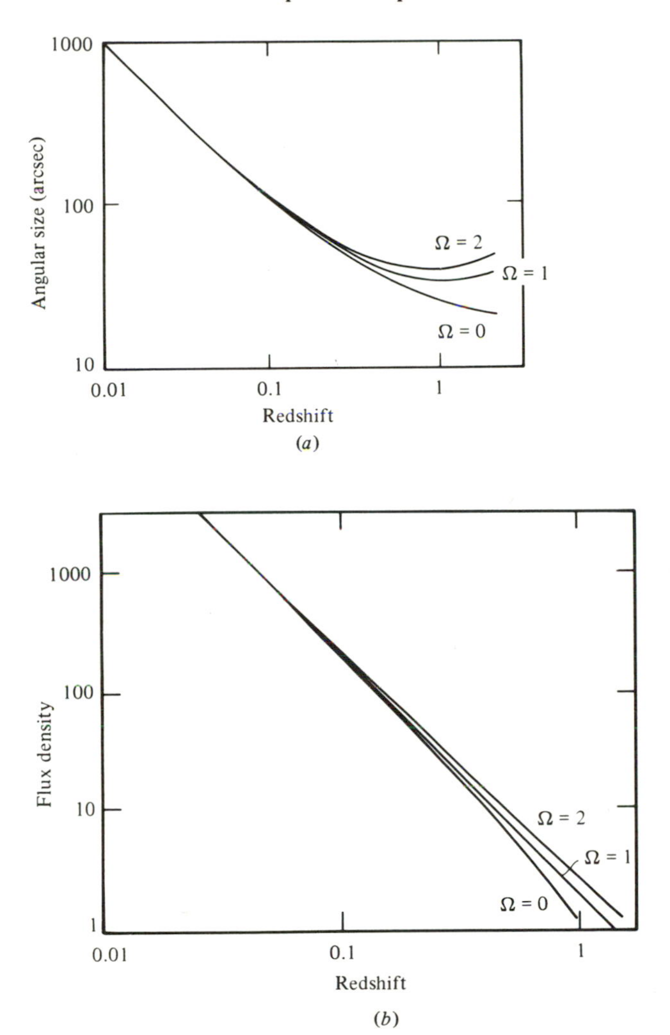

independent of redshift. Again, we can fully expect significant progress in the determination of H_0 over the next 10 years.

15.6.3 The hot big bang model of the Universe

The areas of intensive study have changed from purely geometrical and dynamical questions such as those discussed in Section 15.6.1 and 15.6.2 to astrophysical questions about the evolution of the big bang models. The reason for the name, the *big bang*, is readily appreciated from the following analysis.

So far, we have considered only dust-filled universes in which the dynamics are dominated by the inertial mass of the matter as parameterised by Ω. Let us now look at what happens to the radiation. The relation (15.18) tells us that at earlier epochs, $\nu_1 = \nu_0 R^{-1}$. Therefore, the energy density of radiation varies with cosmic epoch as

$$\begin{aligned}\epsilon(z) &= N(z)h\nu_1 = N_0 R^{-3} h\nu_0 R^{-1} \\ &= N_0 h\nu_0 R^{-4} \\ &= \epsilon_0 (1+z)^4 \end{aligned} \qquad (15.50)$$

where $N(z)$ is the number density of photons at redshift z, N_0 is its value at the present epoch and ϵ_0 is the energy density of radiation at the present epoch. However, this energy density is also $\epsilon = aT_r^4$ where T_r is the thermal radiation temperature and hence

$$T_r = T_0 (1+z) \qquad (15.51)$$

where T_0 is the temperature of the microwave background radiation at the present epoch. When we substitute this relation into the formula for black body radiation, we find that

$$\begin{aligned}\epsilon(\nu_1)\,d\nu_1 &= \frac{8\pi h\nu_1^3}{c^3}\,(e^{h\nu_1/kT_1} - 1)^{-1}\,d\nu_1 \\ &= \frac{8\pi h\nu_0^3}{c^3}\,(e^{h\nu_0/kT_0} - 1)^{-1}\,(1+z)^4\,d\nu_0 \end{aligned} \qquad (15.52)$$

i.e. the form of the Planck spectrum is preserved in the expansion whilst the energy density decreases as $(1+z)^4$. The key point is, however, that the energy density in the radiation increases with increasing redshift more rapidly than that in the matter. The inertial mass density in the matter changes with redshift as $\rho_0(1+z)^3$ whereas the mass density in the radiation changes as $\epsilon(z)/c^2 = \epsilon_0(1+z)^4/c^2$. The ratio of the two therefore changes as

$$\frac{\epsilon(z)}{\rho(z)c^2} = \frac{\epsilon_0}{\rho_0 c^2}\,(1+z) \qquad (15.53)$$

We described in Section 15.2.3 how the value of $\epsilon_0/\rho_0 c^2$ is about 10^{-3}–10^{-4} at the present day and hence, by redshifts $z \approx 10^3$–10^4, there is as much inertial mass in the radiation as there is in the matter. At earlier epochs, $z > 10^4$, the dynamics are dominated by the 'mass' of the radiation – the Universe becomes *radiation dominated* rather than *matter dominated*.

We can work out the new dynamics from equation (15.38). We have proved in Section 8.2 that the equation of state of a 'gas' of electromagnetic radiation is $p = \frac{1}{3}\epsilon$ and hence

$$\ddot{R} = -\frac{8\pi GR}{3c^2}\epsilon = -\frac{8\pi G}{3c^2}\epsilon_0 R^{-3} \tag{15.54}$$

Multiplying through by $\dot{R}$ and integrating with respect to t we find that

$$\int \dot{R}\frac{\mathrm{d}}{\mathrm{d}t}\dot{R}\,\mathrm{d}t = -\frac{8\pi G}{3c^2}\epsilon_0\int\frac{\mathrm{d}R}{R^3}$$

$$\dot{R}^2 = \frac{8\pi G}{3c^2}\epsilon_0 R^{-2} + \text{constant} \tag{15.55}$$

which is exactly what we would expect because the dynamics must also satisfy equation (15.39). When we consider small enough values of R, we can neglect the constant in comparison with the first term on the right-hand side and hence we need integrate only

$$\dot{R} = \left(\frac{8\pi G\epsilon_0}{3c^2}\right)^{\frac{1}{2}} R^{-1}$$

Performing this integral, we find that

$$R = \left(\frac{32\pi G\epsilon_0}{3c^2}\right)^{\frac{1}{4}} t^{\frac{1}{2}} \tag{15.56}$$

This describes the dynamics of a radiation dominated Universe. An interesting feature is that, if electromagnetic radiation were the only contributor to the energy density to be included in ϵ_0, we would now know precisely the early dynamics of the Universe since we know ϵ_0.

We can appreciate why the model is called the hot big bang model. In the early stages, the model is radiation dominated with $T_r \propto R^{-1}$ and $T \propto t^{-\frac{1}{2}}$, i.e. it is an adiabatically cooling expansion.

We can now work out the temperature history of the Universe and identify certain significant epochs in its evolution. These epochs are listed in Table 15.1 which shows the scale factor, a rough value for the time after the big bang, the events that occurred then, the thermal background radiation temperature and the density of the *matter* component in the Universe, assuming the value at the present day to be 10^{-30} g cm^{-3} which corresponds roughly to $\Omega = 0.3$.

Table 15.1. *Significant epochs in the Universe*

Scale factor, R	*Time from big bang*	*Events*	*Temperature of radiation* (K)	*Matter density* (g cm^{-3})
1	2×10^{10} years	Now	3	10^{-30}
1/1500	10^{7} years	At this temperature all the neutral hydrogen in the Universe is ionised	4000	10^{-20}
1/1000–1/10,000	$2–20 \times 10^{6}$ years	There are equal amounts of mass in the matter and radiation. At earlier times the Universe is radiation dominated	3000–30 000	10^{-21}–10^{-18}
$1/10^{9}$	10 minutes	The radiation is so hot that the nuclei of atoms dissociate	3×10^{9}	10^{-3}
$1/3 \times 10^{9}$	1 minute	Electron–positron pairs are created from the thermal background radiation	10^{10}	0.03
$1/10^{13}$	$\sim 10^{-5}$ seconds	Proton–antiproton, baryon–antibaryon pairs are created from the thermal background radiation	$\sim 10^{13}$	10^{9}

Going backwards in time from the present the significant epochs are as follows:

Epoch of 'recombination'. When the background radiation temperature rose to about 4000 K, there was sufficient short wavelength radiation present in the high frequency tail of the Planck spectrum to ionise all the neutral hydrogen. Now most of the matter in the Universe is in the form of the lightest element, hydrogen. This means that at earlier times there was no neutral hydrogen – instead, the matter formed a *plasma*, the phase of matter which is found in the ionosphere of the Earth, in the Sun and in plasma physics experiments. This epoch is called the *recombination epoch* because the hydrogen plasma began to recombine to form neutral hydrogen gas at this time.

We have already discussed the slightly earlier epoch when there were equal amounts of mass density in the matter and radiation. It is coincidental that this took place at roughly the same epoch at which recombination took place. Another interesting physical feature of the evolution of the plasma at epochs earlier than this is that, as soon as all the hydrogen was ionised, there was very strong thermal coupling between the radiation and the plasma. This took place because, whereas neutral gas does not scatter radiation, the free electrons scatter the radiation very strongly and thus energy can be transferred between the matter and radiation. This efficient scattering process, Compton scattering, ensured that the radiation and matter remained at the same temperature at all epochs prior to the epoch of recombination.

Epoch of nuclear reactions. When the Universe was squashed by a factor of 10^9, the temperature of the plasma rose to about 3×10^9 K and the thermal spectrum had maximum intensity at γ-ray energies. These waves are so energetic that they can dissociate the nuclei of atoms into their constituent neutrons and protons. We can therefore be certain that, at earlier epochs, there were no atomic nuclei as we know them – they were all broken down into their constituent parts. In other words, at this epoch, the Universe consisted of protons, neutrons, electrons, photons and various forms of neutrino which accompanied earlier interactions.

Electron–positron pair formation. If the Universe were to be squeezed just a little further, the γ-rays would become so energetic that they could collide to form electron–positron pairs. The basic physical principle involved is that, when the colliding γ-rays have total energy greater than twice the rest-mass energy of an electron ($E = m_e c^2$), there is a probability that the pair of γ-rays will be transmuted into an electron and its antiparticle, the positron. This can be written schematically:

$$\gamma + \gamma \rightarrow e^+ + e^-$$

At a temperature of about 10^{10} K this process became feasible and a new equilibrium was set up in which there are roughly equal numbers of electrons, positrons and γ-rays.

You may well ask, 'How on Earth can we be sure that we understand the physics of matter and radiation at such high densities and temperatures?' The answer is interesting: the last column in Table 15.1 shows that, although the temperatures are very high, the density of matter is modest. In fact, when the scale factor has the value $1/3 \times 10^9$, the mass density of ordinary matter is only about 0.03 g cm^{-3} which is 30 times less than the density of water. Thus, although we are talking about high temperatures, the densities are still at values which are similar to those found in terrestrial laboratories.

Proton–antiproton pair production. We can go further back still and ask what happened when the scale factor R was only $\sim 10^{-13}$. Then the temperature was so high that the γ-rays could collide and create protons and their antiparticles, the antiprotons. This is exactly the same process as electron–positron pair production described above but, in this case, because the rest mass of the proton is 1800 times greater than that of the electron, proton–antiproton pair production took place at a temperature 1800 times higher. In fact, above this temperature pair production of all the heavy particles known to elementary particle physicists can take place. These particles are generally known as baryons (meaning 'heavy particles') and, in general, the process is referred to as baryon–antibaryon pair production. The densities at this epoch were very high and approached the densities found in the nuclei of atoms.

For earlier epochs, we run out of secure physics. Up to this point we have been using 'known' physics in the sense that the nuclear interactions have been studied in terrestrial accelerators. At higher energies, the nuclear physics is not yet available on an experimental basis, although many elementary particle physicists have an excellent picture of what they expect to happen. We can be certain that at these very early epochs, all the different types of elementary particle discovered by particle physicists came into equilibrium. The Universe at these very early times bore very little resemblance to our present Universe.

We should note some of the great successes of the hot big bang models. First, it is natural that the microwave background radiation should be isotropic. Second, it is natural that the microwave background radiation should have a pure black body spectrum. In the very early Universe, all the constituents were in thermal equilibrium and so the radiation spectrum would have taken up its equilibrium form which is that of black body radiation. The thermal radiation cooled and preserved its thermal radiation spectrum as we have described above.

15.6.4 The origin of the light elements

The hot big bang model is an attractive picture but it would be reassuring to have some completely independent evidence for the hot early stages of the Universe. Fortunately, we now have evidence for this by a quite remarkable argument. It has always been a problem in astronomy to understand the origin of some of the lightest elements. We are confident that heavy elements like carbon, nitrogen, oxygen, iron and so on are produced by nuclear reactions in the centres of stars. The main problem lies with elements such as the isotopes of helium, helium-3 (^{3}He) and helium-4 (^{4}He), deuterium D and lithium-7 (^{7}Li). The basic problem is that these are all rather fragile elements and if they are mixed into the hot inner regions of stars, they are rapidly destroyed or converted into heavier species.

The mystery is deepened by the fact that these light elements seem to have more or less the same abundance by mass wherever they are observed. Wherever it is possible to observe helium-4, it always turns out to have an abundance of about 23%, or greater, by mass relative to hydrogen. This is much greater than could be created in stars. Likewise, the D/H ratio seems to be roughly constant with a value of about 1.5×10^{-5} by mass wherever we look in the interstellar gas, despite the fact that there are variations in the heavy element abundances along different lines of sight. In very old halo stars, the ^{7}Li abundance seems to remain constant in those stars in which we have good reason to believe that the surface material has not been mixed into the inner regions of the star. Beautiful observations by the French astronomers Spite and Spite[6] show that the material from which these very old stars formed must have already contained about 1 part in 10 000 million by mass of ^{7}Li.

Let us now investigate in more detail what we expect to come out of the hot big bang model. We can start the evolution of the model at a high temperature, $T \sim 10^{11}$ K, at which we expect all the constituents of the Universe that are stable at that temperature to be in equilibrium. Then we can let all the constituents interact and see what elements are produced as the Universe expands and cools. This simulation is carried out in a computer and one must include in the calculation all the possible interactions between the various constituents of the Universe. Computations like these have been carried out by Robert Wagoner of Stanford University and involved large amounts of computer time to follow the evolution of each species. An example of the time evolution of the chemical composition of the Universe is shown in Figure 15.7. It is not appropriate to go into the details here of why the elements evolve in the way they do but we note that there is a profound difference between element building in stars and element building in the hot big bang. In stars, the synthesis of the elements takes place over millions of years and there is time for equilibrium to be attained among the different chemical species. In the hot big bang, everything is highly non-equilibrium so far as element synthesis is concerned. The process of element formation is over in a few minutes as can be seen from Figure 15.7. This is because, by this time, the temperature has fallen below the value at which nuclear interactions can take place. It turns out that the calculations only depend upon the ratio of the number of photons (or particles of the thermal background radiation) to protons (or baryons) in the Universe. Now we know quite accurately the number density of photons in the microwave background radiation and so the results depend only on the present density of matter in the Universe. These results are shown in Figure 15.8.

The first remarkable feature of the diagram is that the elements which are synthesised in the hot big bang are *exactly* those which are difficult to account for by stellar nucleosynthesis, i.e. D, ^4He, ^3He, ^7Li. The ^4He abundance is remarkably independent of the density of the Universe and there are good thermodynamic reasons for this. For all reasonable values of the mean density of matter

Figure 15.7. An example of the time and temperature evolution of the abundances of different light elements in the early evolution of the hot model of the Universe from detailed computer simulations by Dr Robert Wagoner. Before about 10 seconds from the origin of the model, no significant synthesis of the light elements takes place because deuterium ^2H is destroyed by hard γ-rays in the high energy tail of the black body radiation spectrum which has temperature $T >$ 3×10^9 K. As the temperature decreases, more and more deuterium survives and synthesis of heavier light elements becomes possible through reaction chains such as

$$\mathrm{D + D \rightarrow {}^3He + n \qquad {}^3He + n \rightarrow T + p}$$

$$\mathrm{p + n \rightarrow D}$$

$$\mathrm{D + D \rightarrow T + p \qquad T + D \rightarrow {}^4He + n}$$

Notice that the synthesis of elements such as D (= ^2H), ^3He, ^4He, ^7Li and ^7Be is completed after about 15 minutes. (From R.V. Wagoner (1973), *Astrophys. J.*, 179, 343.)

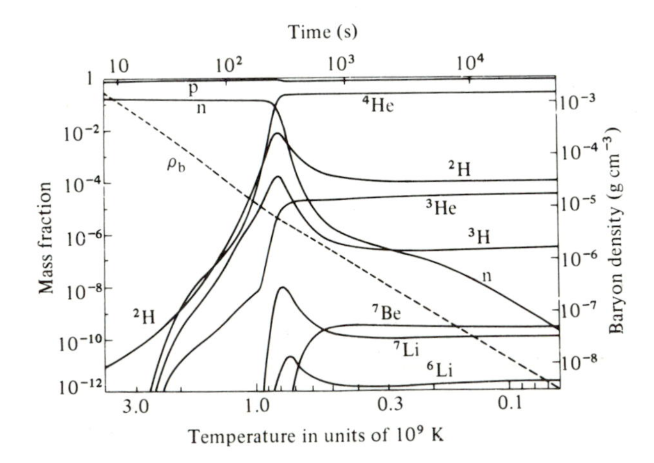

in the Universe, about 25% of ^{4}He is produced, in excellent agreement with observation. Notice, in particular, the strong dependence of the deuterium abundance on density. If the matter density is high, the deuterium is all converted into ^{4}He and so, to obtain the present deuterium abundance, a low value of the mean density of the Universe at the present time is required. Even more important is the fact that we know only ways of destroying deuterium astrophysically and not ways of creating it. Therefore, if some of the deuterium has already been destroyed, its original value must have been greater which drives us to lower values for the mean density of the Universe. On this basis, the density parameter for the baryons (or ordinary matter) Ω must be less than about 0.1, i.e. the Universe must be open. We can, in fact, select a best value for the baryon density of the Universe. We indicate estimates of the abundances of the light elements on Figure 15.8 and it can be seen that a value $\Omega \approx 0.03$ can explain the

Figure 15.8. The predicted primordial abundances of the light elements compared with the observed abundances. The present density of the world model is plotted along the abscissa. The observed abundances are in good agreement with a model having $\rho_0 \approx 3 \times 10^{-31}$ g cm^{-3} corresponding to $\Omega \approx 0.04$ (dotted line). (After R.V. Wagoner, (1973), *Astrophys. J.*, **179**, 343 and J. Audouze (1982), *Astrophysical Cosmology* (eds. H.A. Bruck, G.V. Coyne & M.S. Longair), 395, Pontificia Academia Scientiarum.)

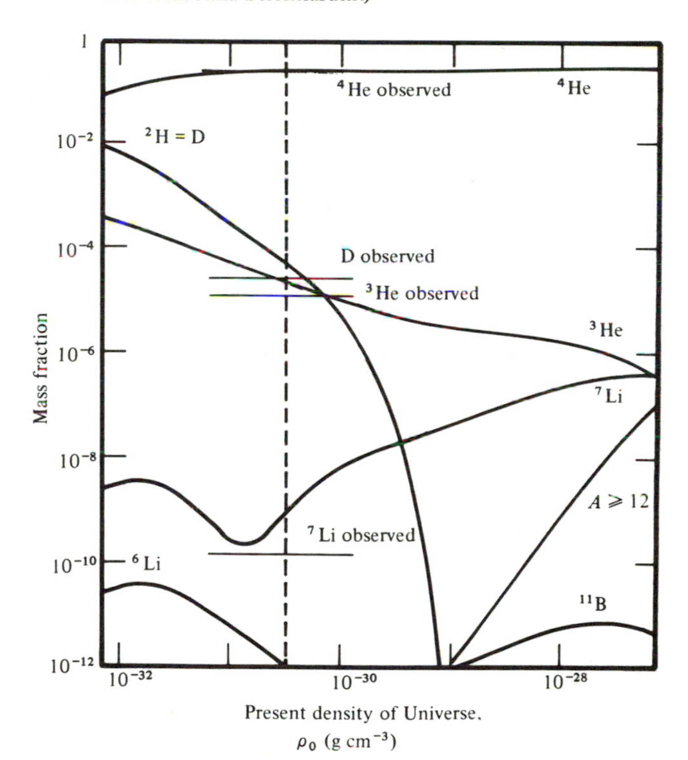

universal abundances of *all* the light elements. It seems to me highly improbable that this could be a result of chance because the computations involve so many different interactions. I interpret these results as *independent* evidence that the Universe went through a very hot dense phase as described by the standard hot big bang model.

15.7 Reflections

I am afraid I have become rather carried away by the physics of the expanding Universe, but, if you have followed the above argument, you may agree with my point of view that the physical content of the theory is very elegant and one can now tell a story which I personally find convincing – we have three independent pieces of evidence which lead naturally to the hot big bang picture, namely (i) the expansion of the Universe as defined by Hubble's law, (ii) the microwave background radiation with its Planckian spectrum, and (iii) the universal abundances of the light elements. We can find a natural explanation for all three in the hot big bang.

Astrophysicists have sufficient confidence in the theory that it has become the standard framework within which to study more profound cosmological questions. Among these are:

(i) How did galaxies and large-scale structures in the Universe come about?

(ii) Why does our Universe seem to consist almost entirely of matter rather than a mixture of matter and antimatter?

(iii) Why is the Universe isotropic?

All three questions are now being tackled vigorously. The first question is related to the fundamental questions of the origin and evolution of galaxies and their contents and is the subject of intense observational and theoretical investigations. The other two questions may well be related to the physics of the very early Universe, much earlier than the earliest epochs discussed in Section 15.6.3. The ultimate solution may involve grand unification theories of elementary particles and an understanding of quantum gravity which must be important in the very earliest stages of the expanding models, $t \approx 10^{-43}$ s.

If these last ideas were to turn out to be correct, we would find the ultimate relation between the physics of elementary particles and the origin of our own Universe. Such a synthesis would be worthy of the intellectual descendants of Newton, Maxwell and Einstein.

APPENDIX TO CHAPTER 15

THE ORIGIN OF THE ROBERTSON–WALKER METRIC IN THE CASE OF AN EMPTY UNIVERSE

The world model containing no matter at all, $\Omega = 0$, is often referred to as the Milne model because it can be developed purely from kinematics. Milne's major contribution was in elucidating precisely the meaning of time and kinematics in cosmology and he developed a particular approach to the construction of cosmological models which are known as kinematic cosmologies. In the empty model there is no influence of gravity upon test particles in the Universe and so they should move apart at constant velocity from $t = 0$ to $t = \infty$. It is possible to derive the Robertson–Walker metric in this special case using only special relativity. This is an instructive exercise since it indicates clearly some of the problems which arise in a more general treatment using general relativity. The origin of the uniform expansion is taken to be [0, 0, 0, 0] and the world lines of particles diverge from this point, each point maintaining constant velocity with respect to the others. The space–time diagram for this case is shown in Figure A15.1. Our own world line is the t axis and that of particle P which has constant velocity v with respect to us is shown.

Figure A15.1. A space–time diagram for an empty universe.

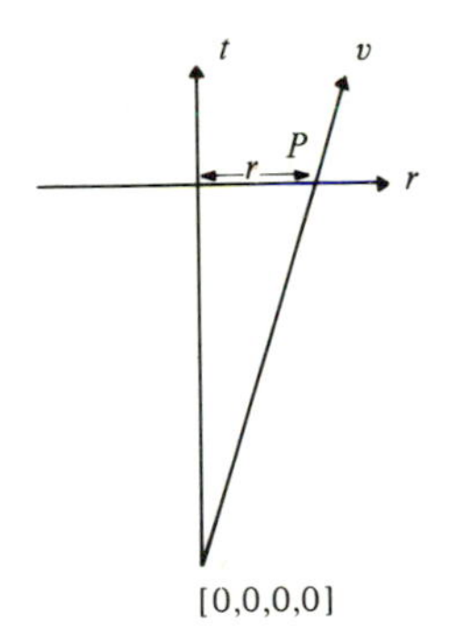

Everything in this picture is entirely special relativistic. The problems become apparent as soon as we attempt to define a suitable *cosmic time* for ourselves and for the fundamental observer moving with the particle P. At time t, his distance from us is r and, since his velocity is constant, $r = vt$ according to us. However, because of the relativity of simultaneity (see Section 13.2), the observer P measures a different time τ on his clock. The relation between t and τ is found from a simple Lorentz transformation

$$\tau = \gamma\left(t - \frac{vr}{c^2}\right); \quad \gamma = (1 - v^2/c^2)^{-\frac{1}{2}}$$

Since $r = vt$, this can be written

$$\tau = t\left[1 - \left(\frac{r}{ct}\right)^2\right]^{\frac{1}{2}} \tag{A15.1}$$

The first problem is now apparent: t is only proper time for the observer at O and for nobody else. We need to be able to define surfaces of constant cosmic time τ, because it is only on these surfaces that we can impose conditions on the large-scale properties of the Universe, in accordance with the cosmological principle. This calculation tells us that the appropriate surface for $\tau =$ constant is given by those points which satisfy

$$\tau = t\left[1 - \left(\frac{r}{ct}\right)^2\right]^{\frac{1}{2}} = \text{constant}$$

Locally, at each point in the space, this surface must be normal to the world line of the particle at that point. This is just another way of saying that the different frames of reference should be related by a simple Lorentz transformation.

The next requirement is to define the local element of radial distance $\mathrm{d}l$ at the point P on the surface $\tau =$ constant. Again, everything is Euclidean and therefore the interval $\mathrm{d}s^2 = \mathrm{d}t^2 - (1/c^2)\,\mathrm{d}r^2$ is an invariant. Over the $\tau =$ constant surface, $\mathrm{d}s^2 = -(1/c^2)\,\mathrm{d}l^2$ and hence

$$\mathrm{d}l^2 = \mathrm{d}r^2 - c^2\,\mathrm{d}t^2 \tag{A15.2}$$

τ and $\mathrm{d}l$ define locally the proper time and proper distance of events at P. We recognise that the coordinates τ and l are exactly equivalent to the cosmic time t and radial distance coordinate x introduced in Section 15.3. This clarifies why the metric of empty space is *not* a simple Euclidean metric. It is only in τ and l coordinates that we can apply the cosmological principle.

Now let us transform from the frame S to that at P by moving at radial velocity v. Distances perpendicular to the radial coordinate remain unaltered under Lorentz transformation and therefore, if in S,

$$\mathrm{d}s^2 = \mathrm{d}t^2 - \frac{1}{c^2}(\mathrm{d}r^2 + r^2\,\mathrm{d}\theta^2) \tag{A15.3}$$

the invariance of ds^2 under radial translation means that

$$dt^2 - \frac{1}{c^2}\, dr^2 = d\tau^2 - \frac{1}{c^2}\, dl^2 \qquad \text{(A15.4)}$$

the perpendicular distance increment $r^2\, d\theta^2$ remaining unaltered. Therefore,

$$ds^2 = d\tau^2 - \frac{1}{c^2}\,(dl^2 + r^2\, d\theta^2) \qquad \text{(A15.5)}$$

Now all we need to do is to express r in terms of l and τ to complete the transformation to τ, l coordinates.

We have already shown that, along the surface of constant τ,

$$dl^2 = dr^2 - c^2\, dt^2$$

In addition, the Lorentz transform of $d\tau$ is

$$d\tau = \gamma \left(dt - \frac{v}{c^2}\, dr\right) = 0 \qquad \text{(A15.6)}$$

and hence,

$$dt^2 = \frac{v^2}{c^4}\, dr^2$$

i.e.

$$dl^2 = dr^2 \left(1 - \frac{v^2}{c^2}\right) \qquad \text{(A15.7)}$$

But $v = r/t$ and hence we need only replace t by τ to find a differential expression for r in terms of l and τ.

$$dl^2 = dr^2 \left(1 - \frac{r^2}{c^2 t^2}\right) \qquad \text{(A15.8)}$$

$$= dr^2 \left(\frac{\tau}{t}\right)^2 \qquad \text{(A15.9)}$$

Now substituting (A15.9) back into (A15.8) to eliminate t, we find that

$$dl = \frac{dr}{\left(1 + \dfrac{r^2}{c^2\tau^2}\right)^{\frac{1}{2}}} \qquad \text{(A15.10)}$$

Integrating, using the substitution $r = \sinh x$, the solution is

$$r = c\tau \sinh\,(l/c\tau) \qquad \text{(A15.11)}$$

The metric (A14.5) can therefore be written

$$ds^2 = d\tau^2 - \frac{1}{c^2}\,[dl^2 + c^2\tau^2 \sinh^2(l/c\tau)\, d\theta^2] \qquad \text{(A15.12)}$$

This corresponds precisely to the expression (15.1) for an isotropic curved space with hyperbolic geometry, the radius of curvature of the geometry $\mathscr{R}$ being $c\tau$. This explains why an empty universe has hyperbolic spatial sections. The conditions (A15.1) and (A15.10) are the key relations which indicate why we can only define a consistent cosmic time and radial distance coordinate in hyperbolic rather than flat space.

16

EPILOGUE

We have come to the end of our story. It is tantalising to leave it there on the brink of modern physics but to take it further would result in a book aimed at a different audience and would need more advanced mathematical tools. How far I have succeeded in the many aims I set myself at the outset is not for me to judge. I can only state that, in preparing these lectures and revising and amplifying them for publication, I have learned a great deal which I only wish I had known many years ago. It is not anybody's fault but it is a pity that courses are nowadays so full of material that it is not possible to digress on some of the fascinating topics discussed in these lectures.

My main impression on rereading the lectures is an enhanced appreciation and admiration for the physicists and mathematicians who worked out the basic laws in the first place. These are outstanding intellectual achievements with the occasional flash of genius which raises the whole subject to a new level of understanding.

In preparing the lectures, I was greatly impressed by the clarity of many of the great papers in the development of classical and modern physics. To be honest, I have found reading the original papers by physicists such as Maxwell, Rayleigh and Einstein easier than many modern text books. In all these great works, I find a clarity of thought and exposition which results from a clear understanding of the relation between our physical world and the mathematics we need to describe it. It is this firm understanding of the basic laws of physics and theoretical physics which acts as the springboard for further advances and discoveries. This understanding cannot be attained by taking short cuts. It requires a great deal of hard work and experience of using the laws before their full content can be appreciated.

Although the working out of any particular problem may be complicated, the basic physical ideas and mathematical structures are simple. Once these are understood, it is basically a matter of technique to apply them to a specific problem. I remember very vividly a story told to me by Peter Scheuer about an incident which took place during his period of army service. Being a physicist, he was put into a section working on radio receivers and one day he asked a sergeant a question about some aspect of one of the circuits. The sergeant's immortal reply is engraved on my memory: 'All you need to know is Ohm's law, but you need to know Ohm's law bloody well!' It is a statement which could be made about any of the laws of physics.

REFERENCES AND FURTHER READING

In preparing my lectures, I found myself making repeated use of the magnificent set of volumes entitled *Dictionary of Scientific Biography* (abbreviated to *DSB* in the following references). The volumes are a mine of authoritative information about all the scientists who appear in this book. The reference is

Dictionary of Scientific Biography Vols. 1–14, 1970. Charles Scribner's Sons, New York.

Chapter 1

1 Dirac, P.A.M., 1977. *History of Twentieth Century Physics*, Proc. International School of Physics 'Enrico Fermi', Course 57, p. 136, Academic Press, New York & London.

2 Dirac, P.A.M., 1977. *op. cit.*, p. 112.

3 Thomson, J.J., 1893. *Notes on Recent Researches in Electricity and Magnetism*, vi, Clarendon Press, Oxford. (Quoted by J.L. Heilbron in ref. 4, p. 42.)

4 Heilbron, J.L., 1977. *History of Twentieth Century Physics*, Proc. International School of Physics 'Enrico Fermi', Course 57, p. 40, Academic Press, New York & London.

5 Heilbron, J.L., 1977. *op. cit.*, p. 43.

6 Heilbron, J.L., 1977. *op. cit.*, p. 43.

Chapter 2

1 Dreyer, J.L.E., 1890. *Tycho Brahe. A Picture of Scientific Life and Work in the Sixteenth Century*, pp. 86–7, Adam & Charles Black, Edinburgh.

2 Christianson, J., 1961. *Scientific American*, **204**, 118, (February issue).

3 Kepler, J., 1609. From *Astronomia Nova.* See *Johannes Kepler Gesammelte Werke*, ed. M. Casper, Vol. III, p. 178, Beck, Munich, 1937.

4 Kepler, J., 1596. From *Mysterium Cosmographorum.* See *Kepleri opera ommia*, ed. C. Frisch, Vol. I, pp. 9ff.

5 Galilei, G., 1630. Quoted by Rupert Hall, A., 1970. *From Galileo to Newton 1630–1720. The Rise of Modern Science 2*, p. 41, Fontana Science, London.

6 Galilei, G., 1632. *Dialogues concerning the Two Chief Systems of the World*, (trans. S. Drake), p. 32, Berkeley 1953.

Chapter 3

1 Maxwell, J.C., 1861–2. These ideas are developed in a series of papers published in *Phil. Mag.* (1861) **21**, 161, 281, 338 and (1862) **23**, 12, 85.

2 Whittaker, E., 1951. *A History of the Theories of Aether and Electricity*, p. 255, Thomas Nelson & Sons Ltd, London.

3 Campbell, L. & Garnett, W., 1882. *The Life of James Clark Maxwell*, p. 342 (Letter of 5 January 1865), MacMillan & Co., London.

4 Hertz, H., 1893. *Electric Waves*, MacMillan & Co., London.

Chapter 4

1 Stratton, J.A., 1941. *Electromagnetic Theory*, McGraw Hill, New York & London.

Chapter 5

1 Goldstein, H., 1950. *Classical Mechanics*, Addison-Wesley, London.

2 Feynman, R.P., 1964. *Lectures on Physics* (eds. R.P. Feynman, R.B. Leighton & M. Sands), Vol. 2, Chapter 19. Addison-Wesley, London.

3 Dirac, P.A.M., 1935. *The Principles of Quantum Mechanics*, Clarendon Press, Oxford.

4 Dirac, P.A.M., 1977. *History of Twentieth Century Physics*, Proc. International School of Physics, 'Enrico Fermi', Course 57, p. 122, Academic Press, New York & London.

5 Landau, L.D. & Lifshitz, E.M., 1960. *Mechanics*, Vol. 1 of *Course of Theoretical Physics*, Pergamon Press, Oxford.

6 Batchelor, G.K., 1967. *An Introduction to Fluid Dynamics*, Cambridge University Press.

7 Landau, L.D. & Lifshitz, E.M., 1959. *Fluid Mechanics*, Vol. 5 of *Course of Theoretical Physics*, Pergamon Press, Oxford.

Chapter 6

1 Harman, P.M., 1982. *Energy, Force and Matter. The Conceptual Development of Nineteenth Century Physics*, Cambridge University Press.

2 Fourier, J.B.J., 1822. *Analytical Theory of Heat*, trans. A. Freeman, reprint edition, New York 1955.

3 Joule, J.P., 1843. See *The Scientific Papers of James Prescott Joule* 2 Volumes, London 1884–1887 (reprint edition London 1963).

4 Pippard, A.B., 1966. *The Elements of Classical Thermodynamics*, Cambridge University Press.

5 Carnot, N.L.S., 1824. *Réflexions sur la Puissance Motrice de Feu et sur les Machines Propres à Developper cette Puissance,* Bachelier, Paris.

6 Feynman, R.P., 1963, *Lectures on Physics* (eds. R.P. Feynman, R.B. Leighton & M. Sands) Vol. 1, 44–4, Addison-Wesley, London.

Chapter 7

1 Clausius, R., 1857. *Annalen der Physik*, **100**, 497 (English translation in S.G. Brush (ed.) *Kinetic Theory*, Vol. 1, p. 111, Pergamon Press, Oxford (1966)).

2 Waterston, J.J., 1843. See article by S.G. Brush, *DSB*, Vol. 14, p. 184. See also *The Collected Scientific Papers of John James Waterston*, (ed. J.G. Haldane), Edinburgh 1928.
3 Rayleigh, Lord, 1892. *Phil. Trans. Roy. Soc.*, **183**, 1.
4 Rayleigh, Lord, 1892. *op. cit.*, 2.
5 Rayleigh, Lord, 1892. *op. cit.*, 3.
6 Maxwell, J.C., 1860. *Phil. Mag., Series 4*, **19**, 19 & **20**, 21. See also *The Scientific Papers of James Clark Maxwell* (ed. W.D. Niven), p. 377, Cambridge University Press, 1890.
7 Everitt, C.W.F., 1970. *DSB*, Vol. 9, p. 218.
8 Maxwell, J.C., 1860. *Report of the British Association for the Advancement of Science*, **28**, pt 2, 16.
9 Maxwell, J.C., 1867. Communication to P.G. Tait quoted by P.M. Harman, *op. cit.*, p. 140.
10 Maxwell, J.C., 1867. Communication to P.G. Tait, quoted by P.M. Harman, *op. cit.*, p. 140.
11 Kittel, C., 1969. *Thermal Physics*, John Wiley & Sons, Chichester.
12 Mandl, F., 1971. *Statistical Physics*, John Wiley & Sons, Chichester.

Chapter 8 (Actually in Introduction to Case Study 5)

1 Klein, M.J., 1977. *History of Twentieth Century Physics*, Proc. International School of Physics, 'Enrico Fermi', Course 57, p. 1, Academic Press, New York & London.

Chapter 9

1 Planck, M., 1950. *Scientific Autobiography and Other Papers*, p. 15, Williams & Norgate, London.
2 Planck, M., 1950. *op. cit.*, p. 18–9.
3 Planck, M., 1896. Quoted by M.J. Klein (1977), *op. cit.*, p. 3.
4 Longair, M.S., 1981. *High Energy Astrophysics*, Cambridge University Press.
5 Planck, M., 1950. *op. cit.*, p. 37–8.
6 Rayleigh, Lord, 1900. *Phil. Mag.*, **49**, 539. See also *Scientific Papers by John William Strutt, Baron Rayleigh Vol. 4, 1892–1901*, p. 483, Cambridge University Press.
7 Planck, M., 1958. *Physikalische Abhandlungen und Vortrage* (*Collected Scientific Papers*), Vol. 1, p. 596, Friedr. Vieweg und Sohn, Braunschweig. (English translation: see Hermann, A., 1971, *The Genesis of Quantum Theory* (*1899–1913*), p. 10, MIT Press, Cambridge, Mass.)
8 Planck, M., 1958. *op. cit.*, Vol. 1, p. 597 (English translation: A. Hermann, 1971, *op. cit.*, p. 10).
9 Planck, M., 1900. *Annalen der Physik*, **1** (**306**), 730. See also *Collected Scientific Papers*, 1958, *op. cit.*, Vol. 1, p. 687. (English translation: *Planck's Original Papers in Quantum Physics*, 1972, annotated by H. Kangro, pp. 35–7, Taylor & Francis, London.)
10 Rayleigh, Lord, 1894. *The Theory of Sound*, 2 volumes, MacMillan, London.
11 Rayleigh, Lord, 1900. *Phil. Mag.*, **49**, 539.

12 Rayleigh, Lord, 1900. Scientific Papers, *op. cit.*, pp. 485.

13 Planck, M., 1900. *Verhandl der Deutschen Physikal. Gessellsch.*, **2**, 202. See also *Collected Scientific Works*, 1958, *op. cit.*, Vol. 1, p. 698. (English translation: *Planck's Original Papers in Quantum Physics*, 1972, *op. cit.*, pp. 38–45.)

14 Planck, M., 1925. *A Survey of Physics*, p. 166, Methuen & Co., London.

Chapter 10

1 Planck, M., 1950. *op. cit.*, p. 41.

2 Planck, M., 1900. See ref. 13, Chapter 9.

3 Planck, M., 1900. Quoted by M.J. Klein, 1977, *op. cit.*, p. 17.

4 Einstein, A., 1912. *The Theory of Radiation and Quanta*, transactions of the First Solvay Conference (eds. P. Langevin & M. de Broglie), p. 115, Gautier-Villars, Paris (translation of quotation: see A. Herman, 1971, *op. cit.*, p. 20).

5 Planck, M., 1950. *op. cit.*, p. 44–5.

6 Planck, M., 1931. Letter from M. Planck to R.W. Wood. See A. Hermann, 1971, *op. cit.*, pp. 23–4, MIT Press, Cambridge, Mass.

Chapter 11

1 Einstein, A., 1905*a*. *Annalen der Physik*, **17**, 549. (English translation: *Investigations on the Theory of the Brownian Movement*, edited with notes by R. Furth, translated by A.D.Cowper, 1926, pp. 1–18, Methuan & Co., London. Reprinted by Dover, London, 1956.)

2 Einstein, A., 1905*b*. *Annalen der Physik*, **17**, 891. (English translation: *The Principle of Relativity* (with notes by A. Sommerfeld), translated by W. Perrett & G.B. Jeffrey, 1923, Methuen & Co., London; republished by Dover, London, 1952.)

3 Einstein, A., 1905*c*. *Annalen der Physik*, **17**, 132. (English translation: A.B. Arons & M.B. Peppard, 1965, *Am. J. Phys.*, **33**, 367.)

4 Einstein, A., 1905*c*. *op. cit.*, 132.

5 Millikan, R.A., 1916. *Phys. Rev.*, 7, 18.

6 Einstein, A., 1906. *Annalen der Physik*, **20**, 199.

7 Einstein, A., 1907. *Annalen der Physik*, **22**, 180.

8 Einstein, A., 1907. *op. cit.*, 183–4.

9 Einstein, A., 1907. *op. cit.*, 184.

Chapter 12

1 Planck, M., 1907. Letter to A. Einstein of 6 July 1907, Einstein Archives, Princeton, N.J., quoted by A. Hermann, 1971, *op. cit.*, p. 56.

2 Lorentz, H.A., 1909. Letter to W. Wien of 12 April 1909, quoted by A. Hermann, 1971, *op. cit.*, p. 56.

3 Einstein, A., 1909. *Phys. Zeitschrift*, **10**, 185.

4 Millikan, R.A., 1916, *Phys. Rev.*, **7**, 355.

5 Heisenberg, W., 1929. 'The Development of the Quantum Theory 1918–1928', *Naturwiss.*, **17**, 491. (See translation of quotation in R.H. Stuewer, *The Compton Effect*, p. 287, Science History Publications, New York.)

Chapter 13

1 Einstein, A., 1905. *Annalen der Physik*, **17**, 891. (English translation: *The Principle of Relativity* (with notes by A. Sommerfeld), translated by W. Perrett & G.B. Jeffery, 1923, Methuen and Co.; republished by Dover, London, 1952.)

2 Rindler, W., 1977. *Essential Relativity – Special, General, and Cosmological*, Springer-Verlag, New York.

3 Einstein, A., 1905. *op. cit.*, English translation p. 37.

Chapter 14

1 Rindler, W., 1977. *Essential Relativity – Special, General, and Cosmological*, Springer-Verlag, New York.

2 Berry, M., 1976. *Principles of Cosmology and Gravitation*, Cambridge University Press.

3 Weinberg, S., 1972. *Gravitation and Cosmology: Principles and Applications of the General Theory of Relativity*, John Wiley & Sons, London.

4 Misner, C.W., Thorne, K. & Wheeler, J.A., 1973. *Gravitation*, W.H. Freeman & Co., San Francisco.

5 Einstein, A., quoted by W. Rindler, 1977, *op. cit.*, p. 18. Einstein's original paper on general relativity is translated into English in *The Principle of Relativity*, p. 111 (see ref. 2 Chapter 11).

6 Berry, M., 1976. *op. cit.*, p. 67 and Appendix B, p. 160.

7 Weinberg, S., 1972. *op. cit.*, p. 9.

8 Rindler, W., 1977. *op. cit.*, p. 126.

9 Hawking, S.W., 1975. *Comm. Math. Phys.*, **43**, 199 and *Quantum Gravity: An Oxford Symposium* ed. C.J. Isham, R. Penrose & D.W. Sciama, p. 219, Oxford University Press.

Chapter 15

1 Bondi, H. & Gold, T., 1948. *Mon. Not. Roy. Astron. Soc.*, **108**, 252. Hoyle, F., 1948. *Mon. Not. Roy. Astron. Soc.*, **108**, 372.

2 Dirac, P.A.M., 1938. *Proc. Roy. Soc.* **A**, **165**, 199.

3 Brans, C. & Dicke, R.H., 1961. *Phys. Rev.* **124**, 925.

4 Einstein, A., 1931. Quoted by A. Pais in *Subtle is the Lord . . . The Science and Life of Albert Einstein*, p. 288, Oxford University Press, 1982. Ref. to *Sitzungsbericht*, p. 235, Prussian Academy of Science, 1931.

5 Sandage, A.R., 1970. *Phys. Today*, **23**, 5.

6 Spite, M. & Spite, F., 1982. *Nature*, **297**, 483.

INDEX

The titles of original papers and books quoted in the main text are shown in italic type.

Those topics which I consider to be particularly valuable for revision purposes are shown in bold type. The page numbers indicated in bold type refer to the principal theoretical discussion of the topic.